Journal of Neural Transmission

Supplement 44

JNT

S. Hoyer, D. Müller, and K. Plaschke (eds.)

Cell and Animal Models in Aging and Dementia Research

Springer-Verlag Wien New York

Prof. Dr. S. Hoyer
Dr. D. Müller
Dr. K. Plaschke
Institute for Pathochemistry and General Neurochemistry,
University of Heidelberg, Heidelberg, Federal Republic of Germany

Typesetting: Best-set Typesetter Ltd, Hong Kong

Printed on acid-free and chlorine-free bleached paper

With 63 Figures

ISSN 0303-6995

ISBN-13: 978-3-211-82549-5 e-ISBN-13: 978-3-7091-9350-1
DOI: 10.1007/978-3-7091-9350-1

Preface

The increasing life expectancy of the population in nearly all highly industrialized countries is not accompanied with physical and/or mental health in any case. With respect to mental function, this is increasingly compromised by abnormal, age-related brain disorders causing a loss in intellectual capacities, i.e., in learning and memory. This dementia state forms one of the greatest burdens in medical and socio-economic terms because most of the early pathogenetic factors of the age-related dementias are still unknown, and, therefore, a successful therapy is not possible at present.

The contribute to the understanding of genetic, molecular, cellular, morphological and behavioral aspects of normal aging and age-related disorders, a workshop "Cell and Animal Models in Aging and Dementia Research" was held in Heidelberg, Federal Republic of Germany, on September 2nd and 3rd, 1993. The workshop was organized in the frame of the IPA Congress in Berlin, and was generously supported by the Hirnliga e. V. (Brain League). The data presented at the workshop deriving from cell and animal models covered many actual aspects of both normal and abnormal brain aging, and its interactions. It is the hope of the authors that the results from these models can help to fill the gap which exists in the understanding of early events of mental aging and age-related (dementing) disorders.

S. Hoyer, Heidelberg

Contents

Listed in Current Contents

J Neural Transm (1994) [Suppl] 44: 1–20

Primary cultures of neurons for testing neuroprotective drug effects*

H. Oberpichler-Schwenk and **J. Krieglstein**

Institut für Pharmakologie und Toxikologie, Philipps-Universität, Marburg, Federal Republic of Germany

Summary. Primary cultures of neurons are widely used for the investigation of pathomechanisms of neuronal damage und for the evaluation of neuroprotective drug effects. The present paper gives a short survey of frequently used primary neuronal culture systems and of experimental measures for the induction of defined neuronal damage with particular respect to the pathomechanisms of cerebral ischemia. Neuroprotective durg effects as achieved under these conditions are reviewed, and the neuroprotective effects of glutamate antagonists, radical scavengers, and neural growth factors are discussed in some more detail.

Introduction

Neurodegeneration is a common feature of disorders such as stroke, multi-infarct dementia, Alzheimer's, Parkinson's and Huntington's disease. Since the number of people affected by these disorders is steadily increasing, there is a great demand for a rational drug therapy. To this end, clarification of the pathophysiological events leading to neurodegeneration is of utmost importance, not only for the understanding of the pathogenesis but also as a prerequisite for the specific development of neuroprotective drugs.

Primary cultures of neurons provide important tools for this task. Neuronal reactions to specific changes of the extracellular milieu can be investigated on the cellular level. Thus, neuronal reactions can be differentiated from non-neuronal reactions. Owing to the absence of the blood-brain barrier, drug effects can be tested without pharmacokinetic considerations. A further general advantage of cultured cells may be seen in the reduction of the number of animal experiments. In this context, primary neuronal cultures, although they still require a certain number of donor

---

* Abbreviations: *AMPA* α-Amino-3-hydroxy-5-methylisoxazole-4-propionate; *APV* 2-Amino-5-phosphonovaleric acid; *CNQX* 6-Cyano-7-nitroquinoxaline-2,3-dione; *DNQX* 6,7-Dinitroquinoxaline-2,3-dione; *LDH* Lactate dehydrogenase; *NBQX* 2,3-Dihydro-6-nitro-7-sulphamoyl-benzo(f)quinoxaline; *NMDA* N-Methyl-D-aspartate

animals, are nevertheless often preferred to neuronal cell lines. While neuronal cell lines contain metabolically deviated cells, primary cultures pertain many morphological and functional features of their in vivo counterparts as will still be discussed.

In the following we shall discuss some primary neuronal culture models commonly used for the investigation of neuroprotective drugs. It will be shown that drug effects evaluated under these in vitro conditions can well be extrapolated to in vivo conditions, and a few more recent concepts of neuroprotection in vitro will be presented.

Neuronal cultures in common use

Like for animal experiments, mice and rats are most frequently used as donor animals for primary neuronal cultures. Cells are derived from nearly all parts of the central nervous system, such as whole cortex (Dichter, 1978; Choi et al., 1987; Prehn et al., 1993a), visual cortex (Huettner and Baughman, 1986), hippocampus (Banker and Cowan, 1977; Rothman, 1984; Yamada et al., 1989; Prehn et al., 1993b), striatum (Weiss et al., 1986), hypothalamus (Yamashita et al., 1992), cerebellum (Messèr, 1977; Balázs et al., 1988; Novelli et al., 1988), and dorsal root ganglia (Scott, 1982). Whereas in most cases the tissue is dissociated into single cells and the cell suspension is seeded onto a prepared surface, a few authors used explants of brain tissue and performed experiments on the manolayer of cells spreading out of the explant (Gähwiler, 1981; Mattson et al., 1988).

Many investigators use nervous tissue from fetal rats or mice for the preparation of their cultures. In this stage of development, differentiation of the neurons is still comparatively poor and the cells presumably withstand the dissociation procedures better than at later stages. Nevertheless, several authors have successfully utilized postnatal donor animals for the preparation of neuronal cultures (Messer, 1977; Huettner and Baughman, 1986; Trussell et al., 1986; Yamada et al., 1989; Zorumski et al., 1989; Prehn et al., 1993a,b). Even 19-day-old rats yielded good cultures of hippocampal neurons (Nakajima et al., 1986).

Neurons in dispersed cell culture start to extend several neurites within one day after plating. These neurites differentiate into an axon and several dendrites (for references see Sargent, 1989) and, depending on the seeding density, can form an extensive network. Synaptic contacts have been shown by electronmicroscopy (Weiss et al., 1986; Harris and Rosenberg, 1993; Ichikawa et al., 1993) and electrophysiology (Dichter, 1978; Rothman and Samaie, 1985). With the formation of neurites, the somata of the neurons take on pyramidal, bipolar or stellate shapes reminiscent of their in situ counterparts. When seeded at high density (i.e. 100,000 to 300,000 cells/cm^2), the neurites often form aggregates extending many neurites. Apart from these morphological characteristics, neurons in culture have been shown to contain neuron-specific markers such as neuron-specific enolase (Söderbäck et al., 1989) and microtubule-associated protein 2 (Mattson et al., 1989) as well as transmitters like glutamate and gamma-aminobutyric acid (Mattson and Kater, 1989).

Besides the neurons, cultures from fetal (15 to 20 days of gestation) or postnatal rat or mouse brain always contain a number of non-neuronal cells, predominantly astrocytes. Since these non-neuronal cells are still mitotic and thus may outgrow the neurons, their proliferation is often halted by addition of an antimitotic agent like cytosine arabinoside or fluorodesoxyuridine after they have become confluent. Hence, these primary neuronal cultures are actually mixed neuronal/glial cultures. A certain proportion of glial cells seems to be necessary for long-term survival of the neurons (Banker and Cowan, 1977; Banker, 1980), perhaps owing to the release of (a) neurotrophic factor(s) (Müller and Seifert, 1982). Some authors have seeded cell suspensions from postnatal nervous tissue (cortex, hippocampus) onto a prepared glia feeder layer (Huettner and Baughman, 1986; Nakajima et al., 1986; Yamada et al., 1989), although postnatal rat cortical and hippocampal neurons could also be cultured successfully without such precautions (Furuya et al., 1989; Jahr and Stevens, 1987; Prehn et al., 1993a,b).

Mixed cultures have been used for numerous neurobiological, physiological and pharmacological investigations and lend themselves to morphological or electrophysiological evaluations. However, in analyses utilizing the whole cell mass, as for example measurement of energy metabolites or protein content, mixed cultures may yield equivocal results, since the non-neuronal cells may react differently to the experimental procedure than the neurons. Thus, for such biochemical measurements pure neuronal cultures are preferable. We have used neuronal cultures from chick embryo telencephalon for various pharmacologic studies (Krieglstein et al., 1988; Ahlemeyer and Krieglstein, 1989; Oberpichler et al., 1990; Peruche et al., 1990; Seif el Nasr et al., 1990; Prehn et al., 1992, 1993b). These neuronal cultures have been first prepared and characterized by Pettmann et al. (1979) and Louis et al. (1981). They are prepared from 7-day-old embryos, i.e. at a developmental stage before gliogenesis has occurred. Proliferation of any glioblasts is hampered by polylysine treatment of the culture flasks and by thorough dissociation of the tissue into single cells. In this way nearly pure neuronal cultures are obtained with about 98% of the cells binding tetanus toxin and less than 0.1% of the cells corresponding to glia (Pettmann et al., 1979). Like in the mixed cultures described above, the neurons form an extensive network of neurites. Membrane thickenings suggestive of synaptic junctions were found three to four days after preparation of the cultures, and characteristic synapses were shown in 6-day-old cultures (Louis et al., 1981). The same authors have shown that the majority of neurons in these cultures are dopaminergic, containing tyrosine hydroxylase and DOPA decarboxylase activity, whereas cholineacetyltransferase and glutamate decarboxylase activity remain rather low. Probably owing to the lack of glial cells, the life span of the neurons is comparatively short, and cultures start to disintegrate after eight to nine days in vitro

(Louis et al., 1981; own unpublished observations). Our experiments with cultured chick neurons were performed on cultures up to seven days in vitro.

Evaluation of neuroprotection in primary cultures

In order to evaluate neuroprotective effects in neuronal cultures, the cultures have to be experimentally damaged in some way. This damage should then be alleviated by the potentially neuroprotective measures. Since our interest lies on the pharmacology of cerebral ischemia, we shall concentrate upon models of experimental neuronal damage with a certain relation to the pathophysiology of cerebral ischemia, although there are also other means to impair cultured neurons, like the intoxication with methylphenyltetrahydropyridine (MPTP).

Neuroprotection under conditions of impaired oxygen availability

In cerebral ischemia, the brain tissue is deprived of both oxygen and glucose. Oxygen deprivation in vitro was used to damage primary cultures of rat hippocampal neurons (Rothman, 1984) or mouse cortical neurons (Goldberg et al., 1987, 1988). Cultures were kept under nominally oxygen-free conditions for several hours and neuronal damage was assessed either directly after the hypoxic incubation (Rothman, 1984) or after 20 to 24 hours of normoxic recovery (Goldberg et al., 1987, 1988). This procedure led to broad neuronal damage, as evidenced microscopically by swelling and vacuolation of the cell somata, loss of phase brightness, and degeneration of neurites. Damaged neurons could also be identified by trypan blue staining. Goldberg et al. (1987) additionally quantified the extent of neuronal damage by measuring the efflux of the cytosolic enzyme lactate dehydrogenase (LDH) into the culture medium over 24 hours after the hypoxic challenge. They found a good correlation between the extent of neuronal loss and the LDH release and could rule out the possibility that a considerable proportion of the LDH had been released by the glia present in their cultures. However, measurement of LDH release may yield equivocal results under conditions where glia are also damaged, since glial cells contain (and, when damaged, release) more LDH activity than neurons (Juurlink and Hertz, 1993).

Neuronal damage by oxygen deprivation required at least 6 hours of nominally oxygen-free conditions. This is due to the facts that 1. in the presence of glucose neurons can survive on minimal oxygen tensions for a rather long time, whereas 2. under the experimental conditions used, by flushing the culture dishes with nitrogen it takes about two hours to lower the oxygen tension in the culture medium to below 1 mmHg (Goldberg et al., 1986). In order to obtain a more defined deprivation of aerobic metabolism, several authors including us used blockers of the respiratory chain like antimycin (Pauwels et al., 1985) or cyanide (Rothman, 1983; Goldberg et al., 1987; Krieglstein et al., 1988). In this so-called cytotoxic hypoxia (Krieglstein et al., 1988) the oxygen tension in the culture medium is not reduced, but the oxygen can not be utilized by the respiratory chain and the energy metabolism of the cells thus is severely compromised. Goldberg et al. (1987) achieved neuronal damage comparable to 6–8 hours of oxygen deprivation by incubation with 20 mM cyanide for only 20 min.

In our laboratory, the effect of cyanide on primary cultures of chick neurons has been studied extensively (Krieglstein et al., 1988; Ahlemeyer and Krieglstein, 1989; Prehn et al., 1992). It could be shown that the cyanide-induced impairment of energy metabolism of the cultures is rapidly reversible, i.e. within few minutes (Krieglstein et al., 1988). But in spite of this rapid recovery the neurons disintegrate over the following hours (Oberpichler et al., 1990), like in the experiments of Rothman (1983) and Goldberg et al. (1987) on murine neurons. For the evaluation of neuroprotective agents we induced neurodegeneration in 4-day-old chick neuronal cultures by 30 min of cytotoxic hypoxia (1 mM sodium cyanide) followed by three days of recovery in cyanide-free culture medium (Ahlemeyer and Krieglstein, 1989). This resulted in a significant reduction of protein content of the cultures accompanied by a decrease in total cell number and in the proportion of viable, i.e. trypan blue excluding neurons. On this model a number of potentially neuroprotective drugs were tested (Ahlemeyer and Krieglstein, 1989; Peruche et al., 1990; Seif el Nasr et al., 1990). Drugs mainly affecting the tone of blood vessels (e.g. bencyclane, dihydroergocristine) remained without effect as had to be expected. On the other hand, centrally depressant drugs (clonazepam, diazepam, lidocaine) and especially NMDA antagonists (dizocilpine, imipramine, ketamine, memantine) proved neuroprotective as evidenced by alleviation of the cyanide-induced reduction in protein content and viability. For the NMDA antagonists, corresponding in vivo effects had been demonstrated by several authors (Simon et al., 1984; Gill et al., 1987; Ozyurt et al., 1988; Karkoutly et al., 1990; Seif el Nasr et al., 1990). The same applies to the nootropic naftidrofuryl, which proved neuroprotective in our chick neuron model (Ahlemeyer and Krieglstein, 1989) as well as in rats subjected to global forebrain ischemia (Krieglstein et al., 1989). Hence, we use the model of cytotoxic hypoxia in cultured chick neurons with long-term (three days) recovery as a reliable screening model for testing neuroprotective drug effects.

Neuroprotection against an excitotoxic challenge — pathophysiological considerations

Another frequently used procedure to damage neurons is the addition of toxic concentrations of glutamate or related excitotoxins. This procedure partly reflects in vivo conditions of cerebral ischemia, since cerebral ischemia leads to strongly elevated extracellular levels of glutamate and aspartate (Benveniste et al., 1984), probably as a result of both increased release and impaired reuptake. These elevated concentrations of excitatory amino acids may become neurotoxic (Olney, 1978) and may underly the neurodegeneration not only after ischemia, but also after hypoxia, prolonged seizures, hypoglycemia and in Huntington's disease (Rothman and Olney, 1986; Choi, 1988). Increased synaptic release of excitatory amino acids was also suggested to be involved in the death of cultured neurons following hypoxia (Rothman, 1984; Goldberg et al., 1988) and may play a role in the cyanide-induced damage of cultured chick neurons, since NMDA antagonists prevented this damage very effectively (see above).

Glutamate as an excitotoxin has been used in primary cultures of murine cortex (Choi et al., 1987; Koh and Choi, 1987), hippocampus (Rothman et al., 1987) and cerebellum (Novelli et al., 1988; Lysko et al., 1989) and more recently also in primary neuronal cultures of chick brain (Prehn and Krieglstein, 1991). In murine neuronal cultures, five minutes of incubation with 500 μM to 1 mM glutamate led to broad neuronal damage as assessed 20 to 24 hours later by trypan blue staining or LDH efflux measurement. In our laboratory, incubation of two-week-old rat hippocampal neuronal cultures with 1 mM glutamate for 30 minutes with subsequent 18 hours of recovery caused reduction of neuronal viability from 93% to 54% (trypan blue exclusion; Lippert et al., 1992). In rat neocortical cultures, 18 to 24 hours of incubation in 1 mM glutamate reduced the proportion of viable cells from about 90% to about 50% (Prehn et al., 1993b). Chick neuronal cultures (six days in vitro) were incubated with Mg^{2+}-free saline including 1 mM glutamate for 60 minutes. After 18 hours of glutamate-free recovery viability was about 70% in control cultures and about 40% in glutamate-treated cultures (Prehn et al., 1992).

The glutamate action on cultured neurons has been shown to have two components (Choi, 1987). Immediately after addition of glutamate, the neurons start to swell, depending on extracellular Na^+ and Cl^-. This is followed by a delayed Ca^{2+}-dependent degeneration of the neurons. Glutamate acts on several receptor subtypes typified after specific agonists. The AMPA-, kainate- and NMDA receptors are ionotropic. Under physiological conditions, activation of the AMPA/kainate receptors leads to increased Na^+ and K^+ conductance (Ascher and Nowak, 1987), whereas activation on NMDA receptors causes a massive Ca^{2+} influx through receptor-operated calcium channels (MacDermott et al., 1986). More recent investigations have revealed that under certain circumstances AMPA/kainate receptor-operated ion channels are also Ca^{2+} permeable (Miller et al., 1992). Apart from these ionotropic receptors glutamate acts on metabotropic receptors specifically activated by quisqualate. Via these receptors glutamate leads to intracellular release of inositol-1,4,5-trisphosphate (IP_3) and diacylglycerol with a subsequent release of Ca^{2+} ions from intracellular stores and activation of protein kinase C (Sladeczek et al., 1985). Thus, glutamate essentially causes an increase in the intracellular free Ca^{2+} level as has been repeatedly shown by microfluorimetry (Connor et al., 1988; Glaum et al., 1990). An increased intracellular free Ca^{2+} level has first been suggested as a cause of cell death by Schanne et al. (1979). Ca^{2+} ions may promote neuronal

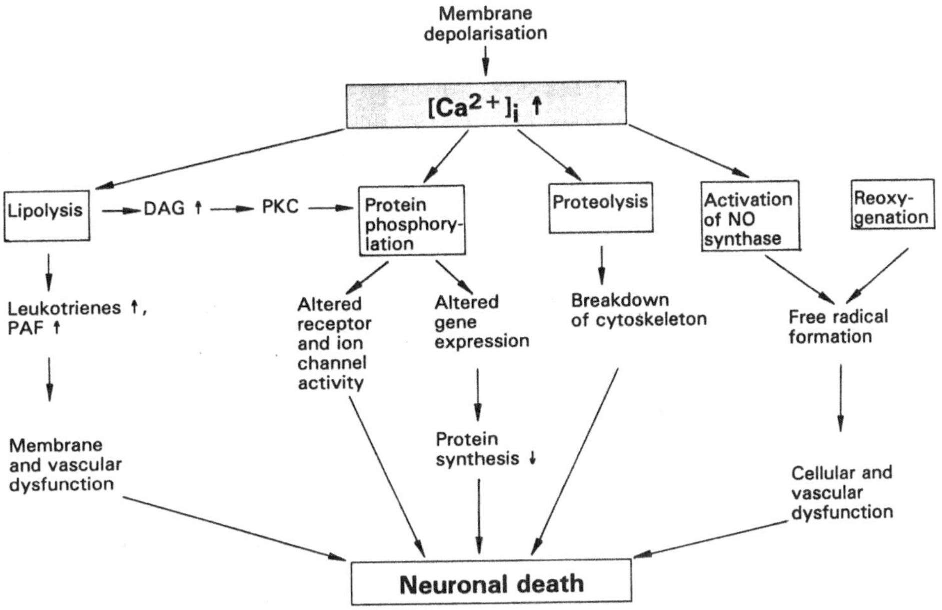

Fig. 1. Ca^{2+}-induced mechanisms leading to neuronal damage

degeneration by several mechanisms (Fig. 1), including activation of lipases, proteases, protein kinases, endonucleases and nitric oxide synthase. This leads to accumulation of free fatty acids and derived mediators as well as to impairment of membrane integrity. Protein phosphorylation and proteolysis cause changes in the activity of receptor and other proteins as well as a degradation of cytoskeletal proteins. Activation of endonucleases and subsequent fragmentation of DNA lead to changes in gene expression. Increased activity of xanthine oxidase and nitric oxide synthase contribute to the generation of free radicals and these subsequently damage various cell components.

Neuroprotection against an excitotoxic challenge — drug effects

These sequelae of glutamate receptor activation can be relieved by glutamate receptor blockade. For several years the NMDA receptor was considered as the major culprit for glutamate-induced neurodegeneration, since NMDA receptor antagonists reduced ischemia-induced neuronal damage in vivo (see above) as well as glutamate-induced neurodegeneration in vitro (Rothman, 1984; Rothman et al., 1987; Choi et al., 1988). This view was supported by the finding that glutamate toxicity was enhanced by addition of glycine (1 µM) or removal of Mg^{2+} ions (Finkbeiner and Stevens, 1988), i.e. conditions which increase NMDA receptor activity (Ascher and Nowak, 1987). With the development of specific non-NMDA antagonists (Honoré et al., 1988) it became possible to investigate the role of the AMPA/kainate receptors, and surprisingly even the postischemic administration of the AMPA antagonist NBQX in gerbils subjected to 5 minutes of global forebrain ischemia yielded a significant neuroprotective effect (Sheardown et al., 1990). Furthermore, the infarct induced by occlusion of the middle cerebral artery in mice was reduced by NBQX (Lippert et al., 1992). In vitro, NBQX dose-dependently reduced the proportion of damaged (trypan blue staining) hippocampal neurons as assessed 18 hours after a 30-min glutamate intoxication (1 mM), although it achieved only about 85% of the possible neuroprotective effect whereas the NMDA antagonist dizocilpine achieved complete neuroprotection (Lippert et al., 1994). One current explanation for the favorable effects of NBQX (or its hydrophilic congeners CNQX and DNQX in vitro) is that via blockade of the AMPA receptor most of the glutamate-induced Na^+ influx is blocked and thus the activation of voltage-operated calcium channels as well as NMDA receptor-operated calcium channels is prevented. On the other hand, Miller et al. (1992) have shown that at least in Purkinje cell cultures part of the cells inherently contain Ca^{2+}-permeable AMPA receptors which thus on their own contribute to the elevation of the intracellular free Ca^{2+} level on the addition of glutamate. It has been shown that after global forebrain ischemia in rats the glutamate receptor subunit gene expression is changed in a way that suggests increased formation of receptor proteins with high Ca^{2+} conductance (Pulsinelli et al., 1992). It is conceivable although not yet proved that such a mechanism is also operant in neuronal cultures subjected to toxic concentrations of glutamate.

In either way, NMDA and non-NMDA antagonists should act synergistically on glutamate-intoxicated neurons. In fact, Frandsen and Schousboe (1992) could show that combined treatment with CNQX and the NMDA antagonist APV completely prevented glutamate-induced increases in intracellular free Ca^{2+} levels in cultured cortical neurons. In our laboratory, dizocilpine and NBQX were tested on cultured rat hippocampal neurons subjected to a 30-min glutamate incubation (1 mM) followed by 18 hours of recovery. The combination of submaximally effective drug concentrations (10^{-8} M each) resulted in early complete neuroprotection, which was significantly more pronounced than the protection by either drug alone. In order to test whether this effect was simply due to addition of two independent neuroprotective mechanisms, we then evaluated concentration-response curves for dizocilpine and NBQX either alone or in the presence of 10^{-8} M of the respective counterpart (which roughly corresponded to the EC_{50} under our experimental conditions). The concentration-response curves thus constructed were compared to calculated concentration-response curves (Pöch and Holzmann, 1980), and in both cases there was a left-hand shift of the measured curve as compared to the theoretical curve (Fig. 2). Thus, the NMDA- and non-NMDA antagonist had an overadditive neuroprotective effect in glutamate-challenged neurons which suggests a functional interaction of the two receptor subtypes. The favorable effect of a combination of NMDA- and non-NMDA antagonist was confirmed in vivo in mice subjected to middle cerebral artery occlusion and in rats subjected to global forebrain

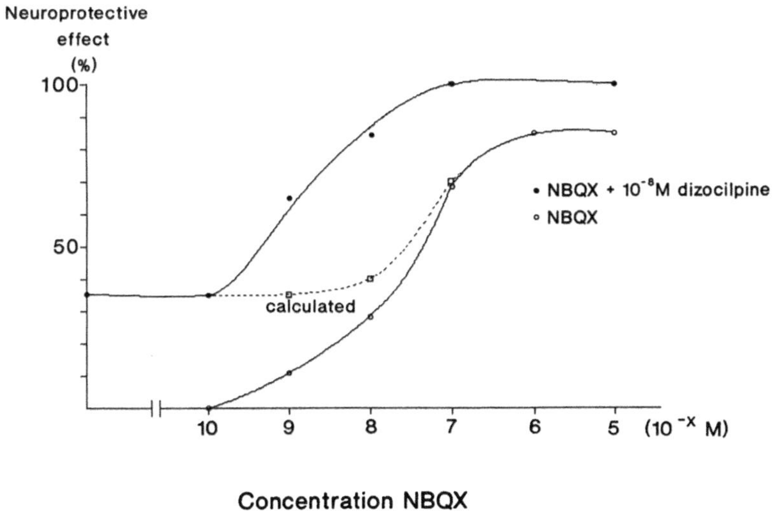

Fig. 2. Overadditive neuroprotective effect of dizocilpine and NBQX in vitro. Primary rat hippocampal cultures (14 days in vitro) were subjected to a 30-min incubation with 1 mM glutamate in the presence of various NBQX concentrations or in the presence of 10^{-8} M dizocilpine and NBQX. Neuronal damage was assessed 18 hours later by trypan blue staining. The neuroprotective effect is expressed as percentage of the difference in neuronal viability between control cultures and untreated glutamate-intoxicated cultures. The dashed line delineates the calculated concentration-response curve for an additive effect of the drug combination. Note the left-hand shift of the measured curve (●) as compared to the calculated curve (□)

ischemia (Lippert et al., 1992, 1994). Kaku et al. (1991) demonstrated an augmentation of the neuroprotective effect of dizocilpine by CNQX in cultured mouse neurons subjected to ischemia-like conditions (combined oxygen and glucose deprivation).

Glutamate via the increase in intracellular free calcium launches a number of potentially deleterious reactions as outlined above. Glutamate intoxication in primary neuronal cultures thus allows to evaluate the importance of these reactions for the final event, the neuronal death. Thus, for example, the extent of neuronal damage in chick neuronal cultures after a 60-min glutamate incubation and 18 hours of recovery was halved when the protein kinase C inhibitor polymyxine B (10^{-6} or 10^{-5} M) was present only during the recovery period (Prehn and Krieglstein, 1991). Ginkgolide A and B, two antagonists of the lipid-derived mediator platelet-activating factor, were similarly active in the same model in concentrations between 10^{-6} and 10^{-4} M (Prehn and Krieglstein, 1993). Only recently it was shown that the endonuclease inhibitor aurintricarboxylic acid (ATA, 10^{-5} M) added to rat hippocampal cultures at the end of a 5-min glutamate incubation (500 μM) completely prevented the glutamate-induced neuronal cell death as assessed 24 hours later by trypan blue staining; the LDH release was reduced by 60% (Samples and Dubinsky, 1993). The evidence concerning a conceivable neuroprotective effect of nitric oxide synthase inhibition is less conclusive in vitro as well as in vivo (Beckman et al., 1992 and references listed therein).

Free radicals as targets for neuroprotection

Accumulation of free radicals is considered as a major cause of postischemic neuronal damage (Siesjö et al., 1989; Peruche and Krieglstein, 1993). Excessive glutamate receptor activation contributes to this accumulation via activation of enzymes involved in free radical generation like phospholipase A (giving rise to release of polyunsaturated fatty acids which may produce free radicals), xanthineoxidase (producing superoxide anion radicals), and nitric oxide synthase (contributing to the generation of hydroxyl radicals) (Choi, 1988; Beckman et al., 1992). Free radicals attack cell components like membrane lipids, proteins, and nucleic acids and thus take part in the neuronal damage. The neuroprotective potency of radical scavengers has been tested repeatedly in vivo (for references see Siesjö et al., 1989; Peruche and Krieglstein, 1993) and protective effects could mainly be shown in focal ischemia. In primary cultures of mouse cortical neurons the glutamate-induced neuronal damage was significantly attenuated by lazaroids, i.e. inhibitors of lipid peroxidation (Monyer et al., 1990).

Table 1. Dihydrolipoate reduces the hypoxic or excitotoxic neuronal damage in cultured chick neurons

[Cyanide] or [Glutamate] (mM)	[DHL] (M)	Cytotoxic hypoxia Protein content (mg/flask)	Glutamate toxicity Damaged neurons (%)
0	0	0.276 ± 0.016	31.4 ± 2.4
0	10^{-9}	0.269 ± 0.026	n.d.
0	10^{-6}	0.183 ± 0.026*	n.d.
0	10^{-4}	0.071 ± 0.032**	54.1 ± 1.4***
1	0	0.173 ± 0.032	59.0 ± 4.2
1	10^{-9}	0.258 ± 0.019**	51.6 ± 1.5
1	10^{-8}	0.250 ± 0.023**	46.6 ± 5.6
1	10^{-7}	0.252 ± 0.025**	41.5 ± 1.0***
1	10^{-6}	0.187 ± 0.037	42.1 ± 2.5***
1	10^{-5}	0.105 ± 0.022**	43.7 ± 2.9**
1	10^{-4}	0.032 ± 0.010**	56.8 ± 2.2

Primary chick neuronal cultures were subjected to either cytotoxic hypoxia (1 mM sodium cyanide for 30 minutes after four days in vitro) or to glutamate toxicity (1 mM sodium glutamate for 60 minutes after six days in vitro). Protein content or cell viability were determined after seven days in vitro. Dihydrolipoate was present in the culture medium from 30 minutes before up to 24 hours (cyanide) or 18 hours (glutamate) after the toxic challenge. Values are given as means ± SD from five (cytotoxic hypoxia) or three to five (glutamate toxicity) experiments. Different from corresponding controls: * $p < 0.05$, ** $p < 0.01$, *** $p < 0.001$, evaluated by one-way-analysis of variance and subsequent Duncan's test (cytotoxic hypoxia) or Scheffé test (glutamate toxicity). *n.d.* not determined

Another agent, dihydrolipoic acid, has been tested for its neuroprotective potency in vitro and in vivo in our laboratory (Prehn et al., 1992). Dihydrolipoate is an endogenous dithiol that acts as a cofactor in several mitochondrial dehydrogenase complexes. It had been shown to have antioxidant activity in microsomal peroxidation (Bast and Haenen, 1988) and to penetrate lipid bilayers (Pagani et al., 1989). Prehn et al. (1992) tested dihydrolipoate and its oxidized form, lipoate, over a dose-range of 10^{-9} to 10^{-4} M in primary chick neuronal cultures subjected to either cytotoxic hypoxia (30-min cyanide incubation followed by three days of recovery) or to 1 mM glutamate for 60 minutes followed by 24 hours of recovery. Dihydrolipoate (10^{-9} to 10^{-7} M) markedly reduced the cyanide-induced loss in protein content when present in the cultures from 30 minutes before until 24 hours after the cyanide incubation (Table 1). Higher concentrations (10^{-6} to 10^{-4} M) proved toxic in normoxic control cultures and thus did not reveal a neuroprotective effect in cyanide-treated cultures. A neuroprotective effect equal to 10^{-9} to 10^{-7} M dihydrolipoate was achieved by the antioxidant dimethylthiourea (10^{-6} M), whereas lipoate remained without effect. The experiments on glutamate-induced neurotoxicity revealed similar results, i.e. dihydrolipoate (although only at 10^{-7} to 10^{-5} M) reduced the number of glutamate-damaged neurons while lipoate was ineffective. Rat hippocampal and cortical cultures subjected to glutamate neurotoxicity also showed a reduced neuronal damage under dihydrolipoate but not lipoate (Müller et

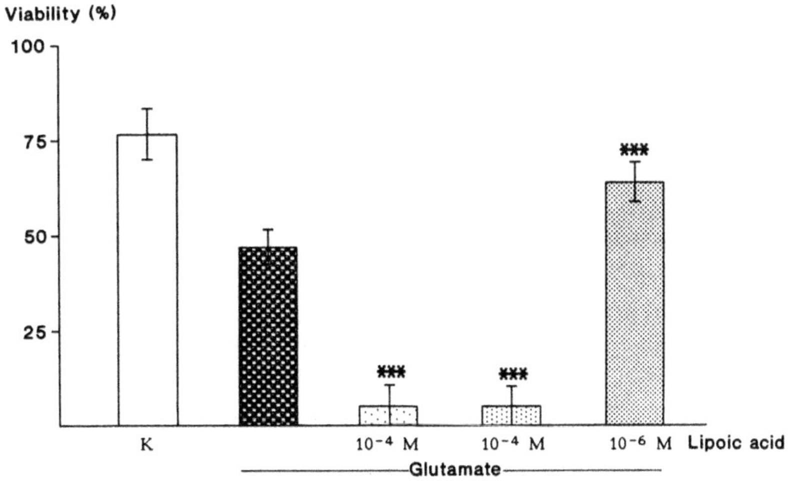

Fig. 3. Chronic pretreatment with lipoic acid reduces glutamate toxicity in hippocampal cultures. Primary rat hippocampal cultures (14 days in vitro) were subjected to a 30-min incubation with 1 mM glutamate. Neuronal viability was assessed by trypan blue staining 18 hours thereafter. Lipoic acid was added to the cultures two hours after plating and again with every medium replacement until the evaluation of neuronal viability. Different from glutamate-treated control: ***$p < 0.001$ (one-way analysis of variance and subsequent Scheffé test)

al., 1993). The neuroprotective effect of dihydrolipoate was confirmed in vivo by a reduction of the infarct size in rats and mice subjected to middle cerebral artery occlusion (Prehn et al., 1992).

In biological systems, dihydrolipoate and lipoate are coupled to each other as an antioxidant system like reduced and oxidized glutathione. Thus, in principle lipoate should be able to contribute to a neuroprotective effect, however, this would require reduction to dihydrolipoate. In the in vitro experiments described above, the drugs had been added to the cultures only 30 minutes before the cyanide or glutamate challenge. This might have been too short a preincubation time to allow for considerable chemical reduction of lipoate. Therefore, chronic pretreatment with lipoate was tried in cultured hippocampal neurons. Lipoate (10^{-6} M) was added to the culture medium two hours after plating and with every medium replacement thereafter. Eighteen hours after a 30-min glutamate incubation in 14-day-old cultures neuronal viability thus was grossly preserved (Fig. 3). A higher concentration of lipoate (10^{-4} M) proved neurotoxic, an effect only seen with dihydrolipoate after acute treatment. These results support the hypothesis that lipoate might exert neuroprotective effects after chronic treatment owing to its endogenous transformation to dihydrolipoate.

Neuroprotection with neural growth factors

A completely different approach to the protection of neurons against ischemia-related damage has evolved recently. Neural growth factors have multiple functions in the brain in the development and maintenance of neurons (Unsicker et al., 1992) and there is growing evidence that the regeneration of lesioned neurons is also influenced by various growth factors. Thus, glutamate-induced neurodegeneration was prevented by basic fibroblast growth factor in cultured rat hippocampal neurons (Mattson et al., 1989) and hypoglycemia-provoked neuronal death was prevented by basic fibroblast factor and nerve growth factor (Cheng and Mattson, 1991) as well as by insulin-like growth factors I and II (Cheng and Mattson, 1992).

The neuroprotective effect of transforming growth factors-β (TGFs-β) has been investigated in detail in our laboratory (Prehn et al., 1993a,b). The TGFs-β are pleiotropic peptide growth factors which act on nearly all cell types (for references see Roberts and Sporn, 1990). Three TGF-β isoforms have been found in the brain of rats and mice (Flanders et al., 1991; Unsicker et al., 1991), but whereas TGF-β_2 and TGF-β_3 exist in a wide range of neuronal and glial populations, TGF-β_1 is confined to meningeal cells and the choroid plexus in the normal, nonlesioned brain. However, a pronounced increase in TGF-β_1 mRNA in rat hippocampus was observed, for example, after an entorhinal cortex lesion (Nichols et al., 1991) as well as after hypoxic ischemia in the infant rat brain (Klempt et al., 1992) and TGF-β_1 thus appears to be an injury-related growth factor. Prehn et al. (1993a,b) evaluated the neuroprotective effects of TGFs-β on chick neuronal cultures and rat neocortical cultures subjected to cytotoxic hypoxia or gluta-

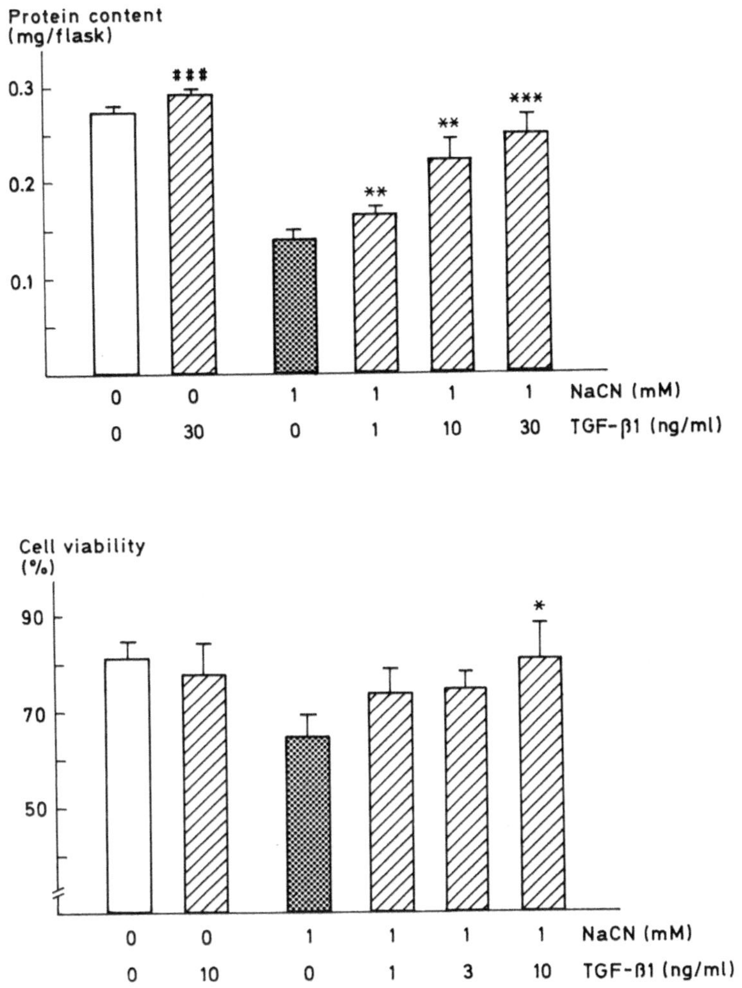

Fig. 4. TGF-β_1 reduces cytotoxic hypoxia-induced neuronal damage in chick neuronal cultures. Primary chick neuronal cultures (four days in vitro) were subjected to cytotoxic hypoxia for 30 minutes by addition of 1 mM sodium cyanide. Either protein content or neuronal viability (trypan blue) were assessed three days later. TGF-β_1 was present in the culture medium from 30 minutes before up to 24 hours after the cyanide incubation. Values are shown as means ± SD from three to five experiments. Different from cyanide-free controls: ###$p < 0.001$ (Student's t test). Different from cyanide-exposed controls: * $p < 0.05$, **$p < 0.01$, ***$p < 0.001$ (one-way analysis of variance and subsequent Duncan's test)

mate neurotoxicity. In primary chick neuronal cultures subjected to cytotoxic hypoxia (30 minutes) followed by three days of recovery, TGF-β_1 in the range of 1 to 30 ng/ml dose-dependently elevated the protein content (Fig. 4). Only the highest dose exerted a certain proliferative effect in slightly elevating the protein content of cyanide-free control cultures. Ten ng/ml TGF-β_1 (not 1 and 3 ng/ml) also normalized the cyanide-dependent reduction of cell viability. TGF-β_2 was nearly ineffective. TGF-β_3 slightly increased the protein content at 0.3 and 1 ng/ml and cell viability at 1 ng/ml in

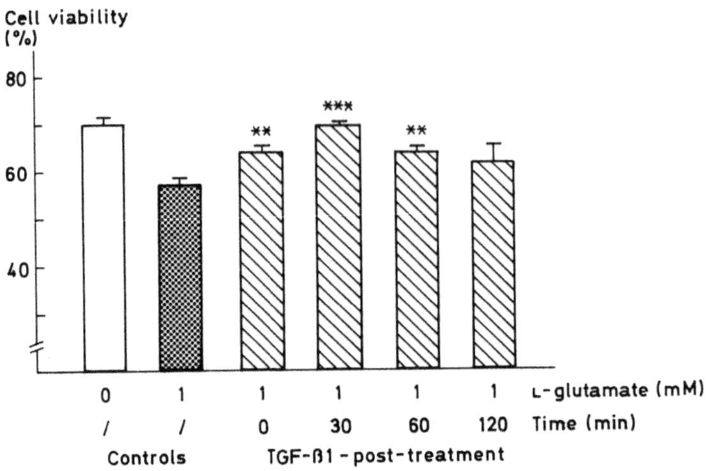

Fig. 5. Delayed administration of TGF-β_1 reduces excitotoxicity in chick neuronal cultures. Primary chick neuronal cultures (five days in vitro) were subjected to a 60-min incubation with 1 mM glutamate in Mg^{2+}-free control saline. Eighteen hours later neuronal viability was determined by trypan blue exclusion. TGF-β_1 (10 ng/ml) was added to the culture medium at various time points after termination of the glutamate exposure. Values are shown as means ± SD from four to five cultures. Different from glutamate-exposed controls: **$p < 0.01$, ***$p < 0.001$ (one-way analysis of variance and subsequent Duncan's test)

cyanide-intoxicated cultures; however, at 10 ng/ml TGF-β_3 proved toxic by a reduction of both protein content and viability in normoxic control cultures. A reduction in viability by TGF-β_3 (1 ng/ml) was also seen in cultured rat neocortical neurons subjected to an 18-hour glutamate incubation in serum-free medium. Only lower doses of TGF-β_3 (0.1 and 0.3 ng/ml) were neuroprotective, whereas TGF-β_1 increased neuronal viability dose-dependently up to 10 ng/ml. The effect of TGF-β_1 was also assessed in chick neuronal cultures subjected to 60 minutes of glutamate incubation (1 mM) at five days in culture followed by 18 hours of recovery. In cultures pretreated with TGF-β_1 since day 1 in culture and treated until the end of the glutamate incubation, TGF-β_1 elevated the cell viability up to control levels at 10 ng/ml. A neuroprotective effect of nearly the same degree could be achieved by adding TGF-β_1 to the culture medium at the end of the glutamate incubation or even up to 60 minutes thereafter, i.e. only during the recovery period (Fig. 5). This suggests that TGF-β_1 can exert a neuroprotective effect by influencing reactions that might be set in motion by the hypoxic or glutamate challenge but grossly proceed during the posthypoxic or post-glutamate period and finally cause neuronal death. The neuroprotective effect of TGF-β_1 may be caused by either direct actions on the neurons or/and by interaction with other neural growth factors. Thus, TGF-β_1 has been shown to stimulate the expression of NGF (as well as its own expression) in cultured rat astrocytes (Lindholm et al., 1990) and to increase the expression of FGF receptor protein in cultured rat auditory neurons

(Lefebvre et al., 1991). TGF-β_1 thus stimulates at least two other neural growth factor systems which are potentially neuroprotective (see above). Whatever the underlying mechanism may be, the neuroprotective effect of TGF-β_1 stated in vitro could be confirmed in vivo (Prehn et al., 1993a). Mice subjected to middle cerebral artery occlusion had smaller infarcts when treated preischemically with TGF-β_1 (1 ng/kg i.c.v.), although this effect was not as pronounced as in vitro, probably owing to pharmacokinetic limitations. Taken together, the results suggest a significant potential of TGF-β_1 to protect neurons against ischemia and other excitotoxicity-related disorders.

Concluding remarks

In conclusion, it shall be stated that primary neuronal cultures are valuable tools for the evaluation of neuroprotective drug effects. As compared to the animal experiment and even the brain slice, they allow quite clear-cut experimental conditions and, with an adequate experimental design, the differentiation between neuronal and glial reactions. To this end, numerous primary neuronal culture systems have been developed. Neuronal damage as a prerequisite for the evaluation of neuroprotective effects has been induced by various manipulations such as oxygen deprivation, cytotoxic hypoxia, combined glucose/oxygen deprivation and, not at least, glutamate intoxication. Pharmacological investigations on these models have contributed considerably to the understanding of the mechanisms leading to neurodegeneration. In consequence, neuroprotection in vitro could be achieved by agents as different as glutamate antagonists, adenosine and serotonin$_{1A}$ agonists (not discussed here), free radical scavengers, antagonists of protein kinase C and endonucleases as well as antagonists of platelet-activating factor. The administration of transforming growth factor-β_1 is a more recent concept of neuroprotection.

References

Ahlemeyer B, Krieglstein J (1989) Testing drug effects against hypoxic damage of cultured neurons during long-term recovery. Life Sci 45: 835–842

Ascher P, Nowak L (1987) Electrophysiological studies of NMDA receptors. Trends Neurosci 10: 284–288

Balázs R, Jorgensen OS, Hack N (1988) N-Methyl-D-aspartate promotes the survival of cerebellar granule cells in culture. Neuroscience 27: 437–451

Banker GA (1980) Trophic interactions between astroglial cells and hippocampal neurons in culture. Science 209: 809–810

Banker GA, Cowan WM (1977) Rat hippocampal neurons in dispersed cell culture. Brain Res 126: 397–425

Bast A, Haenen GRMM (1988) Interplay between lipoic acid and glutathione in the protection against microsomal lipid peroxidation. Biochim Biophys Acta 963: 558–561

Beckman JS, Chen J, Ischiropoulos H, Conger KA (1992) Inhibition of nitric oxide synthesis and cerebral protection. In: Krieglstein J, Oberpichler-Schwenk H (eds) Pharmacology of cerebral ischemia 1992. Wissenschaftliche Verlagsgesellschaft, Stuttgart, pp 383–394

Benveniste H, Drejer J, Schousboe A, Diemer NH (1984) Elevation of the extracellular concentrations of glutamate and aspartate in rat hippocampus during transient cerebral ischemia monitored by intracerebral microdialysis. J Neurochem 43: 1369–1374

Cheng B, Mattson MP (1991) NGF and bFGF protect rat hippocampal and human cortical neurons against hypoglycemic damage by stabilizing calcium homeostasis. Neuron 7: 1031–1041

Cheng B, Mattson MP (1992) IGF-I and IGF-II protect cultured hippocampal and septal neurons against calcium-mediated hypoglycemic damage. J Neurosci 12: 1558–1566

Choi DW (1987) Ionic dependence of glutamate neurotoxicity in cortical cell culture. J Neurosci 7: 369–379

Choi DW (1988) Glutamate neurotoxicity and diseases of the nervous system. Neuron 1: 623–634

Choi DW, Maulucci-Gedde MA, Kriegstein AR (1987) Glutamate neurotoxicity in cortical cell culture. J Neurosci 7: 357–368

Choi DW, Koh JY, Peters S (1988) Pharmacology of glutamate neurotoxicity in cortical cell culture: attenuation by NMDA antagonists. J Neurosci 8: 185–196

Connor JA, Wadman WJ, Hockberger PE, Wong RKS (1988) Sustained dendritic gradients of Ca^{2+} induced by excitatory amino acids in CA1 hippocampal neurons. Science 240: 649–653

Dichter MA (1978) Rat cortical neurons in cell culture: culture methods, cell morphology, electrophysiology, and synapse formation. Brain Res 149: 279–293

Finkbeiner C, Stevens CF (1988) Applications of quantitative measurements for assessing glutamate neurotoxicity. Proc Natl Acad Sci USA 85: 4071–4074

Flanders KC, Lüdecke G, Engels S, Cissel DS, Roberts AB, Kondaiah P, Lafyatis R, Sporn MB, Unsicker K (1991) Localization and actions of transforming growth factor-βs in the embryonic nervous system. Development 113: 183–191

Frandsen A, Schousboe A (1992) Mobilization of dantrolene-sensitive intracellular calcium pool is involved in the cytotoxicity induced by quisqualate and N-methyl-D-aspartate but not by (RS)-2-amino-3-(3-hydroxy-5-methylisoxazol-4-yl) propionate and kainate in cultured cerebral cortical neurons. Proc Natl Acad Sci USA 89: 2590–2594

Furuya S, Ohmori H, Shigemoto T, Sugiyama H (1989) Intracellular calcium mobilization triggered by a glutamate receptor in rat cultured hippocampal cells. J Physiol 414: 539–548

Gähwiler BH (1981) Organotypic monolayer cultures of nervous tissue. J Neurosci Meth 4: 329–342

Gill R, Foster AC, Woodruff GN (1987) Systemic administration of MK-801 protects against ischemia-induced hippocampal neurodegeneration in the gerbil. J Neurosci 7: 3343–3349

Glaum SR, Scholz WK, Miller RJ (1990) Acute and long-term glutamate-mediated regulation of $[Ca^{2+}]_i$ in rat hippocampal pyramidal neurons in vitro. J Pharmacol Exp Ther 253: 1293–1302

Goldberg WJ, Kadingo RM, Barrett JN (1986) Effects of ischemia-like conditions on cultured neurons: protection by low Na^+, low Ca^{2+} solutions. J Neurosci 6: 3144–3151

Goldberg MP, Weiss JH, Pham PC, Choi DW (1987) N-Methyl-D-aspartate receptors mediate hypoxic neuronal injury in cortical culture. J Pharmacol Exp Ther 243: 784–791

Goldberg MP, Viseskul V, Choi DW (1988) Phencyclidine receptor ligands attenuate cortical injury after N-methyl-D-aspartate exposure or hypoxia. J Pharmacol Exp Ther 245: 1081–1087

Harris KM, Rosenberg PA (1993) Localization of synapses in rat cortical cultures. Neuroscience 53: 495–508

Honoré T, Davies SN, Drejer J, Fletcher EJ, Jacobsen P, Lodge D, Nielsen FE (1988) Quinoxalinediones: potent competitive non-NMDA glutamate receptor antagonists. Science 241: 701–703

Huettner JE, Baughman RW (1986) Primary culture of identified neurons from the visual cortex of postnatal rats. J Neurosci 6: 3044–3060

Ichikawa M, Muramoto K, Kobayashi K, Kawahara M, Kuroda Y (1993) Formation and maturation of synapses of rat cerebral cortical cells: an electron microscopic study. Neurosci Res 16: 95–103

Jahr CE, Stevens CF (1987) Glutamate activates multiple single channel conductances in hippocampal neurons. Nature 325: 522–525

Juurlink BHJ, Hertz L (1993) Ischemia-induced death of astrocytes and neurons in primary culture: pitfalls in quantifying neuronal cell death. Dev Brain Res 71: 239–246

Kaku DA, Goldberg MP, Choi DW (1991) Antagonism of non-NMDA receptors augments the neuroprotective effect of NMDA receptor blockade in cortical cultures subjected to prolonged deprivation of oxygen and glucose. Brain Res 554: 344–347

Karkoutly C, Backhauß C, Nuglisch J, Krieglstein J (1990) The measurement of the infarcted area after middle cerebral artery occlusion in the mouse: a screening model. In: Krieglstein J, Oberpichler H (eds) Pharmacology of cerebral ischemia 1990. Wissenschaftliche Verlagsgesellschaft, Stuttgart, pp 63–69

Klempt ND, Sirimanne E, Gunn AJ, Klempt M, Singh K, Williams C, Gluckmann PD (1992) Hypoxia-ischemia induces transforming growth factor β_1 mRNA in the infant rat brain. Mol Brain Res 13: 93–101

Koh JY, Choi DW (1987) Quantitative determination of glutamate mediated cortical neuronal injury in cell culture by lactate dehydrogenase efflux assay. J Neurosci Meth 20: 83–90

Krieglstein J, Brungs H, Peruche B (1988) Cultured neurons for testing cerebroprotective drug effects in vitro. J Pharmcol Meth 20: 39–46

Krieglstein J, Sauer D, Nuglisch J Roßberg C, Beck T, Bielenberg GW, Mennel HD (1989) Naftidrofuryl protects neurons against ischemic damage. Eur Neurol 29: 224–228

Lefebvre PP, Staecker H, Weber T, Vandewater TR, Rogister B, Moonen G (1991) TGF-β_1 modulates bFGF receptor message expression in cultured adult auditory neurons. NeuroReport 2: 305–308

Lindholm D, Hengerer B, Zafra F, Thoenen H (1990) Transforming growth factor-β_1 stimulates expression of nerve growth factor in the rat CNS. NeuroReport 1: 9–12

Lippert K, Welsch M, Krieglstein J (1992) The neuroprotective effect of combined treatment with dizocilpine and NBQX in vitro and in vivo. In: Krieglstein J, Oberpichler-Schwenk H (eds) Pharmacology of cerebral ischemia 1992. Wissenschaftliche Verlagsgesellschaft, Stuttgart, pp 147–153

Lippert K, Welsch M, Krieglstein J (1994) Over-additive protective effect of dizocilpine and NBQX against neuronal damage. Eur J Pharmacol (in press)

Louis JC, Pettmann B, Courageot J, Rumigny JF, Mandel P, Sensenbrenner M (1981) Developmental changes in cultured neurons from chick embryo cerebral hemispheres. Exp Brain Res 42: 63–72

Lysko PG, Cox JA, Vigano MA, Henneberry RC (1989) Excitatory amino acid neurotoxicity at the N-methyl-D-aspartate receptor in cultured neurons: pharmacological characterization. Brain Res 499: 258–266

MacDermott AB, Mayer ML, Westbrook GL, Smith SJ, Barker JL (1986) NMDA receptor activation increases cytoplasmic calcium concentration in cultured spinal cord neurones. Nature 321: 519–522

Mattson MP, Kater SB (1989) Development and selective neurodegeneration in cell cultures from different hippocampal regions. Brain Res 490: 110–125

Mattson MP, Lee RE, Adams ME, Guthrie PB, Kater SB (1988) Interactions between entorhinal axons and target hippocampal neurons: a role for glutamate in the development of hippocampal circuitry. Neuron 1: 865–876

Mattson MP, Murrain M, Guthrie PB, Kater SB (1989) Fibroblast growth factor and glutamate: opposing roles in the generation and degeneration of hippocampal neuroarchitecture. J Neurosci 9: 3728–3740

Messer A (1977) The maintenance and identification of mouse cerebellar granule cells in monolayer culture. Brain Res 130: 1–12

Miller RJ, Brorson JR, Bleakman D, Chard PS (1992) Glutamate receptors in the regulation of neuronal Ca^{2+}. In: Krieglstein J, Oberpichler-Schwenk H (eds) Pharmacology of cerebral ischemia 1992. Wissenschaftliche Verlagsgesellschaft, Stuttgart, pp 3–11

Monyer H, Hartley DM, Choi DW (1990) 21-Aminosteroids attenuate excitotoxic neuronal injury in cortical cell cultures. Neuron 5: 121–126

Müller HW, Seifert W (1982) A neurotrophic factor (NTF) released from primary glial cultures supports survival and fiber outgrowth of cultured hippocampal neurons. J Neurosci Res 8: 195–204

Müller U, Lippert K, Prehn JHM, Krieglstein J (1993) Chronic treatment with thioctic acid reduces excitotoxic neuronal damage in vitro. J Cereb Blood Flow Metab 13 [Suppl] I: S673

Nakajima Y, Nakajima S, Leonard RJ, Yamaguchi K (1986) Acetylcholine raises excitability by inhibiting the fast transient potassium current in cultured hippocampal neurons. Proc Natl Acad Sci USA 83: 3022–3026

Nichols NR, Laping NJ, Day JR, Finch CE (1991) Increases in transforming growth factor-β mRNA in hippocampus during responses to entorhinal cortex lesions in intact and adrenalectomized rats. J Neurosci Res 28: 134–139

Novelli A, Reilly JA, Lysko PG, Henneberry RC (1988) Glutamate becomes neurotoxic via the N-methyl-D-aspartate receptor when intracellular energy levels are reduced. Brain Res 451: 205–212

Oberpichler H, Brungs H, Krieglstein J (1990) Effects of delayed administration of methohexital and ketamine on post-hypoxic cell damage of primary neuronal cultures. Pharmacology 40: 165–173

Olney J (1978) Neurotoxicity of excitatory amino acids. In: McGeer EG, Olney JW, McGeer PL (eds) Kainic acid as a tool in neurobiology. Raven Press, New York, pp 37–70

Ozyurt E, Graham D, Woodruff GN, McCulloch J (1988) Protective effect of the glutamate antagonist, MK-801, in focal cerebral ischemia in the cat. J Cereb Blood Flow Metab 8: 138–143

Pagani S, Iametti S, Cervato G, Cestaro B (1989) Transmembrane mobility of the physiological dithiol D,L-dihydrolipoate. Biochem Int 18: 923–932

Pauwels PJ, Opperdoes FR, Trouet A (1985) Effects of antimycin, glucose deprivation, and serum on cultures of neurons, astrocytes, and neuroblastoma cells. J Neurochem 44: 143–148

Peruche B, Krieglstein J (1993) Mechanisms of drug actions against neuronal damage caused by ischemia — an overview. Prog Neuropsychopharmacol Biol Psychiatry 17: 21–70

Peruche B, Ahlemeyer B, Brungs H, Krieglstein J (1990) Cultured neurons for testing antihypoxic drug effects. J Pharmacol Meth 23: 63–77

Pettmann B, Louis JC, Sensenbrenner M (1979) Morphological and biochemical maturation of neurons cultured in the absence of glial cells. Nature 281: 378–380

Pöch G, Holzmann S (1980) Quantitative estimation of overadditive and underadditive drug effects by means of theoretical, additive dose response curves. J Pharmacol Meth 4: 179–188
Prehn JHM, Krieglstein J (1991) Primary neuronal cultures from chick embryo cerebral hemispheres: a model for studying trophic and toxic effects of excitatory amino acids. J Cereb Blood Flow Metab 11 [Suppl] II: S317
Prehn JHM, Krieglstein J (1993) Platelet-activating factor antagonists reduce excitotoxic damage in cultured neurons from embryonic chick telencephalon and protect the rat hippocampus and neocortex from ischemic injury in vivo. J Neurosci Res 34: 179–188
Prehn JHM, Karkoutly C, Nuglisch J, Peruche B, Krieglstein J (1992) Dihydrolipoate reduces neuronal injury after cerebral ischemia. J Cereb Blood Flow Metab 12: 78–87
Prehn JHM, Backhauß C, Krieglstein J (1993a) Transforming growth factor-β_1 prevents glutamate neurotoxicity in rat neocortical cultures and protects mouse neocortex from ischemic injury in vivo. J Cereb Blood Flow Metab 13: 521–525
Prehn JHM, Peruche B, Unsicker K, Krieglstein J (1993b) Isoform-specific effects of transforming growth factors-β on degeneration of primary neuronal cultures induced by cytotoxic hypoxia or glutamate. J Neurochem 60: 1665–1672
Pulsinelli W, Pellegrini-Giampietro D, Zukin S, Bennett M, Cho S (1992) AMPA receptors: their in vivo role in selective ischemic necrosis of CA1 hippocampal neurons. In: Krieglstein J, Oberpichler-Schwenk H (eds) Pharmacology of cerebral ischemia 1992. Wissenschaftliche Verlagsgesellschaft, Stuttgart, pp 53–58
Roberts AB, Sporn MB (1990) The transforming growth factors-β. In: Sporn MB, Roberts AB (eds) Peptide growth factors and their receptors. Springer, Berlin Heidelberg New York Tokyo, pp 419–472 (Handbook of experimental pharmacology, vol 95)
Rothman SM (1983) Synaptic activity mediates death of hypoxic neurons. Science 220: 536–537
Rothman SM (1984) Synaptic release of excitatory amino acid neurotransmitter mediates anoxic neuronal death. J Neurosci 4: 1884–1891
Rothman SM, Olney JW (1986) Glutamate and the pathophysiology of hypoxic-ischemic brain damage. Ann Neurol 19: 105–111
Rothman SM, Samaie M (1985) Physiology of excitatory synaptic transmission in cultures of dissociated rat hippocampus. J Neurophysiol 54: 701–713
Rothman SM, Thurston JH, Hauhart RE, Clark GD, Solomon JS (1987) Ketamine protects hippocampal neurons from anoxia in vitro. Neuroscience 21: 673–678
Samples SD, Dubinsky JM (1993) Aurintricarboxylic acid protects hippocampal neurons from glutamate excitotoxicity in vitro. J Neurochem 61: 382–385
Sargent PB (1989) What distinguishes axons from dendrites? Neurons know more than we do. Trends Neurosci 12: 203–205
Schanne FAX, Kane AB, Young EE, Farber JL (1979) Calcium dependence of toxic cell death: a final common pathway. Science 206: 700–702
Scott BS (1982) Adult neurons in cell culture: electrophysiological characterization and use in neurobiological research. Progr Neurobiol 19: 187–211
Seif el Nasr M, Peruche B, Roßberg C, Mennel HD, Krieglstein J (1990) Neuroprotective effect of memantine demonstrated in vivo and in vitro. Eur J Pharmacol 185: 19–24
Sheardown MJ, Hansen AJ, Eskesen K, Suzdak P, Diemer NH, Honoré T (1990) Blockade of AMPA receptors in the CA1 region of the hippocampus prevents ischaemia induced cell death. In: Krieglstein J, Oberpichler H (eds) Pharmacology of cerebral ischemia 1990. Wissenschaftliche Verlagsgesellschaft, Stuttgart, pp 245–253
Siesjö BK, Agardh CD, Bengtsson F (1989) Free radicals and brain damage. Cerebrovasc Brain Metab Rev 1: 165–211

Simon RP, Swan JH, Griffiths T, Meldrum BS (1984) Blockade of N-methyl-D-aspartate receptors may protect against ischemic damage in the brain. Science 226: 850–852

Sladeczek F, Pin JP, Recasens M, Bockaert J, Weiss S (1985) Glutamate stimulates inositol phosphate formation in striatal neurones. Nature 317: 717–719

Söderbäck M, Hansson E, Tottmar O, Rönnbäck L (1989) Neurons in primary cultures from five defined rat brain regions: cellular composition and morphological appearance. Cell Mol Biol 35: 1–16

Trussell LO, Thio LL, Zorumski CF, Fischbach GD (1986) Rapid desensitization of glutamate receptors in vertebrate cultured neurons. Proc Natl Acad Sci USA 85: 4562–4566

Unsicker K, Flanders KC, Cissel DS, Lafyatis R, Sporn MB (1991) Transforming growth factor beta isoforms in the adult rat central and peripheral nervous system. Neuroscience 44: 613–625

Unsicker K, Prehn JHM, Backhauß C, Krieglstein K, Otto D, Krieglstein J (1992) Growth factors and their implications for the injured brain: regulation and neuroprotective effects of FGFs and TGF-ßs in ischemic and other lesion paradigms. In: Krieglstein J, Oberpichler-Schwenk H (eds) Pharmacology of cerebral ischemia 1992. Wissenschaftliche Verlagsgesellschaft, Stuttgart, pp 229–238

Weiss S, Pin JP, Sebben M, Kemp DE, Sladeczek F, Gabrion J, Bockaert J (1986) Synaptogenesis of cultured striatal neurons in serum-free medium: a morphological and biochemical study. Proc Natl Acad Sci USA 83: 2238–2242

Yamada KA, Dubinsky JM, Rothman SM (1989) Quantitative physiological characterization of a quinoxalinedione non-NMDA receptor antagonist. J Neurosci 9: 3230–3236

Yamashita N, Nishiyama N, Abe K, Saito H, Fukuda J (1992) Primary culture of postnatal rat hypothalamic neurons in astrocyte-conditioned medium. Brain Res 594: 215–220

Zorumski CF, Thio LL, Clark GD, Clifford DB (1989) Calcium influx through N-methyl-D-aspartate channels activates a potassium current in postnatal rat hippocampal neurons. Neurosci Lett 99: 293–299

Authors' address: Prof. Dr. Dr. J. Krieglstein, Institut für Pharmakologie und Toxikologie, Philipps-Universität, Ketzerbach 63, D-35032 Marburg, Federal Republic of Germany.

J Neural Transm (1994) [Suppl] 44: 21–27

Regulation of proteolytic processing of the amyloid β-protein precursor of Alzheimer's disease in transfected cell lines and in brain slices

R. M. Nitsch[1,2], B. E. Slack[1], S. A. Farber[1], J. G. Schulz[1], M. Deng[1], C. Kim[1], P. R. Borghesani[1], W. Korver[1], R. J. Wurtman[1], and J. H. Growdon[2]

[1] Department of Brain and Cognitive Sciences, Massachusetts Institute of Technology, Cambridge, MA, and

[2] Department of Neurology, Massachusetts General Hospital and Harvard Medical School, Boston, MA, U.S.A.

Summary. βA4 is the principal component of Alzheimer's disease brain amyloid. It is derived from proteolytic processing of amyloid β-protein precursors (APP), a family of transmembrane glycoproteins. Secretion of APP^s, a secreted proteolytic derivative that is cleaved within the βA4 domain of APP, is increased many-fold by the activation of cell-surface receptors, like the muscarinic m1 and m3 receptor subtypes, which are coupled to protein kinase C. Concomitantly, their activation decreases the formation of both secreted soluble βA4 and of endosomal-lysosomal C-terminal APP derivatives. These data suggest that muscarinic m1 and m3 receptors accelerate non-amyloidogenic APP processing and depress the formation of potentially amyloidogenic derivatives. Other receptors that stimulate APP^s secretion include those for bradykinin, vasopressin, and interleukin-1 receptors. A similar control mechanism is present in rat brain tissue slices, in which the release of both APP^s and endogenous neurotransmitters is increased by electrical depolarization. This increase is tetrodotoxin-sensitive and frequency-dependent, suggesting that APP^s release may normally depend on neuronal activity. Taken together, our findings suggest that specific receptor agonists might be effective in reducing the formation of potentially amyloidogenic APP derivatives in vivo.

Introduction

Massive deposits of amyloid in brain tissue are invariant features of Alzheimer's disease and Down syndrome. In lesser amounts, amyloid is also found in brains of neurologically normal, aged human subjects. Amyloid consists of aggregated βA4 peptides, which are 39–43 amino acid residues in length. βA4 is derived from proteolytic processing of larger amyloid β-protein precursors (APP); these are transmembrane glycoproteins with a single membrane spanning domain, a large extracellular N-terminus and a short cytoplasmic C-terminal tail. The N-terminus of the βA4 domain is located in the extracellular portion of APP, 28 amino acids from the membrane, and βA4 extends into the membrane with a hydrophobic region of 11–15 residues. APP exists in various forms generated by alternative splicing of mRNA derived from a single gene on chromosome 21 (for review, see Kosik, 1992). Some forms (e.g. the 695, 751, and 770 residue forms) are expressed at high levels in the brain, as well as in many peripheral tissues. The biological function of APP is uncertain but accumulating evidence suggests that it is involved in cell adhesion (Schubert et al., 1989) and neurite outgrowth (Milward et al., 1992), and recent data indicate that secreted forms of APP exert a protective effect in cell culture by blocking the increase in intracellular calcium caused by excitotoxic concentrations of glutamate or by glucose deprivation (Mattson et al., 1993). Mature APP molecules are rapidly degraded by various alternative proteolytic processing pathways, and proteolytic APP derivatives are secreted into the extracellular space, and are present in human plasma and human cerebrospinal fluid. These secreted derivatives include a large N-terminal cleavage product, termed APP^s, with a C-terminus ending at residue 16 of βA4 (Esch et al., 1990; Sisodia et al., 1990). Because neither this fragment nor the remaining C-terminal stump contains the entire βA4 domain, both products generated by this cleavage event are non-amyloidogenic. Alternatively, APP can be cleaved at both the N- and C-termini of βA4 to yield secreted soluble βA4 molecules (Haass et al., 1992; Shoji et al., 1992) that potentially can aggregate into amyloid. In addition to the secretory processing pathways, full-length APP can be internalized from the cell surface and targeted to the endosomal-lysosomal system (Haass et al., 1992), where multiple additional cleavage products are generated. Some of these contain the intact βA4 domain and thus are also potentially amyloidogenic (Golde et al., 1992; Estus et al., 1992).

βA4 is neurotoxic in some experimental systems (Yankner et al., 1990) and may be involved in apoptosis (Loo et al., 1993). Thus, APP processing pathways yielding either βA4 or APP^s are likely to have distinct cellular consequences: processing events that generate βA4 may be toxic and potentially amyloidogenic, whereas those yielding APP^s may be trophic and non-amyloidogenic. It thus becomes important to understand the cellular mechanisms involved in regulating APP processing pathways.

G protein-coupled cell surface receptors regulate APP processing by activating protein kinase C

APP processing can be regulated by cell-surface receptors coupled to activation of protein kinase C (for review, see Nitsch and Growdon, 1993). To study the effects of receptor activation on APP processing, we used human embryonic kindney (HEK 293) cell lines that were stably transfected with the cDNA encoding for the human muscarinic acetylcholine receptor (mAChR) subtypes m1, m2, m3 and m4. HEK 293 cells endogenously express human APP at high levels and secrete APP^s into the culture medium (Oltersdorf et al., 1990). Stimulation of the m1 and m3 receptor subtypes with the muscarinic agonist carbachol increased APP^s release 6- to 8-fold within minutes (Nitsch et al., 1992; Buxbaum et al., 1992). Concomitantly, levels of cell-associated full-length APP decreased to 50% of control values within 30 minutes of receptor stimulation and continued to drop to undetectable levels within 90 minutes when APP synthesis was blocked by the translation inhibitor cycloheximide. These results demonstrated for the first time that neurotransmitter receptors can regulate APP processing. Control experiments with transfected m2 and m4 receptor subtypes showed that activation of these subtypes did not increase APP^s secretion. Muscarinic m2 and m4 receptor subtypes are coupled to the adenylyl cyclase signal transduction pathway, whereas m1 and m3 receptors activate protein kinase C (PKC) by generating diacylglycerol via phospholipid hydrolysis. Protein kinase inhibitors including chelerythrine chloride, staurosporine and calphostin blocked the receptor-mediated increase in APP^s secretion (Nitsch et al., 1992), and the direct activation of PKC by phorbol esters mimicked the increase in APP^s secretion (Caporaso et al., 1992; Slack et al., 1993), suggesting that the effect of receptor stimulation on APP^s secretion is mediated by activation of PKC. Cell-surface receptors coupled to PKC activation as part of their signal transduction pathways include a large list of neurotransmitter and neuromodulator receptors. We are currently testing which of these can regulate APP processing. Among the receptors that increase APP^s secretion are bradykinin and vasopressin receptors.

Cellular mechanism of receptor-mediated APP^s secretion

Receptor agonists stimulated APP^s secretion even when de novo synthesis of APP was blocked by the translation inhibitor cycloheximide indicating that preexisting full-length APP molecules are cleaved in response to receptor activation. Furthermore, inhibition of vesicular transport with colchicine, an agent that inhibits the polymerization of tubulin did not modify receptor-mediated APP^s release, suggesting that APP cleavage occurs close to, or directly at, the plasma membrane. This interpretation is compatible with the suggestion that a plasma membrane-bound endoprotease may cleave APP at a distance of 12 residues from the extracellular face of the membrane (Sisodia, 1992). This protease, which is frequently referred to as α-secretase, has not yet been isolated and its cellular mechanism of activation is unknown. It is, however, possible that this protease is activated by PKC-mediated phosphorylation, either directly or by a protein kinase downstream to PKC. Furthermore, APP may be phosphorylated in response to receptor activation.

Initial experiments that addressed the action of cell-surface receptors on intracellular APP derivatives showed that the decrease in full-length APP was paralleled by decreases in the levels of endosomal-lysosomal C-terminal APP derivatives. These preliminary results suggest that the endosomal-lysosomal system degrades full-length APP molecules that were not previously cleaved by the α-secretase pathway.

Muscarinic receptor activation blocks βA4 secretion

Unstimulated cells in culture normally secrete soluble βA4 (Haass et al., 1992; Shoji et al., 1992), and similar, if not identical peptides are detectable in human cerebrospinal fluid (Seubert et al., 1992). It was thus suggested that normal cellular metabolism generates and secretes APP derivatives that are potentially amyloidogenic. In order to test whether the production and secretion of soluble βA4 can also be regulated by cell-surface receptors, HEK 293 cells that stably express the muscarinic m1 and m3 receptors were cotransfected with the cDNA encoding for APP695. Activation of the receptors with carbachol caused a 60% decrease in the secretion of soluble βA4, indicating that cell-surface receptors can modulate APP processing pathways in a manner that increases APP^s secretion and concomitantly depresses βA4 secretion (Hung et al., 1993). It is thus possible that the pathway generating APP^s occurs in a cellular compartment proximal to the cleavage event which yields soluble βA4, and that receptor activation thus decreases the availability of substrate for generating soluble βA4 molecules. Alternatively, the proteases involved in the generation of either APP^s or βA4 may be regulated differentially by phosphorylation.

Cellular secretion of βA4 in cell culture is increased 4- to 5-fold by the double mutation found in the APP gene of a Swedish familial Alzheimer's disease kindred (Citron et al., 1992; Cai et al., 1993), suggesting that this mutation may cause increased amyloid formation in Alzheimer's disease patients carrying this mutation. Transfection experiments with cDNA encoding the m1 or m3 receptors, and with an additional expression construct encoding for APP695 with the Swedish double mutation, showed that the increased secretion of soluble βA4 caused by the Swedish double mutation was blocked by receptor activation (Hung et al., 1993). These data show that the cleavage events involved in generating soluble βA4 can be inhibited by receptor activation independently of the presence of the Swedish double mutation. The results demonstrate that both normal and pathologically-high βA4 secretion can be suppressed by cell-surface receptors, and suggest that specific receptor agonists might be useful for inhibiting βA4 formation.

Electrical depolarization of mammalian brain slices modulates APP processing

In order to establish whether neurotransmitter receptors can also regulate APP processing in the mammalian brain, we prepared brain tissue slices from the hippocampus, cortex, striatum, and the cerebellum of rats, and incubated them in superfusion chambers equipped with field stimulation electrodes (Nitsch et al., 1993). Electrical stimulation with 10 to 30 Hz caused 3- to 10-fold increases in the release of endogenous neurotransmitters including glutamate and acetylcholine. In order to control for cellular damage during electrical stimulation, we measured the release of lactate dehydrogenase; this was unchanged during electrical stimulation. The increased release of endogenous neurotransmitters was paralleled by an averaged 2-fold increase in the release of APP^s during a 30 Hz stimulation period of 50 minutes' duration. Individual brain areas varied in terms of the magnitude of the increase in APP^s formation: in brain cortex stimulation increased APP^s secretion 2.5-fold, in striatum 1.7-fold, in cerebellum 1.4-fold and in hippocampus 1.9-fold. The depolarization-induced increase in APP^s secretion was blocked by the sodium channel blocker tetrodotoxin, indicating that action potentials directly affect APP processing in mammalian brain. The effect of electrical stimulation on APP^s secretion from hippocampal slices was frequency-dependent in the range of 0 to 30 Hz, and reached its maximum at 30 Hz. These data also support the suggestion that the rate of APP^s secretion is a function of neuronal activity.

Implications for Alzheimer's disease

Cell-surface receptors coupled to activation of PKC can regulate APP processing by increasing the secretion of APP^s (which is non-amyloidogenic and presumably subserves both neurotrophic and neuroprotective biological functions), and concomitantly decreasing the formation of potentially amyloidogenic and neurotoxic βA4 peptides. Receptors for which a stimulatory effect on APP processing has already been demonstrated include the muscarinic receptor subtypes m1 and m3, and those for bradykinin, vasopressin, and interleukin-1. These and other receptors may thus be potential targets for pharmacological treatments designed to modulate APP processing. If the assumption is correct that APP^s is a neurotrophic, and possibly even a neuroprotective agent, and that βA4 is both neurotoxic and amyloidogenic, pharmacologically mediated increases in APP^s and concomitant decreases in βA4 produced by low molecular mass agonists to receptors coupled to PKC may constitute a rational therapy for Alzheimer's disease, perhaps inhibiting amyloidogenesis, suppressing βA4 neurotoxicity and enhancing the neurotrophic actions of APP^s. Human studies will show whether this treatment approach can also ameliorate the clinical symptoms of Alzheimer's disease.

Acknowledgments

This work was supported by the NIA, NIMH, the Center for Brain Sciences and Metabolism Charitable Trust, and the Hoffman Fellowship in Alzheimer's disease.

References

Buxbaum JD, Oishi M, Chen HI, Pinkas-Kramarski R, Jaffe EA, Gandy SE, Greengard P (1992) Cholinergic agonists and interleukin 1 regulate processing and secretion of the Alzheimer β/A4 amyloid protein precursor. Proc Natl Acad Sci USA 89: 10075–10078

Cai X-D, Golde TE, Younkin SG (1993) Release of excess amyloid β-protein from a mutant amyloid β-protein precursor. Science 259: 514–516

Caporaso GL, Gandy SE, Buxbaum JD, Ramabhadran TV, Greengard P (1992) Protein phosphorylation regulates secretion of Alzheimer β/A4 amyloid precursor protein. Proc Natl Acad Sci USA 89: 3055–3059

Citron M, Oltersdorf T, Haass C, McConlogue L, Hung AY, Seubert P, Vigo-Pelfrey C, Lieberburg I, Selkoe DJ (1992) Mutation of the β-amyloid precursor protein in familial Alzheimer's disease increases β-protein production. Nature 360: 672–674

Esch FS, Keim PS, Beattie EC, Blacher RW, Culwell AR, Oltersdorf T, McClure D, Ward P (1990) Cleavage of amyloid β-peptide during constitutive processing of its precursor. Science 248: 1122–1124

Estus S, Golde TE, Kunishita T, Blades D, Lowery D, Eisen M, Usiak M, Qu X, Tabira T, Greenberg BD, Younkin SG (1992) Potentially amyloidogenic, carboxyl-terminal derivatives of the amyloid protein precursor. Science 255: 726–728

Golde TE, Estus S, Younkin LH, Selkoe DJ, Younkin SG (1992) Processing of the amyloid protein precursor to potentially amyloidogenic derivatives. Science 255: 728–730

Haass C, Koo EH, Mellon A, Hung AY, Selkoe DJ (1992) Targeting of cell-surface β-amyloid precursor protein to lysosomes: alternative processing into amyloid-bearing fragments. Nature 357: 500–503

Haass C, Schlossmacher MG, Hung AY, Vigo-Pelfrey C, Mellon A, Ostaszewski BL, Lieberburg I, Koo EH, Schenk D, Teplow DB, Selkoe DJ (1992) Amyloid β-peptide is produced by cultured cells during normal metabolism. Nature 359: 322–325

Hung AY, Haass C, Nitsch RM, Qiu WQ, Citron M, Wurtman RJ, Growdon JH, Selkoe DJ (1993) Activation of protein kinase C inhibits cellular production of the amyloid β-protein. J Biol Chem (in press)

Kosik KS (1992) Alzheimer's disease: a cell biological perspective. Science 256: 780–783

Loo DT, Copani A, Pike CJ, Whittemore ER, Walencewicz AJ, Cotman CW (1993) Apoptosis is induced by β-amyloid in cultured central nervous system neurons. Proc Natl Acad Sci USA: 90: 7951–7955

Mattson MP, Cheng B, Culwell AR, Esch FS, Lieberburg I, Rydel RE (1993) Evidence for excitoprotective and intraneuronal calcium-regulating roles for secreted forms of the β-amyloid precursor protein. Neuron 10: 243–254

Milward EA, Papadopoulos R, Fuller SJ, Moir RD, Small D, Beyreuther K, Masters CL (1992) The amyloid protein precursor of Alzheimer's disease is a mediator of the effects of nerve growth factor on neurite outgrowth. Neuron 9: 129–137

Nitsch RM, Farber SA, Growdon JH, Wurtman RJ (1993) Release of amyloid β-protein precursor derivatives by electrical depolarization of rat hippocampal slices. Proc Natl Acad Sci USA 90: 5191–5193

Nitsch RM, Growdon JH (1994) The role of neurotransmission in the regulation of amyloid β-protein precursor processing. Biochem Pharmacol (in press)

Nitsch RM, Slack BE, Wurtman RJ, Growdon JH (1992) Release of Alzheimer amyloid precursor derivatives stimulated by activation of muscarinic acetylcholine receptors. Science 258: 304–307

Oltersdorf T, Ward PJ, Henriksson R, Beattie EC, Neve R, Lieberburg I, Fritz LC (1990) The Alzheimer amyloid precursor protein. Identification of a stable intermediate in the biosynthetic/degradative pathway. J Biol Chem 265: 4492–4497

Schubert D, Jin L-W, Saitoh T, Cole G (1989) The regulation of amyloid β precursor secretion and its modulatory role in cell adhesion. Neuron 3: 689–694
Seubert P, Vigo-Pelfrey C, Esch F, Lee M, Dovey H, Davis D, Sinha S, Schlossmacher M, Whaley J, Swindlehurst C, McCormack R, Wolfert R, Selkoe D, Lieberburg I, Schenk D (1992) Isolation and quantification of soluble Alzheimer's β-peptide from biological fluids. Nature 359: 325–327
Shoji M, Golde TE, Ghiso J, Cheung TT, Estus S, Shaffer LM, Cai X-D, McKay DM, Tintner R, Frangione B, Younkin SG (1992) Production of the Alzheimer amyloid β-protein by normal proteolytic processing. Science 258: 126–129
Sisodia SS (1992) β-Amyloid precursor protein cleavage by a membrane-bound protease. Proc Natl Acad Sci USA 89: 6075–6079
Sisodia SS, Koo EH, Beyreuther K, Unterbeck A, Price DL (1990) Evidence that β-amyloid protein in Alzheimer's disease is not derived by normal processing. Science 248: 492–495
Slack BE, Nitsch RM, Livneh E, Kunz Jr GM, Breu J, Eldar H, Wurtman RJ (1993) Regulation by phorbol esters of amyloid precursor protein release from Swiss 3T3 fibroblasts overexpressing protein kinase Cα. J Biol Chem 268: 21097–21101
Yankner BA, Duffy LK, Kirschner DA (1990) Neurotrophic and neurotoxic effects of amyloid beta protein: reversal by tachikinin neuropeptides. Science 250: 279–282

Authors' address: Dr. R.M. Nitsch, Department of Brain and Cognitive Sciences, Massachussetts Institute of Technology, E25-604, Cambridge, MA 02139, U.S.A.

J Neural Transm (1994) [Suppl] 44: 29–45

Glutamate, beta-amyloid precursor proteins, and calcium mediated neurofibrillary degeneration

V. L. Smith-Swintosky and **M. P. Mattson**

Sanders-Brown Research Center on Aging and Department of Anatomy and Neurobiology, University of Kentucky, Lexington, KY, U.S.A.

Summary. In this article we present evidence supporting the interaction between excitotoxicity, βAPP mismetabolism, metabolic compromise and intracellular calcium destabilization in the process of neurodegeneration associated with Alzheimer's disease (AD). AD is characterized by the presence of neurofibrillary tangles and amyloid-containing plaques in specific regions of the brain. There appear to be several processes which contribute to the neurodegeneration associated with AD. Although AD has been linked to genetic mutations on chromosomes 21, 19 and 14, there are sporadic forms of AD that have no known genetic mutation involved. Aging is the major risk factor for AD. During the course of normal aging several metabolic compromises may occur in the brain. Both decreased glucose transport and utilization, and increased glucocorticoid levels are known to occur with aging and may lead to decreased energy supplies, ATP depletion, failure of Ca^{2+} buffering systems, excess glutamate release and activation of glutamate receptors. In addition, a reduction in antioxidant enzymes and consequently an increase in free radicals has also been associated with aging. Each of the preceeding alterations would lead to an increase in neuronal $[Ca^{2+}]_i$. Elevated calcium could then activate calcium-dependent proteases which degrade particular cytoskeletal proteins, and lipases which generate free radicals resulting in membrane damage and possible cell death. In this article we provide evidence that amyloid β-peptide (Aβ), the substance which accumulates in AD plaques, exacerbates excitotoxic and metabolic compromises to neurons resulting in changes in the cytoskeleton which resemble those seen in the neurofibrillary tangles of AD. We also provide evidence that secreted forms of β-amyloid precursor protein (βAPP) are neuroprotective against excitotoxic insults. Recent findings concerning the normal function of βAPP and the mechanism of Aβ toxicity place βAPP at the center of changes leading to neuronal degeneration in AD.

Introduction

Two prominent theories concerning the mechansims underlying brain aging and neurodegenerative disorders are the calcium hypothesis (Khachaturian,

1989; Mattson, 1992) and the free radical hypothesis (Harmon, 1956, 1987). Calcium is critically involved in regulating dynamic aspects of neuronal structure such as axon polarity, outgrowth regulation and synaptogenesis (Kater et al., 1988; Mattson, 1992). It is therefore not surprising that neurons possess highly elaborate systems for regulating $[Ca^{2+}]_i$ and responding to changes in $[Ca^{2+}]_i$. For example, neurons possess voltage-gated and ligand-gated calcium channels which allow calcium into the cell; Ca^{2+}-ATPase and Na^+-Ca^{2+} membrane exchangers for moving calcium out of the cell; and organelles and calcium-binding proteins which sequester free cytosolic calcium. It appears that calcium regulation is disrupted in aging. For example, recent evidence indicates that aging leads to prolongation of calcium currents possibly leading to an increase in $[Ca^{2+}]_i$ (Landfield, 1987). A sustained increase in $[Ca^{2+}]_i$ can lead to the activation of calcium-dependent proteases and free radicals which destabilize the neuronal cytoarchitecture leading to cell damage and eventual death (Mattson et al., 1988a; Siman et al., 1989; Yanagihara et al., 1990; Johnson et al., 1991).

Free radicals have been implicated in the process of aging since the 1950's. This hypothesis has been modified to state that free radicals and free radical reactions are involved in the etiology and development of a number diseases, especially those that increase mortality (Harmon, 1987; Pryor, 1987). Several studies indicate that there are reductions in antioxidant enzymes and increased levels of free radicals in the aging brain and in AD (Carney et al., 1991; Smith et al., 1991). For example, aged gerbils have significantly higher levels of oxidized protein and decreased glutamine synthetase and neutral protease activites as compared to young adult gerbils. Moreover, treatment with N-tert-butyl-a-phenylnitrone (PBN), a spin-trapping compound, decreased the amount of oxidation occurring in the brains of aged gerbils and improved their cognitive performance on radial arm maze tasks (Carney et al., 1991). In addition, evidence suggests that dietary supplementation with antioxidants (i.e. vitamin E) can decrease free radical accumulation and increase longevity (Koistinaho et al., 1990). There is evidence suggesting the impairment in brain function with age may result from damage to brain DNA (Goldgaber et al., 1987). For example, an age-associated progressive impairment of mitochondrial DNA in the cerebral cortices of primates was reported (Bowling et al., 1993). If DNA is compromised with aging, its organic components are susceptible to hydroxyl radical attack. The free radical-induced DNA aberrations would result in a decrease in messenger RNA (Clark and Parhad, 1989). Lowered transcription rates could result in a loss of control for various biochemical processes including those for intracellular iron-handling and intracellular calcium homeostasis. Such alterations could lead to devastating structural and functional changes in neurons. Abnormalities in iron-handling and calcium homeostasis have been implicated in AD (Richardson et al., 1990; Subbarao et al., 1990; Andorn et al., 1990; Zhang et al., 1993; Goodman et al., 1993). The beneficial effects of deprenyl and the 21-aminosteroid antioxidant U-74500A in AD that have been reported (Tariot et al., 1987, 1988; Subbarao et al., 1990) lend credence to the causal role of free radicals in the pathogenesis of AD.

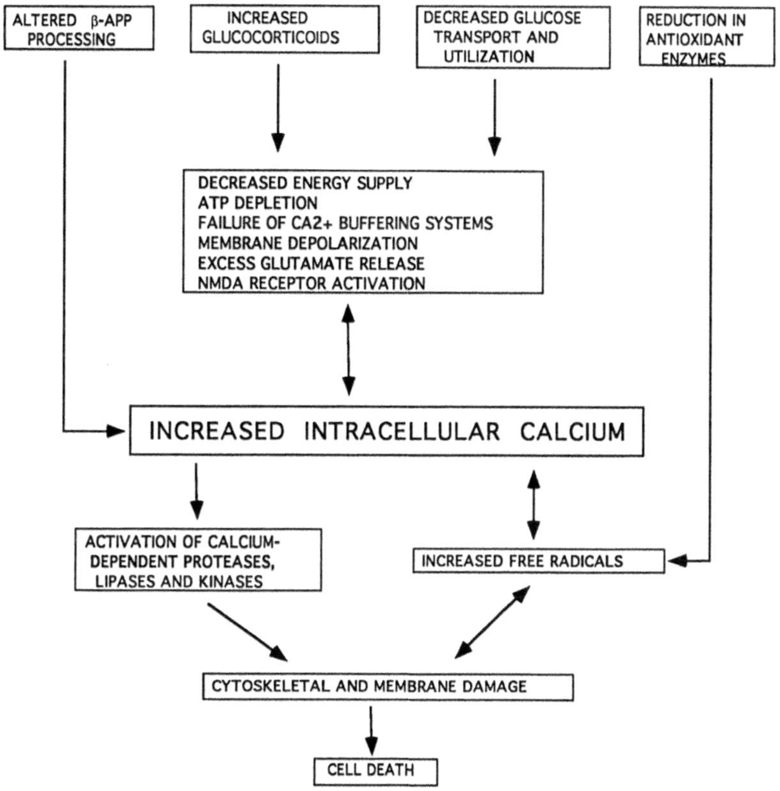

Fig. 1. Metabolic compromises which occur during normal aging. Several metabolic compromises are known to occur during normal aging. Altered βAPP metabolism, increased glucocorticoids, decreased glucose transport and utilization and a reduction of antioxidant enzymes may all contribute to a rise in neuronal $[Ca^{2+}]_i$. Elevations in $[Ca^{2+}]_i$ can lead to the activation of calcium-dependent proteases, kinases and lipases and the formation of free radicals which degrade cytoskeletal proteins and membranes resulting in cell damage and death

During normal human aging several changes in the brain occur which contribute to the metabolic compromise of neurons (Fig. 1). For example, aging is often accompanied by atherosclerosis which can lead to a decreased blood supply to the brain parenchyma. A reduction in glucose transport and utilization may occur with aging and to a greater extent in AD (Hoyer, 1988; Hoyer et al., 1988; Kalaria and Harik, 1989). Reduced glucose availability results in ATP depletion, failure of Ca^{2+} extrusion/buffering systems, membrane depolarization, excess glutamate release, and glutamate receptor activation. Increased levels of glucocorticoids that occur with age (Sapolsky, 1992) are believed to reduce glucose uptake in neurons and thereby increase their vulnerability to damage due to overactivation of excitatory amino acid receptors (Sapolsky, 1985). Recent evidence demonstrated that glucocorticoids and physiological stress can exacerbate excitotoxic damage and associated neurofibrillary tangle-like antigenic changes and spectrin proteolysis in hippocampal neurons in vivo (Elliott et al., 1993; Stein-Behrens et al., 1994). Taken together with previous findings (Greenamyre and Young,

1989), these data suggest a mechanism of neurofibrillary degeneration in AD involving metabolic impairment and a loss of calcium homeostasis. Several of the metabolic compromises that occur in normal aging appear to be exacerbated by neurodegenerative conditions such as AD, Huntington's disease, status epilepticus and ischemia.

AD is characterized histopathologically by the presence of numerous amyloid-containing plaques and neurofibrillary tangles in vulnerable brain regions such as the entorhinal cortex, hippocampus and basal forebrain. β-amyloid precursor protein (βAPP), a large transmembrane glycoprotein, is the source of amyloid β-peptide (Aβ) which accumulates as diffuse and aggregated plaques in the brains of AD victims. Aβ is a 40–42 amino acid segment of βAPP that lies partially extracellularly and partially within the plasma membrane (Kang et al., 1987). βAPP is metabolized via several pathways that give rise to different products that are secreted from cells. An enzymatic cleavage occurs near the cell surface (Esch et al., 1990; Sisodia et al., 1990) that liberates secreted forms of βAPP (APP^{s}s) into the extracellular milieu; this cleavage is within the Aβ sequence (α-secretase) and precludes deposition of Aβ. Another enzymatic cleavage occurs at the N-terminus of Aβ (β-secretase); this cleavage leaves a potentially amyloidogenic C-terminal fragment in the plasma membrane. Alternative processing of such C-terminal fragments or full length βAPP can result in the liberation of intact Aβ from cells (Golde et al., 1992; Haass et al., 1992; Seubert et al., 1992; Shoji et al., 1992). Release of Aβ from cells occurs during normal metabolism of βAPP, and Aβ circulates at low levels (nM) within the cerebrospinal fluid and blood.

It is believed that altered expression and/or processing of βAPP contributes to the pathogenesis of AD (Selkoe, 1991). The gene for βAPP has been mapped to chromosome 21, suggesting that the βAPP overexpression believed to occur in Down syndrome (trisomy 21) is causally involved in the development of AD-like tangles and plaques in essentially all persons with Down syndrome. More recently it has been shown that mutations in βAPP underlie at least some cases of inherited AD. Several different βAPP mutations have been linked to AD including a conversion of valine to isoleucine, phenylalanine or glycine at codon 717 just C-terminal to the Aβ sequence (Chartier-Harlin et al., 1991; Goate et al., 1991; Murrell et al., 1991). A double mutation in βAPP involving lysine to asparagine and methionine to leucine conversions at codons 670 and 671, respectively, just N-terminal to Aβ was also recently reported in a family with AD (Mullan et al., 1992). It appears that cells expressing βAPP containing the 670–671 double mutation show increased Aβ production (Cai et al., 1993; Citron et al., 1993). Other cases of familial AD have been linked to mutations in chromosomes 14 or 19 although the altered genes have not been identified. Two possibilities in the case of chromosome 19 are genes that encode an amyloid precursor-like protein (APLP1; Wasco et al., 1993) and apolipoprotein E (ApoE; Strittmatter et al., 1993a). APLP1 is structurally smilar to βAPP but lacks the Aβ sequence. The ApoE gene and the suspect region in AD are in the same place on chromosome 19. A recent study indicated that persons with multiple copies of the ApoE4 allele were at greater risk for developing AD than those with no copies of this allele (Corder et al., 1993). Moreover, ApoE4 binds β-amyloid, but it is unknown whether this binding is beneficial or contributes to the neurodegeneration associated with AD (Wisniewski et al., 1992; Strittmater et al., 1993b). The genes affected in other AD families such as those mapping to chromosome 14 (Schellenberg et al., 1992) remain to be determined.

In this article we have examined the interactive effects of excitatory amino acids, metabolic compromise and APP metabolites on neuronal calcium homeostasis and survival in both in vitro and in vivo models of neurodegeneration.

Methods

Details of all methods referred to here can be found in previous studies as follows: dissociated cell culture of rat hippocampal, septal, cortical and human cortical neurons (Mattson and Kater, 1988; Mattson and Rychlik, 1990); glutamate toxicity assays (Mattson et al., 1989); glucose deprivation (Cheng and Mattson, 1991); assessment of neuronal injury (Mattson et al., 1989); immunocytochemistry and Western blots (Mattson et al., 1991b); fluorescence ratio imaging of intracellular free calcium levels using the calcium indicator dye fura-2 (Mattson et al., 1989); adrenalectomy, corticosterone administration and kainic acid injection in vivo (Elliott et al., 1993); and four vessel occlusion model of global transient ischemia in vivo (Smith-Swintosky et al., 1993).

Results and discussion

The role of glutamate in neural development and degeneration

Glutamate, the predominant excitatory neurotransmitter in the brain, is essential in neuronal development and synaptic plasticity. It can stabilize the outgrowth of dendrites and promote synapse formation in developing hippocampal neurons (Mattson et al., 1988a,b), and plays a major role in the activity-dependent refinement of synaptic connections in the visual system (Cline et al., 1990). Moreover, glutamate is intimately involved with the process of long-term potentiation (LTP) which is one mechanism associated with learning and memory in man (Collingridge and Bliss, 1987). However, overactivation of glutamate receptors can cause maladaptive structural changes in neurons such as dendritic pruning and cell death (Rothman and Olney, 1986; Mattson et al., 1988a). Activation of glutamate receptors is involved in the pathogenesis of neuronal injury in stroke since glutamate receptor antagonists can reduce neuronal damage induced by ischemia in vivo or hypoxia/hypoglycemia in vitro (Simon et al., 1984; Siesjo et al., 1989; Cheng and Mattson, 1992a). An excitotoxic mechanism of neuronal injury is also implicated in the pathogenesis of a number of chronic neurodegenerative disorders including AD, amyotrophic lateral sclerosis, Huntington's disease, and Parkinson's disease although the evidence is less exacting. Excitatory damage is mediated in large part by uncontrolled Ca^{2+} influx resulting in sustained elevation of $[Ca^{2+}]_i$ (Mattson et al., 1988a, 1989; Siesjo et al., 1989). Neurons that bear glutamate receptors are particularly vulnerable to excitotoxicity when energy supplies (glucose and oxygen) are low (e.g., hypoglycemia), probably because the removal of Ca^{2+} from the cytoplasm requires ATP (Novelli et al., 1988; Siesjo et al., 1989). Glutamate toxicity leads to antigenic changes in the neuronal cytoarchitecture resembling the changes seen in AD. Neuronal degeneration in AD manifests itself as neurofibrillary tangles (NFT's) which are characterized by several structural and antigenic features including the accumulation of straight and paired-helical filaments and increased immunoreactivity with tau and ubiquitin antibodies (Wisniewski et al., 1979; Selkoe, 1989). Cultured rat hippocampal and human cortical neurons exposed to toxic levels of glutamate exhibited increased immunostaining for both tau and ubiquitin resembling the antigenic changes seen in NFT's; these alterations were dependent on calcium influx (Mattson, 1990; Mattson et al., 1991a). Biochemical studies also revealed NFT-like changes in tau induced by glutamate (Mattson et al., 1991a; Sautiere et al., 1992). Thus it appears that glutamate and a concomittant loss of calcium homeostasis may be involved in the formation of NFT in AD. Further evidence supporting a role for elevated intracellular calcium levels comes from studies showing that tau is excessively phosphorylated in AD NFT (Grundke-Iqbal et al., 1986). In addition, it was recently reported that transglutaminase promotes the formation of tau polymers with structural and antigenic properties similar to those in NFT (Dudek and Johnson, 1993). Transglutaminase is a calcium-activated enzyme that catalyzes covalent association of certain proteins to form insoluble polymers (Greenberg et al., 1991).

Recent in vivo studies also support the notion of an excitotoxic mechanism of NFT formation. Kainic acid injection into the hippocampus of rats elicited a rapid (hrs) but transient appearance of immunoreactivity to tau antibodies Alz-50 and 5E2 in CA3 neurons. The degradation of spectrin, a cytoskeletal protein sensitive to calcium-mediated proteolysis, was also induced by kainate. The neuronal damage, appearance of tau immunoreactivity and spectrin breakdown caused by kainate were exacerbated by glucocorticoids (Elliott et al., 1993) and physiological stress (Stein-Behrens et al., 1994). These findings are consistent with an excitotoxic mechanism of neuronal degeneration in AD since glucocorticoids are known to metabolically compromise neurons and increase their vulnerability to excitotoxicity (Elliott and Sapolsky, 1992; Stein-Behrens and Sapolsky, 1992). Moreover, there is evidence of altered glucocorticoid regulation (Greenwald et al., 1986) and reduced energy availability to neurons (Hoyer, 1988; Hoyer et al., 1988; Kalaria and Harik, 1989) in AD.

Hypoglycemia may also play a role in the neuronal damage that occurs in AD due to the marked reduction in glucose metabolism and glucose transport seen in patients with early-onset AD (Hoyer, 1988; Hoyer et al., 1988; Kalaria and Harik, 1989). When neuronal cultures are deprived of glucose they die over a period of 16–24 hours. Within 4–12 hours after the onset of glucose-deprivation the $[Ca^{2+}]_i$ begins to rise, during which time the neurons appear to be undamaged. However, by 12–16 hours there is a marked increase in $[Ca^{2+}]_i$ and the neurons die rapidly within the next 4 hours (Cheng and Mattson, 1991). We found that neurons deprived of glucose exhibited similar antigenic changes in tau and ubiquitin as seen in glutamate toxicity and the NFT of AD (Cheng and Mattson, 1992b). Growth factor administration (i.e. bFGF, NGF and IGF's) significantly protected neurons against glucose deprivation-induced damage and glutamate toxicity (Mattson et al., 1989; Cheng and Mattson, 1991, 1992a). Moreover, these growth factors also prevented the loss of calcium homeostasis that normally mediates glucose deprivation-induced damage in vitro, and inhibited antigenic changes in tau (Cheng and Mattson, 1991, 1992b). Subsequently, these findings were confirmed, and extended to in vivo models of ischemic brain injury. For example, bFGF (Nozaki et al., 1993), NGF (Shigeno et al., 1991; Frim et al., 1993) and IGF-1 (Gluckman et al., 1992) protected hippocampal and/or striatal neurons against ischemic/hypoxic injury in vivo. Thus endogenous growth factors appear to be an important line of defense for neurons against metabolic and excitotoxic insults.

Altered processing of βAPP may disrupt neuronal metabolism of calcium and free radicals

The major risk factor for AD is aging. One of the key histopathological characteristics of AD is the deposition of Aβ in senile plaques. It has been hypothesized that the accumulation of Aβ is causally involved in neuronal degeneration. Isolation of cerebrovascular amyloid and identification of the amino acid sequence of Aβ (Glenner and Wong, 1984) allowed the synthesis of pure preparations of Aβ which could then be administered to neuronal cultures. Aβ has been shown to have direct toxic effects on neuronal cells (Yankner et al., 1990; Mattson et al., 1993a; Pike et al., 1993) and can potentiate glutamate toxicity (Koh et al., 1990; Mattson et al., 1993a,b). Several reports have demonstrated that Aβ must be aggregated in order to exert its neurotoxic effects (Busciglio et al., 1992; Pike et al., 1991; Mattson et al., 1993a). Increased aggregation of Aβ, and possibly increased β-pleated sheet conformation, correlates well with increasing intracellular calcium levels and cell death (Mattson et al., 1993a). Exposure of cultured human cerebral cortical or rat hippocampal cells to Aβ results in a progressive elevation of $[Ca^{2+}]_i$, the time course and magnitude of which depends on the degree of aggregation of Aβ at the time of exposure to the cultures (Mattson et al., 1993a). The greater the aggregation, the greater the increase in $[Ca^{2+}]_i$. With the use of immunocytochemistry, we discovered that the greater the amount of exogenous Aβ associated with the cell surface, the greater the level of $[Ca^{2+}]_i$. Moreover, neurons exposed to Aβ exhibit enhanced $[Ca^{2+}]_i$ responses to excitatory stimuli (e.g. glutamate and KCl) and potentiation of excitotoxicity (Mattson et al., 1993a,b). Aβ also renders

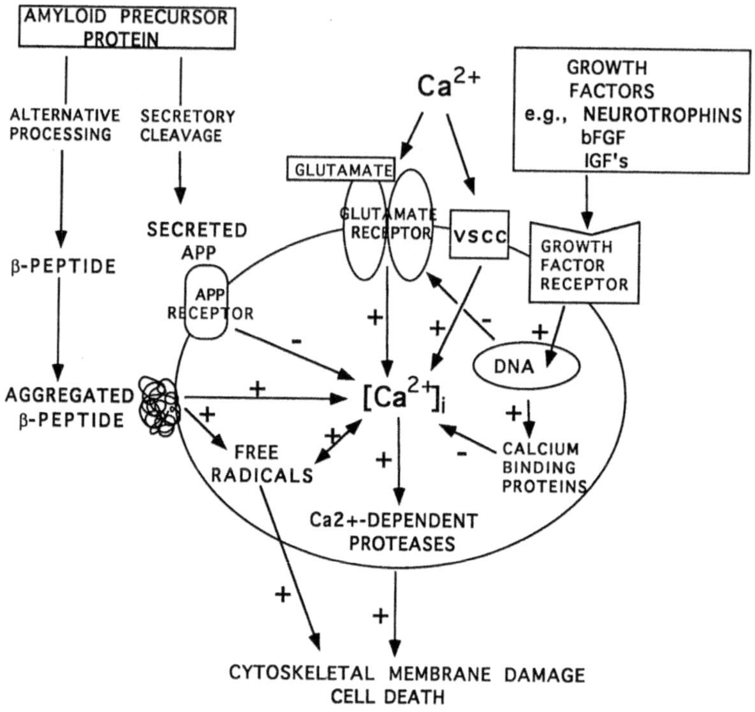

Fig. 2. Possible mechanisms contributing to perturbated calcium and free radical metabolism in AD. Schematic diagram of the proposed interactions between βAPP, glutamate, growth factors and intracellular calcium with respect to cell degeneration and subsequent death. Amyloid precursor protein (βAPP) can be cleaved within the Aβ region by an α-secretase to generate secreted forms of APP. Alternative processing of βAPP liberates Aβ from cells. Under certain environmental conditions (e.g., altered pH or presence of free radicals) Aβ forms insoluble aggregates that can accumulate at the plasma membrane of neurons and disrupt $[Ca^{2+}]_i$ regulation. Aβ alone increases $[Ca^{2+}]_i$ and exacerbates glutamate receptor activated $[Ca^{2+}]_i$ increases. Administration of APP^ss leads to a rapid decrease in $[Ca^{2+}]_i$, suggesting that APP^ss bind to a cell surface receptor. Growth factors have also been shown to prevent increases in $[Ca^{2+}]_i$ and subsequent cytoskeletal changes otherwise induced by glutamate or hypoglycemia. Evidence suggests that growth factors can exert their protective effects by increasing calcium binding protein expression and reducing glutamate receptor expression. If mechanisms for stabilizing $[Ca^{2+}]_i$ are jeopardized, $[Ca^{2+}]_i$ may rise to neurotoxic levels (μM) and activate calcium-dependent proteases (e.g. calpain) which may contribute to neuronal injury and death. **VSCC** voltage-sensitive calcium channels

neurons more vulnerable to hypoglycemic injury (Copani et al., 1991). These and other findings are consistent with the fact that most of the neuronal populations affected by AD possess glutamate receptors. Aβ appears to exert its effects at the cell surface since aggregates of Aβ accumulate at the plasma membrane and because Aβ toxicity is attenuated when Ca^{2+} influx is experimentally reduced (Mattson et al., 1993a). Clearly, these data support a role for Aβ in an excitotoxic mechanism of neuronal degeneration in AD (Fig. 2).

Recently, we found that pretreatment of hippocampal and cortical cultures with bFGF protects neurons from Aβ toxicity (Mattson et al., 1993a). Basic FGF reduced the rise in $[Ca^{2+}]_i$ associated with Aβ toxicity. We concluded that endogenous growth factors may normally protect neurons against toxicity associated with Aβ accumulation. The ways in which growth factors stabilize $[Ca^{2+}]_i$ and protect against excitotoxic insults are beginning to be understood. Recent findings suggest that growth factors can stabilize $[Ca^{2+}]_i$ by influencing specific $[Ca^{2+}]_i$-regulating proteins. For example, in hippocampal cell cultures bFGF and NT-3 increased the expression of the 28 kDa calcium-binding protein calbindin (Collazo et al., 1992; M.P. Mattson and S. Christakos, unpublished data). Neurons expressing calbindin are better able to handle toxic increases in $[Ca^{2+}]_i$ following an insult and are relatively resistant to excitotoxicity (Mattson et al., 1991b). In addition, we recently demonstrated that bFGF attenuates the expression of a 71 kDa NMDA receptor protein that mediates calcium influx and excitotoxicity in cultured hippocampal neurons (Mattson et al., 1993c). Thus it appears that growth factors play a role in protecting neurons against potentially damaging consequences of excitatory stimulation. The fact that growth factor expression can be modulated by electrical activity (Ballarin et al., 1988; Gall and Isackson, 1989) supports the latter hypothesis.

In addition to stabilizing $[Ca^{2+}]_i$, growth factors appear to regulate free radical levels. For example, NGF has been shown to protect PC12 cells from hydrogen peroxide damage (Jackson et al., 1990). In addition, NGF, bFGF and IGF's attenuated iron-induced oxidative damage and death in cultured rat hippocampal and human cortical neurons (Zhang et al., 1993). One way growth factors are thought to protect against free radical damage is by their ability to increase the expression of free radical scavenging proteins such as catalase, superoxide dismutase and glutathione peroxidase (Nistico et al., 1992). Thus, endogenous growth factors help protect neurons against the relationship between $[Ca^{2+}]_i$ and free radicals. For example, increased $[Ca^{2+}]_i$ levels can lead to the generation of free radicals; likewise, a free radical accumulation can lead to a rise in $[Ca^{2+}]_i$.

Normal functions of β-amyloid precursor protein in neurons: regulation of $[Ca^{2+}]_i$?

While the role of Aβ has been intensely studied due to its potentially neurotoxic role in AD, comparatively little is known about the function of βAPP in neurons. A major processing pathway for βAPP involves an enzymatic cleavage in the middle of Aβ resulting in the release of a soluble amino terminal portion of the APP molecule into the extracellular milieu. These secreted APPs (APP^ss) include forms lacking (APP^s695 and APP^s714) or containing (APP^s751 and APP^s770) a Kunitz-type protease inhibitor region. APP^ss have been shown to affect outgrowth and survival of developing neurons in culture (Araki et al., 1992; Milward et al., 1992; Mattson et al., 1993d). APPs are axonally transported (Koo et al., 1990), and stimulation of neurons by excitatory transmitters and electrical activity induces secretion of APP^{s}s (Nitsch et al., 1992, 1993). Presumably APP^{s}s are released from synaptic terminals and are therefore poised to serve a neuromodulatory role. Consistent with such a role for APP^{s}s are the findings that APP^{s}s can modify neuronal responses to glutamate (Mattson et al., 1993d).

Recent studies suggest that APP^{s}s may play a role in the brain's response to injury. For example, βAPP levels are increased by growth factors, cytokines and brain injury (Mobley et al., 1988; Goldgaber et al., 1989; Siman et al., 1989; Quon et al., 1990; Abe et al., 1991; Roberts et al., 1991; Wallace et al., 1991; Nakamura et al., 1992; Stephenson et al., 1992; Wakita et al., 1992). Recently, we demonstrated that exogenously applied APP^{s}695 and APP^{s}751 (1–100 nM) protect cultured hippocampal and cortical neurons against glucose deprivation-induced injury and excitoxicity in a concentration-dependent manner (Mattson et al., 1993d). In addition, we found that APP^{s}s attenuate the rise in $[Ca^{2+}]_i$ that accompanies these environmental insults thereby protecting the neurons from damage. Recent studies suggest that the cyclic GMP second messenger pathway may be involved in APP^{s}s neuroprotective and $[Ca^{2+}]_i$ lowering effects. For example, in cultured rat hippocampal neurons: APP^{s}s elevated cyclic GMP levels; membrane-permeant cyclic GMP analogs lowered $[Ca^{2+}]_i$; and cyclic GMP antagonists blocked the $[Ca^{2+}]_i$-lowering effects of APP^{s}s (Barger et al., 1993).

We recently extended the in vitro studies of APP^{s}s neuroprotective effects to a four-vessel occlusion model of transient global ischemia in the rat. We found that APP^{s}695 and APP^{s}751 significantly increased cell survival within the vulnerable CA1 region of the hippocampus. Moreover, ^{14}C-leucine autoradiography demonstrated that these neurons were not only structurally intact but functional (Smith-Swintosky et al., 1993). Previous studies demonstrated a reduction of MAP-2 in region CA1 of rats 72 hrs following transient global forebrain ischemia (Tomioka et al., 1992). The loss of MAP-2 immunoreactivity is believed to result from glutamate receptor stimulation in dendrites of pyramidal neurons which leads to Ca^{2+} influx and activation of Ca^{2+}-dependent proteases (e.g. calpain) (Mattson et al., 1988b; Siman et al., 1989; Yanagihara et al., 1990; Johnson et al., 1991; Saito et al., 1993). We found that APP^{s}751 largely prevented the ischemia-induced decrease of MAP-2 immunoreactivity demonstrating preservation of pyramidal neuron dendrites. Transient global ischemia is known to induce other antigenic changes in hippocampal neurons similar to those seen in AD. Both tau and ubiquitin are increased early after the ischemic event, however by 72 hr the immunostaining is greatly reduced or disappears altogether (C. Schwab, personal communication). The loss of immunoreactivity for tau and ubiquitin is presumeably due to cell death. Recent evidence demonstrated that calpain immunoreactivity is strongly associated with tangle formation and senile plaques in AD patients compared to age-matched controls (Iwamoto et al., 1991). These data suggest that calpain may be an important factor in the abnormal proteolysis underlying plaque and tangle formation in AD.

Conclusion

Taken together, the data presented here suggest that energy reduction, excitatory amino acid receptor overactivation, and abnormal βAPP processing leading to a destabilization of intracellular calcium homeostasis are intimately involved in neurodegeneration and cell death associated with Alzheimer's disease (Fig. 2). Combined activation of glutamate receptors and the accumulation of Aβ at the cell surface may increase $[Ca^{2+}]_i$ levels to a point at which the cell's normal compensatory mechanisms for regulating fluctuations in calcium levels are unable to handle the large sustained increase. Consequently, calcium-dependent proteases, kinases, and lipases may be activated which promote degradation of cytoskeletal proteins, destabilizing the cytoarchitecture and possibly leading to cell death. Endogenous growth factors and APP^ss appear to serve as powerful neuroprotective agents which can rescue neurons from injury. Growth factors exert their influence, in part, by upregulating the expression of calcium-buffering proteins; however, the mechanism of action of APP^ss is still unknown. Cumulative evidence suggests that altered processing of βAPP may be the major factor contributing to neuronal injury in AD. Several inherited forms of AD have been linked to mutations in βAPP (Goate, 1991; Murrell et al., 1991; Mullan et al., 1992), and at least some of these mutations may result in increased Aβ production (Cai et al., 1993; Citron et al., 1993), and perhaps reduced levels of APP^ss. We propose that in AD altered processing of βAPP exacerbates metabolic compromises which normally accompany aging (i.e. reduced glucose supply, reduced antioxidant enzymes, increased glucocorticoids) and may compromise the neuroprotective function of APP^ss which would lead to increased neuronal injury and death (Figs. 1 and 2).

Acknowledgements

We would like to thank our collaborators S. W. Barger, B. Behrens, B. Cheng, J. M. Carney, S. Christakos, E. M. Elliott, K. Kumar, I. Lieberburg, R. Mark, E. K. Michaelis, J. A. O'Keefe, L. C. Pettigrew, R. E. Rydel and R. M. Sapolsky, for their invaluable contributions to the research. We also appreciate the excellent technical assistance of M. M. Barger, S. Bose, Y. Goodman, J. J. Joseph, S. E. Loughran, J. Mattson, J. Oeltgen and Y. Zhang. Research in the authors laboratory was supported by grants to M.P.M. from the NIH, The Metropolitan Life Foundation, The Alzheimer's Association, The French Foundation for Alzheimer's Research, and by a fellowship to V.L.S. from the Kentucky Research Foundation.

References

Abe K, Tanzi RE, Kogure K (1991) Selective induction of Kunitz-type protease inhibitor domain-containing amyloid precursor protein mRNA after persistent focal ischemia in rat cerebral cortex. Neurosci Lett 125: 172–174

Andorn AC, Britton RS, Bacon RR (1990) Evidence that lipid peroxidation and total iron are increased in Alzheimer's brain. Neurobiol Aging 11: 316

Araki W, Kitaguchi N, Tokushima Y, Ishii K, Aratake H, Shimohana S, Nakamura S, Kimura J (1991) Trophic effects of β-amyloid precursor protein on cerebral cortical neurons in culture. Biochem Biophys Res Commun 181: 265–271

Ballarin M, Ernfors P, Lindefors N, Persson H (1988) Hippocampal damage and kainic acid injection induce a rapid increase in mRNA for BDNF and NGF in the rat brain. Exp Neurol 114: 35–43

Barger SW, Fiscus RR, Rydel RE, Culwell AR, Lieberburg I, Mattson MP (1993) Mechanistic studies of the depression of intraneuronal calcium concentration by secreted forms of β-amyloid precursor protein. Soc Neurosci Abstr 19: 1252

Bowling CA, Mutisya ME, Walker CL, Price LD, Cork CL, Beal FM (1993) Age dependent impairment of mitochondrial function in primate brain. J Neurochem 60: 1964–1967

Busciglio J, Lorenzo A, Yankner BA (1992) Methodological variables in the assessment of beta amyloid neurotoxicity. Neurobiol Aging 13: 609–612

Cai X, Golde T, Younkin S (1993) Release of excess amyloid β protein from a mutant amyloid β protein precursor. Science 259: 514–516

Carney JM, Starke-Reed PE, Oliver CN, Landum RW, Cheng MS, Wu JF, Floyd RA (1991) Reversal of age-related increase in brain protein oxidation, decrease in enzyme activity, and loss in temporal and spatial memory with chronic administration of the spin-trapping compound N-tert-butyl-alpha-phenylnitrone. Proc Natl Acad Sci 88: 3633–3636

Chartier-Harlin M, Crawford F, Houlden H, Warren A, Hughes D, Fidani L, Goate A, Rossor M, Roques P, Hardy J, Mullan M (1991) Early-onset Alzheimer's disease caused by mutations at codon 717 of the beta-amyloid precursor protein gene. Nature 353: 844–846

Cheng B, Mattson MP (1991) NGF and bFGF protect rat hippocampal and human cortical neurons against hypoglycemic damage by stabilizing calcium homeostasis. Neuron 7: 1031–1041

Cheng B, Mattson MP (1992a) IGF-I and IGF-II protect cultured hippocampal and septal neurons against calcium-mediated hypoglycemic damage. J Neurosci 12: 1558–1566

Cheng B, Mattson MP (1992b) Glucose deprivation elicits neurofibrillary tangle-like antigenic changes in hippocampal neurons: Prevention by NGF and bFGF. Exp Neurol 117: 114–123

Citron M, Oltersdorf T, Haass C, McConlogue L, Hung AY, Seubert P, Vigo-Pelfrey C, Lieberburg I, Selkoe DJ (1993) Mutation of the beta-amyloid precursor protein in familial Alzheimer's disease increases beta-protein production. Nature 360: 672–674

Clark AW, Parhad IM (1989) Expression of neuronal mRNAs in Alzheimer's type degeneration of the nervous system. Can J Neurol Sci 16: 477–482

Cline HT, Constantine-Paton M (1990) NMDA receptor agonist and antagonists alter retinal ganglion cell arbor structure in the developing frog retinotectal projection. J Neurosci 10: 1197–1216

Collazo D, Takahashi H, McKay RDG (1992) Cellular targets and trophic functions of neurotrophin-3 in the developing rat hippocampus. Neuron 9: 643–656

Collinridge GL, Bliss TVP (1987) NMDA receptors: Their role in long-term potentiation. Trends Neurosci 10: 288–293

Copani A, Koh J-Y, Cotman CW (1991) β-amyloid increases neuronal susceptibility to injury by glucose deprivation. Neuro Report 2: 763–765

Corder EH, Saunders AM, Strittmatter WJ, Schmechel DE, Gaskell PC, Small GW, Roses AD, Haines JL, Pericak-Vance MA (1993) Gene dose of apolipoprotein E type, 4 allele and the risk of Alzheimer's disease in late onset families. Science 261: 921–923

Dudek SM, Johnson GVW (1993) Transglutaminase catalyzes the formation of sodium dodecyl sulfate-insoluble, Alz-50-reactive polymers of tau. J Neurochem 61: 1159–1162

Elliott EM, Sapolsky RM (1992) Corticosterone enhances kainic acid-induced calcium elevation in cultured hippocampal neurons. J Neurochem 59: 1033–1040
Elliott EM, Mattson MP, Vanderkish P, Lynch G, Chang I, Sapolsky RM (1993) Corticosterone exacerbates kainate-induced alterations in hippocampal tau immunoreactivity and spectrin proteolysis in vivo. J Neurochem 61: 57–67
Esch FS, Keim PS, Beattie EC, Blacher RW, Culwell AR, Oltersdorf T, McClure D, Ward PJ (1990) Cleavage of amyloid β peptide during constitutive processing of its precursor. Science 248: 1122–1124
Frim MD, Short PM, Rosenberg SW, Simpson J, Xandra AB, Breakfield O, Isacson O (1993) Local protective effects of nerve growth factor-secreting fibroblasts against excitotoxic lesions in the rat striatum. J Neurosurg 78: 267–273
Gall C, Isackson PJ (1989) Limbic seizures increase neuronal production of mRNA for nerve growth factor. Science 245: 758–761
Glenner GG, Wong CW (1984) Alzheimer's disease: initial report of the purification and characterization of a novel cerebrovascular amyloid protein. Biochem Biophys Res Commun 120: 885–890
Gluckman P, Klempt N, Guan J, Mallard C, Sirimanne E, Dragunow M, Klempt M, Singh K, Williams C, Nikolics K (1992) A role for IGF-1 in the rescue of CNS neurons following hypoxic-ischemic injury. Biochem Biophys Res Commun 182: 593–599
Goate A, Chartler-Harlin MC, Mullan M, Brown J, Crawford F, Fidani L, Gluffra L, Haynest A, Irving N, James L, Mant R, Newton P, Rooke K, Roques P, Talbot C, Williamson R, Rossor M, Owen M, Hardy J (1991) Segregation of a missense mutation in the amyloid precursor protein gene with familial Alzheimer's disease. Nature 349: 704–707
Golde TE, Estus S, Younkin LH, Selkoe DJ, Younkin SG (1992) Processing of the amyloid protein precursor to potentially amyloidogenic derivatives. Science 255: 728–730
Goldgaber D, Harris HW, Hla T, Maciag T, Donnelly RJ, Jacobsen JS, Vitek MP, Gajdusek DC (1989) Interleukin 1 regulates synthesis of amyloid β-protein precursor mRNA in human endothelial cells. Proc Natl Acad Sci USA 86: 7606–7610
Goldgaber D, Lerman MI, McBridge OW, Saffiotti U, Gajdusek DC (1987) Characterization and chromosomal localization of a cDNA encoding brain amyloid of Alzheimer's disease. Science 235: 877–890
Goodman Y, Rydel RE, Culwell AR, Mattson MP (1993) Secreted forms of βAPP protect hippocampal neurons against Aβ toxicity and oxidative injury. Soc Neurosci Abstr 19: 1251
Greenamyre JT, Young AB (1989) Excitatory amino acids and Alzheimer's disease. Neurobiol Aging 10: 593–602
Greenberg CS, Birckbichler PJ, Rice RH (1991) Transglutaminases: multifunctional cross-linking enzymes that stabilize tissue. FASEB J 5: 3071–3077
Greenwald BS, Mathe AA, Mohs RC, Levy MI, Johns CA, Davis KL (1986) Cortisol and Alzheimer's disease. II. Dexamethasone suppression, dementia severity, and affective symptoms. Am J Psychiatry 143: 442–446
Grundke-Iqbal I, Iqbal K, Turg Y-C, Quinlan M, Wisniewski HM, Binder LI (1986) Abnormal phosphorylation of the microtubule-associated protein tau in Alzheimer cytoskeletal pathology. Proc Natl Acad Sci USA 83: 4913–4917
Haass C, Schlossmacher MG, Hung AY, Vigo-Pelfrey C, Mellon A, Ostaszewski BL, Lieberburg I, Koo EH, Schenk D, Teplow DB, Selkoe DJ (1992) Amyloid beta-peptide is produced by cultured cells during normal metabolism. Nature 359: 322–325
Harmon D (1956) Aging: a theory based on free radical and radiation chemistry. J Gerontol 11: 298–300
Harmon D (1987) Modern biological theories of aging. Raven Press, New York, p 81
Hoyer S (1988) Glucose and related brain metabolism in dementia of Alzheimer type and its morphological significance. Age 11: 158–166

Hoyer S, Oesterreich K, Wagner O (1988) Glucose metabolism as the site of the primary abnormality in early-onset dementia of Alzheimer type? J Neurol 235: 143–148

Iwamoto N, Thangnipon W, Crawford C, Emson PC (1991) Localization of calpain immunoreactivity in the senile plaques and in neurones undergoing neurofibrillary degeneration in Alzheimer's disease. Brain Res 561: 177–180

Jackson GR, Apfell L, Werrbach-Perez K, Perez-Polo JR (1990) Role of nerve growth factor in oxidant-antioxidant balance and neuronal injury. I. Stimulation of hydrogen peroxide resistance. J Neurosci Res 25: 360–368

Johnson GV, Litersky JM, Jope RS (1991) Degradation of microtubule-associated protein 2 and brain spectrin by calpain: a comparative study. J Neurochem 56: 1630–1638

Kalaria RN, Harik SI (1989) Reduced glucose transporter at the blood-brain barrier and in cerebral cortex in Alzheimer's disease. J Neurochem 53: 1083–1088

Kang J, Lemaire H-G, Unterbeck A, Salbaum JM, Masters CL, Grzesckik K-H, Multhaup G, Beyreuther K, Muller-Hill B (1987) The precursor of Alzheimer's disease amyloid A4 protein resembles a cell-surface receptor. Nature 325: 733–736

Kater SB, Mattson MP, Cohan CS, Connor JA (1988) Calcium regulation of the neuronal growth cone. Trends Neurosci 11: 315–321

Khacharturian ZS (1989) The role of calcium regulation in brain aging: reexamination of a hypothesis. Aging 1: 17–34

Koh J-Y, Yang LL, Cotman CW (1990) β-amyloid protein increases the vulnerability of cultured cortical neurons to excitotoxic damage. Brain Res 533: 315–320

Koo EH, Sisodia SS, Archer DR, Martin LJ, Weidemann A, Beyreuther K, Fischer P, Masters CL, Price DL (1990) Precursor of amyloid protein in Alzheimer disease undergoes fast anterograde axonal transport. Proc Natl Acad Sci USA 87: 1561–1565

Koistinaho J, Alho H, Hervonen A (1990) Effect of vitamin E and selenium on the aging peripheral neurons of the male Sprague-Dawley rat. Mech Aging Dev 51: 63–72

Landfield PW (1987) Increased calcium current hypothesis of brain aging. Neurobiol Aging 8: 346–347

Mattson MP (1990) Antigenic changes similiar to those seen in neurofibrillary tangles are elicited by glutamate and calcium influx in cultured hippocampal neurons. Neuron 4: 105–117

Mattson MP (1992) Calcium as sculptor and destroyer of neural circuitry. Exp Gerontol 27: 29–49

Mattson MP, Kater SB (1987) Calcium regulation of neurite elongation and growth cone motility. J Neurosci 7: 4034–4043

Mattson MP, Kater SB (1988) Isolated hippocampal neurons in cryopreserved long term culture: development of neuroarchitecture and sensitivity to NMDA. Int J Dev Neurosci 6: 439–452

Mattson MP, Dou P, Kater SB (1988a) Outgrowth-regulating actions of glutamate in isolated hippocampal pyramidal neurons. J Neurosci 8: 2087–2100

Mattson MP, Lee RE, Adams ME, Guthrie PB, Kater SB (1988b) Interactions between entorhinal axons and target hippocampal neurons: a role for glutamate in the development of hippocampal circuitry. Neuron 1: 865–876

Mattson MP, Kater SB (1989) Excitatory and inhibitory neurotransmitters in the generation and degeneration of hippocampal neuroarchitecture. Brain Res 478: 337–348

Mattson MP, Murrain M, Guthrie PB, Kater SB (1989) Fibroblast growth factor and glutamate: opposing roles in the generation and degeneration of hippocampal neuroarchitecture. J Neurosci 9: 3728–3740

Mattson MP, Rychlik B (1990) Cell culture of cryopreserved human fetal cerebral cortical and hippocampal neurons: neuronal development and responses to trophic factors. Brain Res 522: 204–214

Mattson MP, Engle MG, Rychlik B (1991a) Effects of elevated intracellular calcium levels on the cytoskeleton and tau in cultured human cortical neurons. Mol Chem Neuropathol 15: 117–142
Mattson MP, Rychlik B, Chu C, Christakos S (1991b) Evidence for calcium-reducing and excitoprotective roles for the calcium-binding protein calbindin-D28k in cultured hippocampal neurons. Neuron 6: 41–51
Mattson MP, Tomaselli KJ, Rydel RE (1993a) Calcium-destabilizing and neurodegenerative effects of aggregated β-amyloid peptide are attenuated by basic FGF. Brain Res 621: 35–49
Mattson MP, Cheng B, Davis D, Bryant K, Lieberberg I, Rydel RE (1993b) β-amyloid peptides destabilize calcium homeostasis and render human cortical neurons vulnerable to excitotoxcity. J Neurosci 12: 379–389
Mattson MP, Kumar K, Wang H, Cheng B, Michaelis EK (1993c) Basic FGF regulates the expression of a functional 71 kDa NMDA receptor protein that mediates calcium influx and neurotoxicity in hippocampal neurons. J Neurosci 13: 4575–4588
Mattson MP, Cheng B, Culwell AR, Esch FS, Lieberberg I, Rydel RE (1993d) Evidence for excitoprotective and intraneuronal calcium-regulating roles for secreted forms of β-amyloid precursor protein. Neuron 10: 243–254
Milward EA, Papadoopoulos R, Fuller SJ, Moir RD, Small D, Beyreuter K, Masters C (1992) The amyloid protein precursor of Alzheimer's disease is a mediator of the effects of nerve growth factor on neurite outgrowth. Neuron 9: 129–137
Mobley WC, Neve RL, Prusiner SB, McKinley MP (1988) Nerve growth factor increases mRNA levels for the prion protein and the beta-amyloid protein precursor in developing hamster brain. Proc Natl Acad Sci USA 85: 9811–9815
Mullan M, Crawford F, Axelman K, Houlden H, Lilius L, Winblad B, Lannfelt L (1992) A pathogenic mutation of probable Alzheimer's disease in the APP gene at the N-terminus of β-amyloid. Nature Genet 1: 345–347
Murrell J, Farlow M, Ghetti B, Benson MD (1991) A mutation in the amyloid precursor protein associated with hereditary Alzheimer's disease. Science 254: 97–99
Nakamura K, Wu G, Ledeen RW (1992) Protection of neuro-2a cells against calcium ionophore cytotoxicity by gangliosides. J Neurosci Res 31: 245–253
Nistico G, Ciriolo MR, Fiskin K, Iannone M, De Martino A, Rotilio G (1992) NGF restores decrease in catalase activity and increases superoxide dismutase and glutathione peroxidase activity in the brain of aged rats. Free Rad Biol Med 12: 177–181
Nitsch RM, Farber AS, Growdon HJ, Wurtman JR (1993) Release of amyloid β-protein precursor derivatives by electrical depolarization of rat hippocampal slices. Proc Natl Acad Sci USA 90: 5191–5193
Nitsch RM, Slack BE, Wurtman RJ, Growdon JH (1992) Release of Alzheimer amyloid precursor derivatives stimulated by activation of muscarinic acetylcholine receptors. Science 258: 304–307
Novelli A, Reilly JA, Lyska PG, Henneberry RC (1988) Glutamate becomes neurotoxic via the N-methyl-D-asparate receptor when intracellular energy levels are reduced. Brain Res 451: 205–212
Nozaki K, Finklestein SP, Beal MF (1993) Basic fibroblast growth factor protects against hypoxia/ischemia and NMDA neurotoxicity in neonatal rats. J Cereb Blood Flow Metab 13: 221–228
Pike C, Burdick D, Walencewicz A, Glabe C, Cotman C (1993) Neurodegeneration induced by β-amyloid peptides in vitro: the role of peptide assembly state. J Neurosci 13: 1676–1686
Pike CJ, Walencewicz AJ, Glabe CG, Cotman CW (1991) In vitro aging of β-amyloid protein causes peptide aggregation and neurotoxicity. Brain Res 563: 311–314
Pryor WA (1987) Modern biological theories of aging. Raven Press, New York, p 89
Quon D, Catalano R, Cordell B (1990) Fibroblast growth factor induces beta amyloid precursor mRNA in glial but not neuronal cultured cells. Biochem Biophys Res Commun 167: 96–102

Richardson JS, Subbarao KV, Ang LC (1990) Biochemical indices of peroxidation in Alzheimer's and control brains. Trans Am Soc Neurochem 21: 113
Roberts GW, Gentleman SM, Lynch A, Graham DI (1991) βA4 amyloid protein deposition in brain after head injury. Lancet 338: 1422–1423
Rothman SM, Olney JW (1986) Glutamate and the pathophysiology of hypoxic-ischemic brain damage. Ann Neurol 19: 105–111
Saito K-I, Elce JS, Hamos JE, Nixon RA (1993) Widespread activation of calcium-activated neutral proteinase (calpain) in the brain of Alzheimer disease: a potential molecular basis for neuronal degeneration. Proc Natl Acad Sci USA 90: 2628–2632
Sapolsky RM (1985) Glucocorticoid toxicity in the hippocampus: temporal aspects of neuronal vulnerability. Brain Res 359: 300–305
Sapolsky RM (1992) Do glucocorticoid concentrations rise with age in the rat? Neurobiol Aging 13: 171–174
Sautiere P-E, Sindou P, Couratier P, Hugon J, Wattez A, Delacourte A (1992) Tau antigenic changes induced by glutamate in rat primary culture model: a biochemical approach. Neurosci Lett 140: 206–210
Schellenberg GD, Bird TD, Wijsman EM, Orr HT, Anderson L, Nemens E, White JA, Bonnycastle L, Weber JL, Alonso ME, Potter H, Heston LL, Martin J (1992) Genetic linkage evidence for a familial Alzheimer's disease locus on chromosome 14. Science 258: 668–671
Selkoe DJ (1989) Biochemistry of altered brain proteins in Alzheimer's disease. Ann Rev Neurosci 12: 463–490
Selkoe DJ (1991) The molecular pathology of Alzheimer's disease. Neuron 6: 487–498
Seubert P, Vigo-Pelfrey C, Esch F, Lee M, Dovey H, Davis D, Sinha S, Schlossmacher M, Whaley J, Swindlehurst C, McCormack R, Wolfert R, Selkoe D, Lieberburg I, Schenk D (1992) Isolation and quantification of soluble Alzheimer's beta-peptide from biological fluids. Nature 359: 325–327
Seubert P, Olterdorf T, Lee MG, Barbour R, Blomquist C, Davis DL, Bryant K, Fritz LC, Galasko D, Thal LJ, Lieberburg I, Schenk DB (1993) Secretion of beta-amyloid precursor protein cleaved at the amino terminus of the beta-amyloid peptide. Nature 361: 260–263
Shigeno T, Mima T, Takakura K, Graham DI, Kato G, Hashimoto Y, Furukawa S (1991) Amelioration of delayed neuronal death in the hippocampus by nerve growth factor. J Neurosci 11: 2914–2919
Shoji M, Golde TE, Ghiso J, Cheung TT, Estus S, Shaffer LM, Cai X-D, McKay DM, Tintner R, Frangione B, Younkin SG (1992) Production of the Alzheimer amyloid beta protein by normal proteolytic processing. Science 258: 126–129
Siesjo BK, Bengtsson F, Grampp W, Theander S (1989) Calcium, excitotoxins, and neuronal death in the brain. Ann NY Acad Sci 568: 234–251
Siman R, Card JP, Nelson RB, Davis LG (1989) Expression of β-amyloid precursor protein in reactive astrocytes following neuronal damage. Neuron 3: 275–285
Siman R, Noszek JC, Kegerise C (1989) Calpain I activation is specifically related to excitatory amino acid induction of hippocampal damage. J Neurosci 9: 1579–1590
Simon RP, Swan JH, Griffiths T, Meldrum BS (1984) Blockade of N-methyl-D-aspartate receptors may protect against ischemic damage in the brain. Science 226: 850–852
Sisodia SS, Koo EH, Beyreuther K, Unterbeck A (1990) Evidence that beta-amyloid protein in Alzheimer's disease is not derived by normal processing. Science 248: 492–495
Smith CD, Carney JM, Starke-Reed PE, Oliver CN, Stadtman ER, Floyd RA, Markesbery WR (1991) Excess brain protein oxidation and enzyme dysfunction in normal aging and in Alzheimer disease. Proc Natl Acad Sci USA 88: 10540–10543
Smith-Swintosky VL, Pettigrew LC, Craddock SD, Culwell A, Rydel RE, Mattson MP (1993) Secreted forms of β-amyloid precursor protein protect against ischemic brain injury. Soc Neurosci Abstr 19: 1251
Stein-Behrens BA, Sapolsky RM (1992) Stress, glucocorticoids, and aging. Aging (Milano) 4: 197–210

Stein-Behrens B, Mattson MP, Chang I, Yeh M, Sapolsky RM (1994) Stress exacerbates neuron loss and cytoskeletal pathology in the hippocampus. J Neurosci 14 (in press)

Stephenson DT, Rash K, Clemens JA (1992) Amyloid precursor protein accumulates in regions of neurodegeneration following focal cerebral ischemia in the rat. Brain Res 593: 128–135

Strittmater WJ, Saunders AM, Schmechel, Pericak-Vance M, Enghild J, Salvesen GS, Roses AD (1993a) Apolipoprotein E: High aviditiy binding to β-amyloid and increased frequency of type 4 allele in late-onset familial Alzheimer disease. Proc Natl Acad Sci USA 90: 1977–1981

Strittmater WJ, Huang DY, Bhasin R, Roses AD, Goldlaber D (1993b) Avid binding of β amyloid to its own precursor. Exp Neurol 122: 327–334

Subbarao KV, Richardson JS, Ang LC (1990) Autopsy samples of Alzheimer's cortex show increased peroxidation in vitro. J Neurochem 55: 342–345

Tariot PM, Cohen RM, Sunderland T, Newhouse PA, Yount D, Mellow AM, Weingartner H, Meuller EA, Murphy DL (1987) L-deprenyl in Alzheimer's disease. Arch Gen Psychiatry 44: 427–433

Tariot PM, Sunderland T, Cohen RM, Newhouse PA, Meuller EA, Murphy DL (1988) Tranylcypromine compared with L-deprenyl in Alzheimer's disease. J Clin Psychopharm 8: 23–27

Tomioka C, Nishioka K, Kogure K (1992) Absence of early destructive changes of cytoskeletal proteins after transient ischemia in the rat. Brain Res 586: 352–357

Wakita H, Tomimoto H, Akiguchi I, Ohnishi K, Nakamura S, Kimura J (1992) Regional accumulation of amyloid beta/A4 protein precursor in the gerbil brain following transient cerebral ischemia. Neurosci Lett 146: 135–138

Wallace WC, Bragin V, Robakis NK, Sambamurti K, VanderPutten D, Merril CR, Davis KL, Santucci AC, Haroutunian V (1991) Increased biosynthesis of Alzheimer amyloid precursor protein in the cerebral cortex of rat with lesions of the nucleus basalis of Meynert. Brain Res Mol Brain Res 10: 173–178

Wasco W, Brook JD, Tanzi RE (1993) The amyloid precursor-like protein (APLP) gene maps to the long arm of human chromosome 19. Genetics 15: 237–239

Wisniewski K, Jervis GA, Moretz RC, Wisniewski HM (1979) Alzheimer neurofibrillary tangles in diseases other than senile and presenile dementia. Ann Neurol 5: 288–294

Wisniewski T, Frangione B (1992) Apolipoprotein E: a pathological chaperone protein in patients with cerebral and systemic amyloid. Neurosci Lett 135: 235–238

Yanagihara T, Brengman JM, Mushynski WE (1990) Differential vulnerability of microtubule components in cerebral ischemia. Acta Neuropathol 80: 499–505

Yankner BA, Duffy LK, Kirschner DA (1990) Neurotrophic and neurotoxic effects of amyloid β protein: reversal by tachykinin neuropeptides. Science 250: 279–282

Zhang Y, Tatsuno T, Carney J, Mattson MP (1993) Basic FGF, NGF, and IGFs protect hippocampal neurons against iron-induced degeneration. J Cereb Blood Flow Metab 13: 378–388

Authors' address: Dr. V.L. Smith-Swintosky, 211 Sanders-Brown Building, University of Kentucky, Lexington, KY 40536-0230, U.S.A.

J Neural Transm (1994) [Suppl] 44: 47–60

An aggregate brain cell culture model for studying neuronal degeneration and regeneration

S. S. Chatterjee and **M. Nöldner**

Department of Pharmacology, Research Laboratories, Dr. Willmar Schwabe Arzneimittel, Karlsruhe-Durlach, Federal Republic of Germany

Summary. Rotation-mediated aggregating cell cultures from fetal rat telencephalons containing glial and neuronal cells mature in a fashion comparable to that known to occur in brain in vivo. Large aggregates of 300–500 μM diameters can now reproducibly be cultivated and maintained for more than 40 days in a well defined serum free medium. Validity of the use of such cultures for in vitro studies of various physiological, pharmacological and toxicological phenomenon has already been demonstrated. In this communication some observations suggesting the usefulness of such cultures for pharmacological studies clarifying the possible effects of drugs and other agents on excitatory amino acid induced pathological processes will be presented. The advantages and limitations of the use of aggregated brain cell culture based models for the development of agents potentially useful for the treatment of aging and dementia will also be discussed.

Introduction

It is now more than thirty years that Muscona (1960, 1961) observed that dispersed immature cells of any fetal organ — including those from fetal brain — reassociate spontaneously in vitro, giving rise to three dimensional organotypic cultures. Since then simplified methodologies for obtaining aggregating brain cell cultures derived from various parts of fetal mammalian brain, including those from human fetus (Pulliam et al., 1988), have been described by several groups (DeLong, 1970; Garber and Muscona, 1972a,b; Garber, 1977; Honegger and Richelson, 1976; Honegger et al., 1979; Honegger, 1985; Muscona, 1965). Numerous morphological and biochemical investigations have demonstrated (DeLong, 1970; Garber and Muscona, 1972a,b; Honegger, 1985; Seeds, 1973) that such aggregating brain cell cultures are able to mimic several morphogenetic events like cell migration, synaptogenesis and myelination and that they in vitro eventually attain a high degree of cellular differentiation and maturation similar to that observed in the tissue in vivo. Indeed, through the use of such cultures valuable data on the developmental processes of mammalian brain have been generated (Kozak et al., 1977; Matthieus et al., 1978; Romijn et al., 1981; Trapp et al., 1979) and their usefulness for testing potential neurotoxicities of drugs and chemicals has also been suggested (Halks-Miller, 1991; Honegger and Werffeli, 1988).

The rotation-mediated aggregating brain cell cultures provide thus an interesting alternative between the monolayer cell cultures which are easy to manipulate but very restricted in cell-cell and cell-matrix interactions, and explant cultures which are truely organotypic but limited in quantity and reproducibility which make them (Laerium, 1985) unsuitable for routine applications or for multidisciplinary studies. However, despite their unique qualities, such brain cell cultures have only rarely (Atterwill and Meakin, 1990; Halks-Miller et al., 1986, 1991) been used for studies related to aging, dementia or neurodegeneration and till now no attempts have been made to develop pharmacological screening models based on the use of rotation-mediated aggregating brain cell cultures for potential neuroprotective or antidementia agents. Some essential prerequisit for developing such pharmacological models are that the neurodegenerative and other processes associated with aging and dementia are inducible and reproducibly quantifiable in such cultures, and that the relevant effects of known neuroprotective and other CNS-active agents are detectable in rotation-mediated aggregates of the neuronal tissues.

Survey of the available information on the aggregating brain cell cultures maintained in serum free medium clearly indicates that depending on the age and part of the fetal brain tissue and on the cultivation condition used, the morphological and biochemical characteristics of the aggregates can vary. Thus, modulation of thyroid hormone or nerve growth factor (NGF) concentration in the medium influences the number of cholinergic neurons in the aggregates (Honegger and Lenoir, 1980, 1982) and the process of myelination in such cultures can be drastically influenced by modifying the culturing conditions (Matthieu et al., 1980). In addition, reaggregated brain cell cultures derived from different parts of rat brain markedly differ in their densities of specific neuronal populations (Garber and Muscona, 1972a). Such observations clearly indicate that until and unless the culture conditions are appropriately defined and standardized, no reproducible in vitro pharmacological studies can be conducted on such cultures, and that the optimal use of brain cell reaggregates might also depend on the type of culturing conditions used. Thus, the flexibility of rotation-mediated reaggregating brain cell culture systems could be an advantage for some types of studies but at the same time could be the reason for incoherent, contradictory or irrelevant (for the in vivo situation) observations.

In this communication, some observations made during studies directed towards standardizing an in vitro pharmacological screening model for detecting the potential effects of agents on excitatory amino acid mediated processes will be presented and some potential usefulness and limitations of the aggregate brain cell cultures for in vitro studies will be discussed.

Table 1. Biochemical methods used

Parameter measured	Units	Ref.
Protein (Prot.)	mg/flask	Lowry et al. (1951)
Desoxyribonucleic acid (DNA)	µg/mg Prot.	Labarca and Paigen (1980)
Acetylcholine-esterase (AChE)	nM/min/mg Prot.	Ehrenpreis et al. (1967), Reed et al. (1966), Wilson et al. (1972)
Choline-acetyl-transferase (ChAT)	pM/min/mg Prot.	Schrier and Schuster (1967), Wilson et al. (1972)
Glutamin-synthetase (GS)	nM/min/mg Prot.	Patel et al. (1982), Pishak and Phillips (1979)
Glutamic-acid-decarboxylase (GAD)	pM/min/mg Prot.	Wilson et al. (1972), Wingo and Awapara (1950)
2′-3′-Cyclic-nucleotide-3′-phosphohydrolase (CNPase)	nM/min/mg Prot.	Sogin (1976)
Lactate-dehydrogenase (LDH)	nM/min/flask	Korzeniewski and Callewaert (1983)

Materials and methods

Rotation-mediated aggregating cell cultures were obtained from cells of mechanically dissociated fetal (15–16 d gestation) rat telencephalon using the method described by Honegger (Honegger and Richelson, 1976; Honegger, 1985). After repeated washing and centrifugation, 4 ml aliquotes of the cell suspension (1 * 10–7 cells/ml) were seeded into 25 ml DeLong flasks and kept under constant rotation (68 rpm) at 37°C in a humidified atmosphere of 10% CO_2 in air. After two days the cell aggregates were transfered to 50 ml flasks and the volume of the culture medium adjusted to 8 ml. The medium was changed (exchange of 5 ml) every 3 days until day 11 and every other day afterwards. Due to the observed increase in the metabolic rate of the cultures with progressing maturation, the cultures were split at day 19 by redistributing the aggregates of each flask into two separate cultures. This technique has permitted us to maintain the cultures for more than 40 days in a highly differentiated state.

The details of the different incubation procedures used in the present study are described in the result and discussion section. For biochemical studies the cultures were harvested and washed with cold phosphate buffered saline (PBS) and homogenized. The sonicated homogenates were then stored frozen (−80°C) until analysis. The enzymatic and other methods used in this study are summarized in Table 1. For each experimental observation at least triplicates of the aggregates from the same batch of culture were used and each determination was conducted at least in duplicate samples.

Results and discussions

The importance of excitatory amino acid mediated processes for the normal functioning of human brain and in several neurodegenerative and other pathological phenomenon associated with aging are now well recognized (Whetsell and Shapira, 1993). Various experimental models for studies related to such processes have also been proposed (Wilmot, 1989). To our knowledge, till now only two reports have, however, appeared (Halks-

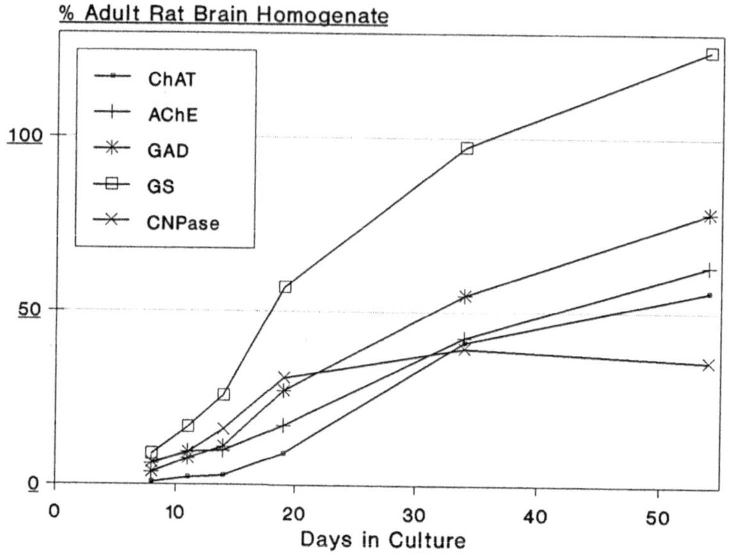

Fig. 1. Neuronal and glial marker enzyme levels in serum-free aggregating fetal-rat telencephalon cultures

Miller et al., 1989; Honegger and Richelson, 1977) which describe the effects of excitatory amino acids on aggregate brain cell cultures. In one of these studies (Honegger and Richelson, 1977) the excitotoxic effects of kainic acid were found to be more prominent in older — i.e. in more differentiated — cultures, and that as compared to the aggregates derived from other fetal rat brain regions the telencephalon derived cultures were more susceptible to the neurotoxin. In the other study (Halks-Miller et al., 1989), although the types and ages of the cell cultures used are not specified, the aggregated cultures were shown to contain high concentrations of NMDA-receptors and NMDA was shown to be toxic to such cultures. These reports encouraged us to compare the effects of the three excitatory amino acid ligands glutamic acid, NMDA and kainic acid on mature and well differentiated aggregate cell cultures from rat telencephalons, so that a versatile in vitro pharmacological screening model for potential neuroprotective agents could eventually be developed and the effects of drugs and other agents on the excitatory amino acid system could be clarified in details.

The aggregate cell cultures from rat telencephalon used in the studies contained astrocytes, oligodendrocytes and neurons arranged in a pattern similar to that reported by others (Honegger and Richelson, 1977; Honegger, 1985). The developmental changes in the marker enzyme profile of our aggregate cell cultures — shown in Fig. 1 — were similar to those reported by Honegger (1985) and others. After the first seven days in culture the levels of all the marker enzymes studied increased gradually. However, compared to the neuronal marker enzymes GAD, ChAT and AChE the developmental patterns of the two glial enzymes GS and CNPase were distinctly different. Such data indicate that under the culturing conditions

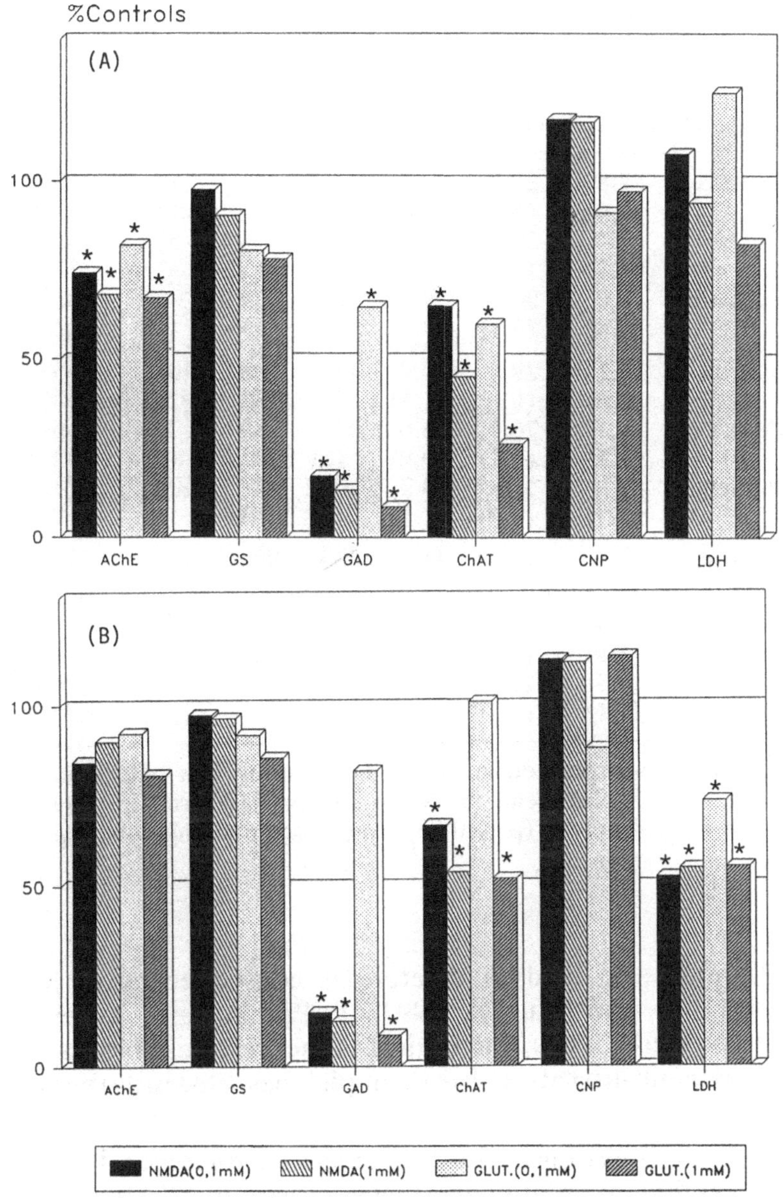

Fig. 2. Effects of NMDA and glutamic acid (0.1 and 1 mM) on aggregate cell cultures maintained in vitro for 15 (*A*) or 22 (*B*) days. The values significantly different from corresponding controls ($p < 0.05$) are marked with *

used, the aggregates contain disproportionately higher numbers of astrocytes and lower numbers of oligodendrocytes and that after 10–15 days the ratio of the numbers of glial cells to neurons in the aggregates continuously increase as a function of days the cultures are maintained in vitro.

It is now well established that cell-cell interactions between neurons and glial cells are vital for mutual differentiation, development and functioning of the central nervous system and that preturbation or damage of non-

Table 2. Effects of NMDA or glutamic acid (100 μM) on the aggregated rat brain cell cultures maintained in vitro for different periods

Days (in vitro)	Treatment	Protein	DNA	AChE	GS	GAD	ChAT	CNPase	LDH
15	Controls	7.8	28.0	11.6	41	431	80	1634	5322
		(0.4)	(4.3)	(1.0)	(2)	(183)	(15)	(18)	(535)
	NMDA	7.4	29.0	8.6*	40	74*	52*	1924	5732
		(1.2)	(3.4)	(1.2)	(3)	(11)	(7)	(379)	(1211)
	Glutamat	8.0	30.2	9.5*	33	279*	48*	1487	6653
		(0.8)	(3.1)	(0.1)	(4)	(46)	(9)	(128)	(1288)
22	Controls	10.1	29.5	19.7	110	598	108	2551	12871
		(0.8)	(6)	(1.5)	(6)	(85)	(26)	(535)	(1879)
	NMDA	7.1	26.7	16.6	107	89*	72*	2869	6734*
		(1.1)	(5.7)	(0.9)	(8)	(35)	(3)	(322)	(1726)
	Glutamat	7.8	21.8	18.2	101	489	109	2242	9438*
		(1.3)	(4.7)	(1.7)	(12)	(53)	(13)	(690)	(2374)
30	Controls	5.7	11.6	41.2	80	902	129	1710	—
		(1.2)	(0.7)	(4.6)	(25)	(328)	(60)	(690)	
	NMDA	4.4	12.4	42.1	98	354*	89*	1610	—
		(1.0)	(1.1)	(7.8)	(28)	(212)	(14)	(260)	
	Glutamat	4.7	11.5	42.5	89	782	159	2240	—
		(1.1)	(0.7)	(4.5)	(11)	(331)	(36)	(1320)	

The amino acids were incorporated in the medium during 24 h before harvesting the aggregates. Each value is the mean (± SD) of three flasks run in parallel. Values statistically significantly different ($p < 0.05$) from corresponding controls are marked with *

neuronal cells may also lead to deleterious consequences in juxtaposed neurons (Eddleston and Mucke, 1993; LoPachin and Aschner, 1993; Schousboe et al., 1993). Thus, the critical influence of astrocytes on excitatory amino acid neurotoxicities, for example, has elegantly been demonstrated by Rosenberg and Aizenman (1989). They report that cortical neurons cocultured in presence of astrocyte-poor supplements were ca. 100 fold more susceptible to glutamate toxicity than in those cocultured in presence of astrocyte-rich supplements. Such observations could suggest that astrocyte enriched aggregate cultures could as well be less susceptible to glutamate or NMDA inducible neurotoxicities and that depending on the days the aggregates are maintained in vitro, the excitotoxicities of the amino acids will vary.

In an initial study, therefore, the effects of 100 μM NMDA or glutamic acid on the aggregate cell cultures maintained in vitro for 15, 22 or 30 days were studied. The results of this study, summarized in Table 2, clearly demonstrate that both the amino acids only specifically reduced the neuronal marker enzyme levels (i.e. those of AChE, ChAT and GAD) in the aggregates and did not significantly alter the glial marker enzyme levels (GS and CNPase) or protein and DNA contents. LDH levels in the aggregates

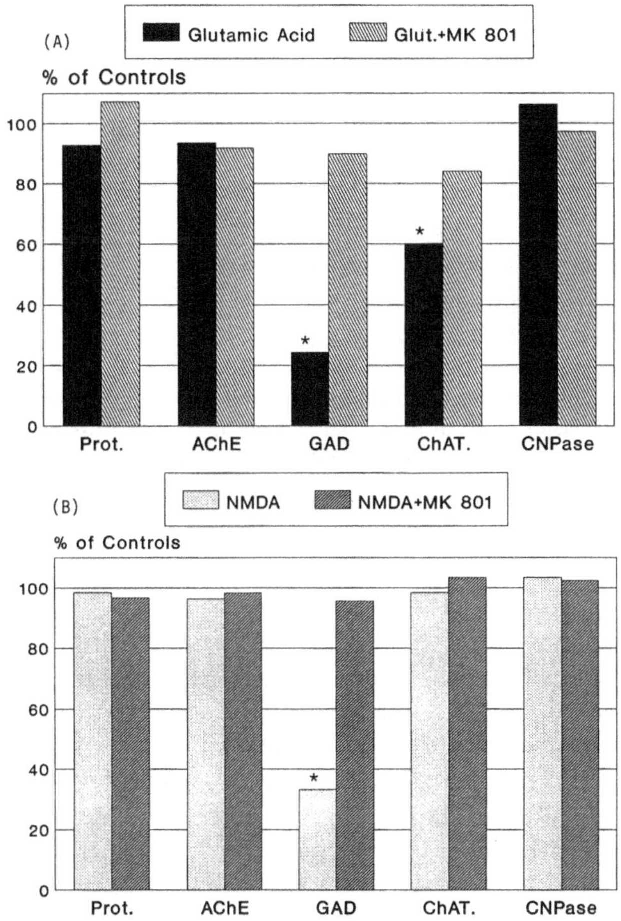

Fig. 3. Inhibition of the effects of 1 mM glutamic acid (*A*) or 10 μM NMDA (*B*) on 20 day old cultures by MK 801 (1 μM). The only values statistically significantly different from controls ($p < 0.05$) are marked with *

were also reduced by the excitotoxic amino acids. Quantitatively, however, the neurotoxicities of NMDA and glutamic acid were dependent on the maturation and/or differentiation status of the cultures, and in equimolar concentrations, NMDA was always more neurotoxic in the aggregates than glutamic acid. Further, in contrast to the 15 day old aggregates, no evidence of glutamic acid toxicity was observable in 22 or 30 day old cultures and the effects of NMDA was more pronounced in 15 or 22 day old cultures than in cultures maintained in vitro for 30 days.

Even after higher concentrations of glutamic acid or NMDA the profile of their toxic effects on the aggregates did not qualitatively change drastically from that observed after 100 μM of the amino acids. Quantitatively, however, as shown in Fig. 2, the effects of both the amino acids were more prominent after higher concentration and statistically highly significant effects of 1 mM glutamic acid could easily be quantified in 22 day old aggregate cultures. Such observations add support to the suggestion (Rosen-

Table 3. Effects of kainic acid (1 mM), on aggregated rat brain cell cultures maintained in vitro for 21 days

Treatment	Prot.	DNA	AChE	GS	GAD	ChAT	CNP	LDH
Controls	9.6 (1.4)	21.1 (1.4)	19.4 (2.3)	110 (15)	604 (47)	102 (13)	1814 (162)	12789 (998)
Kainate	6.5* (0.3)	26.0 (4.0)	12.3* (0.9)	118 (17)	86* (98)	32* (5)	1249* (216)	5084* (279)
Kainate + MK 801	6.1* (1.4)	24.3 (6.7)	11.6* (1.9)	92 (19)	102* (123)	29* (7)	1614 (363)	4814* (633)

Kainic acid was incorporated in the medium during 24 h, thereafter the cultures were maintained for two further days in standard medium. In the kainate + MK 801 group, the antagonist (1 μM) was added together with kainic acid and was also incorporated in the medium after kainate was removed. Each value is the mean (± SD) of four flasks run in parallel. Values statistically significantly different ($p < 0.05$) from corresponding controls are marked with *

berg and Aizenman, 1989) that neurotoxicities of excitatory amino acids are effectively modulated by the glial cells of the cerebral tissue, and indicate further that the neurotoxicity of glutamic acid might, as well, be mainly due to its effects on NMDA-type excitatory amino acid receptor complexes.

This last suggestion is also in agreement with the results of studies on interaction with MK 801 (a specific NMDA-receptor antagonist) and NMDA or glutamate; and those conducted with kainic acid — an excitatory amino acid neurotoxin with negligible effects on NMDA-receptor complex. As shown in Fig. 3, the neurotoxic effects of NMDA or glutamic acid could be completely eliminated by co-incubating the aggregates with 1 μM MK 801. No excitatory amino acid induced LDH release was observed in the presence of the antagonist. In this concentration of MK 801 it had no effects on the measured parameters (results not shown).

In some preliminary experiments, no reproducible toxic effects of kainic acid on the aggregate brain cell cultures of different ages could be detected after incubations for 24 h. Such observations are in accordance with those reported by Honegger and Richelson (1977). However, as shown in Table 3, when the aggregates were maintained in culture for two further days (without kainic acid) after an initial 24 h incubation period in presence of the acid, the levels of the measured marker enzymes and other cellular parameters (except those of DNA and GS) were reduced. A drastic release of LDH in the medium was also observed. The observed effects of kainic acid were not influenced at all by co-incubation of the cultures with MK 801 (Table 3). Taken together, these observations strongly indicate that differences in the toxicity of NMDA and kainic acid can be demonstrated in aggregate brain cell cultures. By co-incubation with MK 801 the involvement of the NMDA-receptor complex in various neurotoxic phenomenon can also be estimated in these cultures.

If, however, the observed neurotoxic effects of NMDA and glutamic acid were indeed due to their effects on NMDA-receptor complex and that of kainic acid was mediated via other type(s) of excitatory amino acid receptor(s), different effects of magnesium and glycine concentrations in the medium on the two types of toxicities should be expected. Indeed, in some preliminary studies (Nöldner and Chatterjee, 1993) the effects of magnesium and glycine on the influences of NMDA and glutamic acid on the cultures were speculated. Quantitatively, however, the magnitude of the effects of magnesium or glycine observed were not comparable to those reported by others in different in vitro neuronal cell culture systems.

Due to the nature of the aggregate brain cell culture system and the long periods of incubation (24 h), the concentrations of glycine, magnesium and other constituents in the medium can not be exactly predicted or controlled in these studies. Till now we have attempted to resolve such problems by two different approaches. The first was to shorten the duration of incubation with the excitotoxins, and the other was to maintain the aggregates for a longer period before the experiments in a medium in which the concentrations of magnesium and glycine were deliberately altered.

Indeed, as shown in Table 4, the incubation period necessary to observe and quantify the neurotoxicity of NMDA in aggregate brain cell cultures (in our standard medium) could be reduced to 15 min (instead of 3 or 24 h). In addition, it could also be demonstrated that maintenance of cultures for 24 or 48 h after treating with NMDA for 15 min does not modulate (quantitatively and qualitatively) the observed effects. Similar results were also achieved in the case of glutamic acid. Till now we do not have experimental

Table 4. Effects of NMDA (100 μM) on aggregated rat brain cell cultures maintained in vitro for 20 days

Incub. Period	Harvest after	Group	Prot.	DNA	AChE	GS	GAD	ChAT	CNP	LDH
15 min	24 h	Cont.	8.8	23.3	38.3	99	498	81	3890	13456
			(0.7)	(0.8)	(4.8)	(8)	(134)	(29)	(396)	(2752)
		NMDA	6.6*	27.1	37.2	108	91*	49	4594*	7849*
			(0.7)	(1.1)	(4.1)	(8)	(35)	(15)	(381)	(989)
15 min	70 h	Cont.	6.8	26.2	49.7	153	659	99	4775	12722
			(0.6)	(2.5)	(2.7)	(14)	(161)	(49)	(587)	(1144)
		NMDA	5.4*	28.0	40.3	218*	37*	71	6438*	7654*
			(0.3)	(3.5)	(10.8)	(8)	(27)	(8)	(411)	(940)
3 h	24 h	Cont.	8.9	25.4	41.0	112	468	83	4189	10340
			(0.4)	(4.7)	(5.1)	(8)	(129)	(36)	(357)	(1687)
		NMDA	7.0*	26.7	36.8	120	78*	53	5093*	8639
			(0.7)	(1.9)	(4.9)	(21)	(43)	(24)	(347)	(1422)

The amino acid was incorporated in the medium for 15 min or 3 h and the aggregates were harvested after a further 24 or 70 h. Each value is the mean (± SD) of at least three flasks run in parallel. Values statistically significantly different ($p < 0.05$) from corresponding controls are marked with *

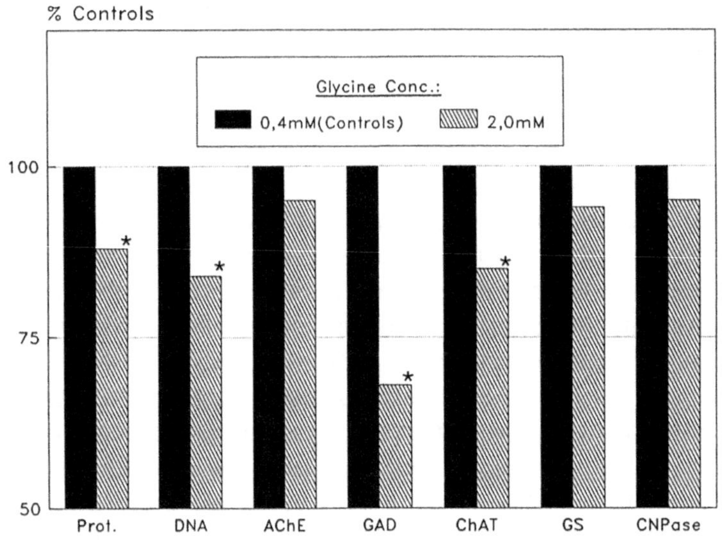

Fig. 4. Comparison of the spectrum of marker enzymes and other parameters in the aggregate cell cultures maintained in standard medium (0.4 mM glycine) for 21 days and in cultures in which glycine concentration was increased to 2 mM for 24 h during day 20–21. The significant changes ($p < 0.05$) are marked with *

data on incubation periods shorter than 15 min or maintenance periods less than 24 h.

Attempts to maintain the aggregates in media with altered magnesium or glycine concentrations led us to some unexpected observations which possibly could explain the quantitative discrepancies mentioned above. Increasing the glycine concentration in the medium of 20 days old cultures from 0.4 to 10 mM for 24 h effected the marker enzyme levels and other parameters in a way comparable to those observed after incubation of similar cultures in standard medium with 10 µM NMDA for 24 h. Significant amounts of LDH are also released in the medium when the aggregates are maintained for 24 h in a medium with 10 mM glycine. Although, increasing glycine concentrations from 0.4 to 2 mM in the medium of 20 days old cultures for 24 h did not release any measurable LDH in the medium, the marker enzyme levels and other measured parameters in the aggregates were significantly altered (Fig. 4). Such observations indicate that glycine itself can induce toxicities in aggregate brain cell cultures and that this amino acid could have modulating effects on differentiation and maturation of the brain cells. In a detailed study on the amino acid concentrations in the medium of the rotation-mediated aggregating brain cell cultures Honegger (1985) also pointed out the regulating role of glycine on in vitro cellular development.

The results of the study in which magnesium concentration in the medium was varied between 0.8 (concentration in the standard medium) to 0 mM from day 8 to 30 are shown in Fig. 5. Reduction of magnesium increased concentration dependently the DNA:protein ratio and glial

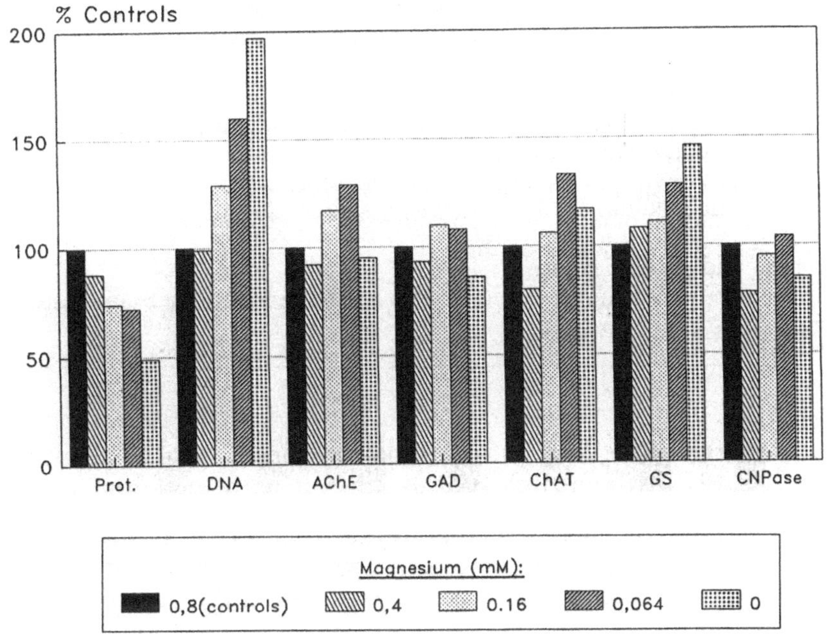

Fig. 5. Effects of changing in magnesium concentrations in medium from day 8 to 30. Magnesium concentration in the standard medium is 0.8 mM

marker enzyme levels in the aggregates and distorted the relative ratios of other marker enzyme levels in an unsystematic manner. No LDH release were detected in any of these cultures. Such observations indicate only that, like glycine, magnesium ion concentration in the medium plays a decisive role in the differentiation, maturation or integrity of the cells in aggregates, and that reduction of magnesium concentration in the medium suitable induces glial proliferation or neuronal loss in the aggregates. Further studies are needed to clarify the situation.

In any case, the observations presented and cited in this communication clearly demonstrate that the rotation-mediated aggregating brain cell cultures offer a versatile and uniquely useful in vitro system suitable for studies of age related and other types of neurodegenerative disorders. The main advantages of these cell culture system compared to other in vitro models are summarized in Table 5. To our judgement, for several neurological problems related to aging, myelination and demyelination etc., the aggregate cell culture offers some unique possibilities and allows to conduct studies which could not be (otherwise) performed in till now known in vitro systems. It can not be overemphasized, however, that our understanding of the control mechanisms and other processes involved in aggregation, maturation, differentiation etc. in aggregate cell cultures is limited, and that the in vivo relevance of some observations made in such cultures have also not yet been clarified. Experience in several laboratories clearly demonstrate that under strictly controlled culture conditions, very reproducible

Table 5. Some characteristics of aggregate brain cell cultures which could of special value for aging and dementia research

1.	Histotypic arrangements of different cell types
2.	Aggregation, differentiation, maturity can be controlled and quantified easily
3.	High yields, stability and reproducibility
4.	Numerous types of studies possible even in the same cultures
5.	Totally synthetic medium can be used which allows easy manipulations
6.	Time and age related studies possible
7.	Myelination and synaptogenesis
8.	Species specificity of complex processes can be studied (human brain tissue aggregate cultures are possible)

results can be generated in different laboratories. Since several in vivo neurological processes can also be reproduced with ease and reliability in this in vitro system, it can be speculated that together with the advancement of our knowledge on the etiology and the pathogenesis of aging related dementia, suitably controlled and relevant modulations of this system should eventually be possible.

We conclude, therefore, that in vitro models based on aggregated brain cell cultures are not only suitable for neurotoxicological and neurophysiological studies but also for developing pharmacological screening methods in the field of aging and dementia. Further studies, using aggregate brain cell cultures from other fetal brain tissues and with other excitatory amino acid receptor agonists and antagonists as well as neurotoxic, neuroprotective and CNS-active agents should, therefore, be conducted to fully evaluate the potential usefulness of this brain culture system. We must be aware, however, that experimental conditions suitable for other neuronal cell culture systems are not necessarily valid for rotation-mediated aggregating brain cell cultures and that till better knowledge on such culture systems and on the aging and dementia related neurological processes are acquired, proper controls and precautions are needed for meaningful use of such cultures.

Acknowledgements

Thanks are due to Dr. Honegger for helping us establishing the aggregating brain cell culture system in our laboratories and Mrs. C. Weibezahn for skillful technical assistance.

References

Atterwill CK, Meakin JM (1990) Delayed treatment with Nerve Growth Factor (NGF) reverses ECMA-induced cholinergic lesions in rat brain reaggregate cultures. Biochem Pharmacol 39: 2073–2076

DeLong GR (1970) Histogenesis of fetal mouse isocortex and hippocampus in reaggregating cell cultures. Dev Biol 22: 563–583

Eddleston M, Mucke L (1993) Molecular profile of reactive astrocytes — implications for their role in neurologic disease. Neuroscience 54: 15–36

Ehrenpreis S, Chiesa A, Bigo-Gullino M, Patrick P (1967) Correlation between acetylcholin (ACH) potentiation and cholinesterase (CHE) inhibition by diisopropylfluorophosphate (DFP) using a simplified radiometric assay for the enzyme. Fed Proc 26: 296

Garber BB, Muscona AA (1972a) Reconstruction of brain tissue from cell suspensions. I. Aggregation patterns of cells dissociated from different regions of the developing brain. Dev Biol 27: 217–234

Garber BB, Muscona AA (1972b) Reconstruction of brain tissue from cell suspension. Specific enhancement of aggregation of embryonic cerebral cells by supernatants from homologous cell cultures. Dev Biol 27: 235–243

Garber BB (1977) Cell aggregation and recognition in the self-assembly of brain tissue. In: Fedoroff S, Herz L (eds) Cell, tissue and organ cultures in neurobiology. Academic Press, New York, pp 515–537

Halks-Miller M, Henderson M, Eng LF (1986) Alpha-tocopherol decreases lipid peroxidation, neuronal necrosis, and reactive gliosis in reaggregate cultures of fetal rat brain. J Neuropathol Exp Neurol 45: 471–484

Halks-Miller M, Toll L, Davies MF (1989) Excitotoxins and antioxidants in reaggregate cultures of developing rat brain. Neurosci Abstr 15: 764

Halks-Miller M, Fedor V, Tyson CA (1991) Overview of approaches to in vitro neurotoxicity testing. J Am Coll Toxicol 10: 727–736

Honegger P, Richelson E (1976) Biochemical differentiation of mechanically dissociated mammalian brain in aggregating cell culture. Brain Res 109: 335–354

Honegger P, Richelson E (1977) Kainic acid alters neurochemical development in fetal rat brain aggregating cell cultures. Brain Res 138: 580–584

Honegger P, Lenoir D, Favrod P (1979) Growth and differentiation of aggregating fetal brain cells in serum-free defined medium. Nature 82: 305–308

Honegger P, Lenoir D (1980) Triiodothronine enhancement of neuronal differentiation in aggregating fetal rat brain cells cultured in a chemically defined medium. Brain Res 99: 425–434

Honegger P, Lenoir D (1982) Nerve growth factor (NGF) stimulation of cholinergic and tencephalic neurons in aggregating cell cultures. Dev Brain Res 3: 229–238

Honegger P (1985) Biochemical differentiation in serum-free aggregating brain cell cultures. In: Bottenstein JE, Sato G (eds) Cell culture in the neurosciences. Plenum, New York, pp 223–243

Honegger P, Werffeli P (1988) Use of aggregating cell cultures for toxicological studies. Experientia 44: 817–823

Korzeniewski C, Callewaert DM (1983) An enzyme-release assay for natural cytotoxicity. J Immunol Meth 64: 313–320

Kozak LP, Eppig JJ, Dhal D, Bignami A (1977) Ultrastructural and immunohistological characterization of a cell culture model for the study of neuronal-glial interactions. Dev Biol 59: 206–227

Labarca C, Paigen K (1980) A simple, rapid and sensitive DNA assay procedure. Anal Biochem 102: 344–352

Laerium OD, Steinsvag S, Bjerkvig R (1985) Cell the tissue culture of the central nervous system: Recent developments and current applications. Acta Neurol Scand 72: 529

LoPachin RM, Aschner M (1993) Glial-neuronal interactions: relevance to neurotoxic mechanisms. Toxicol Appl Pharmacol 118: 141–158

Lowry OH, Rosebrough NJ, Farr AL, Randall RJ (1951) Protein measurement with the Folin phenol reagent. J Biol Chem 193: 265–275

Matthieu JM, Honegger P, Favrod P, Poduslo JF, Constatino-Ceccarini E, Krstic R (1980) Myelination and demyelination in aggregating cultures of rat brain cells. In: Giacobinin E, et al (eds) Tissue culture in neurobiology. Raven Press, New York, pp 441–459

Matthieus JM, Honegger P, Trapp BD, Cohen SR, Webster HDeF (1978) Myelination in rat brain aggregating cell cultures. Neuroscience 3: 565–572
Muscona AA (1960) Patterns and mechanisms of tissue reconstruction from dissociated cells. In: Rudnick D (ed) Developing cell systems and their control. Ronald Press, New York, pp 45–70
Muscona AA (1961) Rotation-mediated histogenic aggregation of dissociated cells: a quantifiable approach to cell interactions in vitro. Exp Cell Res 22: 455–475
Muscona AA (1965) Recombination of dissociated cells and the development of cell aggregates. In: Willmer E (ed) Cells and tissue culture, vol 1. Academic Press, New York, pp 489–529
Nöldner M, Chatterjee SS (1993) Excitatory amino acid induced toxicity in the aggregate cell cultures from fetal rat brain. Naunyn Schmiedebergs Arch Pharmacol [Suppl] 347: R133
Patel AJ, Hunt A, Gordon RD, Balazs R (1982) The activities in different neural cell types of certain enzymes associated with the metabolic compartmentation glutamate. Dev Brain Res 4: 3–11
Pishak MR, Phillips AT (1979) A modified radioisotopic assay for measuring glutamine synthetase activity in tissue extracts. Anal Biochem 94: 82–88
Pulliam L, Berens ME, Rosenbloom ML (1988) A normal human brain cell aggregate model for neurobiological studies. J Neurosci Res 21: 521–530
Reed DJ, Goto K, Wang CH (1966) A direct radioisotopic assay for acetylcholinesterase. Anal Biochem 16: 59–64
Romijn HJ, Habets AMMC, Mus MT, Wolters PS (1981) A quantitative electron microscopic study of synaps formation in dissociated fetal rat cerebral cortex in vitro. Dev Brain Res 1: 591–605
Rosenberg PA, Aizenman E (1989) Hundred-fold increase in neuronal vulnerability to glutamate toxicity in astrocyte-poor cultures of rat cerebral cortex. Neurosci Lett 103: 162–168
Schousboe A, Westergaard N, Hertz L (1993) Neuronal-astrocytic interactions in glutamate metabolism. Biochem Soc Trans 21: 49–53
Schrier BK, Schuster L (1967) A simplified radiochemical assay for choline acetyltransferase. J Neurochem 14: 977–985
Seeds NW (1973) Differentiation of aggregating brain cell cultures. In: Sato G (ed) Tissue culture of the nervous system. Plenum, New York, pp 35–53
Sogin DC (1976) 2′,3′-cyclic nucleotide 3′-phosphohydrolase. J Neurochem 27: 1333–1337
Trapp BD, Honegger P, Richelson E, Webster H (1979) Morphological differentiation of mechanically dissociated fetal rat brain in aggregating cell cultures. Brain Res 160: 117–130
Whetsell WO, Shapira NA (1993) Biology of diseases: neuroexcitation, excitotoxicity and human neurological diseases. Lab Invest 68: 372–387
Wilmot CA (1989) Excitatory amino acid antagonists: behavioral and biochemical approaches for the development of new central nervous system therapeutic agents. Drug Dev Res 17: 339–365
Wilson SH, Schrier BK, Farber JL, Thompson EJ, Rosenberg RN, Blume AJ, Nirenberg MW (1972) Markers for gene expression in cultured cells from nervous system. J Biol Chem 247: 3159–3169
Wingo WJ, Awapara J (1950) Decarboxylation of L-glutamic acid by brain. J Biol Chem 187: 267–271

Authors' address: Dr. S. S. Chatterjee, Department of Pharmacology, Research Laboratories, Dr. Willmar Schwabe Arzneimittel, D-76227 Karlsruhe-Durlach, Federal Republic of Germany.

J Neural Transm (1994) [Suppl] 44: 61–71

The organotypic entorhinal-hippocampal complex slice culture of adolescent rats. A model to study transcellular changes in a circuit particularly vulnerable in neurodegenerative disorders

S. Diekmann, R. Nitsch, and **T. G. Ohm**

Zentrum der Morphologie, Johann Wolfgang-Goethe-Universität, Frankfurt am Main, Federal Republic of Germany

Summary. The entorhinal-hippocampal system is severely altered in many neurodegenerative disorders with mnemonic malfunction, e.g. Alzheimer's, Parkinson's and Huntington's disease. The present approach characterizes an organotypic complex slice culture comprising both the entorhinal cortex and the hippocampal formation in order to establish a tool for experimental studies of the entorhinal-hippocampal interaction and its presumed neurodegenerative alterations *in vitro*.

Slices were obtained from rats at about postnatal day 15 and maintained in culture using the interface technique. Thus, also structures known to be developed gradually during the first weeks postnatally are in accord to structures seen in adult rats. After two-three weeks *in vitro*, slices in the culture dish still revealed the typical morphological features of the entorhinal-hippocampal formation as visible with the dissecting microscope. Biocytin, which is taken up by and transported within living cells, labeled typical cell bodies, dendrites and axons of stellate neurons in layer II and pyramidal cells in layer III when applied to the outer layers of the entorhinal cortex. Small injections of biocytin within the dentate gyrus displayed living granule cells and the maintenance of their projection to the pyramidal cells in CA3, i.e., a typical suprapyramidal plexus of mossy fibers. The presence of axons of entorhinal neurons traveling towards the hippocampus and growth cones traversing the deep layers of the entorhinal cortex indicate that both brain regions are still interacting. Immunocytochemistry for calbindin D-28K revealed labeled neurons in layer II of the entorhinal cortex and dentate granule cells which are known to contain this calcium-binding protein.

Introduction

The entorhinal-hippocampal system appears to be a very early and major target for neurodegeneration within mnemonic disorders such as Alzheimer's disease. Furthermore, recent studies suggest its involvement in cognitive

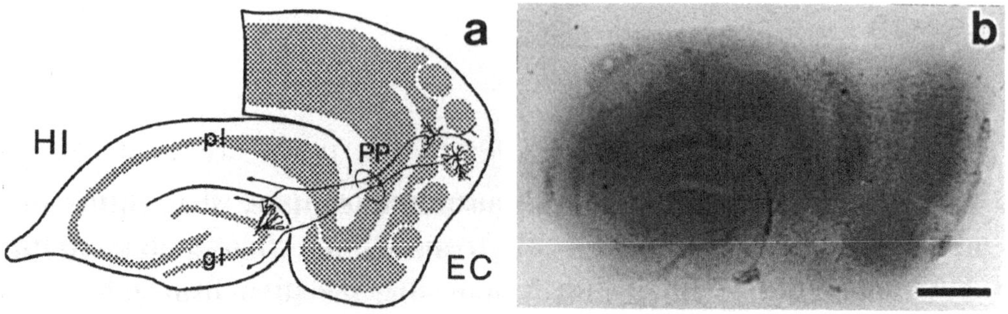

Fig. 1. **a** Schematic drawing of the entorhinal-hippocampal complex. Stellate neurons in layer II clustered in cell islands and medium sized pyramidal cells in layer III of the entorhinal cortex (*EC*) give rise to the perforant path (*PP*) which projects to the hippocampal formation (*HI*). There, the afferents terminate in the middle and outer molecular layer of the dentate gyrus and the stratum lacunosum-moleculare of the Ammon's horn and form synapses with distal dendrites of hippocampal neurons, i.e., dentate granule cells and, to a lesser extent, CA1 and CA3 pyramidal neurons (not illustrated). **b** Entorhinal-hippocampal complex slice culture of a 15 day old rat maintained in vitro for three weeks as observed in the culture dish under the dissecting microscope. *gl* granule cell layer; *pl* pyramidal cell layer. Scale bar in **b** = 1 mm

malfunction within Parkinson's and Huntington's disease (Braak and Braak, 1993). Pre-processed data from the association cortices converge via the entorhinal cortex into the hippocampus both in rodents and primates (Steward, 1976; Witter et al., 1989; Braak and Braak, 1993; c.f. Witter, 1993). Stellate neurons in layer II (pre-alpha in human) which are characteristically clustered in cell islands in primates (see Braak, 1980) and medium sized pyramidal cells in layer III (pre-beta in human) of the entorhinal cortex (see Lorente de Nó, 1934) give rise to the perforant path (Fig. 1a). This presumably excitatory projection (Storm-Mathisen and Ottersen, 1984; Grandes and Streit, 1991) terminates in the middle and outer molecular layer of the dentate gyrus and the stratum lacunosum-moleculare of the ammon's horn forming synapses with distal dendrites of principal neurons, i.e., dentate granule cells and, to a lessser extent, CA1 and CA3 pyramidal neurons. (Nafstad, 1967; Matthews et al., 1976; Witter et al., 1989) (see Fig. 1a). Furthermore, distal dendrites of non-principal cells, among them the GABAergic inhibitory neurons which control the discharge of principal cells within the so-called "trisynaptic pathway" of the hippocampus (see Frotscher, 1988), are targets of these excitatory afferents (Zipp et al., 1989). Granule cell axons terminate on dendrites of CA3 pyramids and their axons eventually on dendrites of CA1 pyramidal cells (Frotscher, 1988). In turn, the pyramidal cells in CA1 and in the subiculum project to the deep layers of the entorhinal cortex (Witter, 1993), from which fibers originate which form synapses with neurons in layers II and III (Jones and Heinemann, 1988). This way, a powerful synaptic chain is established, which is regarded to play a crucial role in higher cognitive function such as learning and memory (Bachvalier and Beauregard, 1993; Jaffard and Meunier, 1993).

There is evidence that the transentorhinal/entorhinal area is the first brain region to be histopathologically affected in Alzheimer's disease (Braak and Braak, 1991). The progressive accumulation of the intraneuronal pathological neurofilaments eventually leads to a deafferentiation of the hippocampus (Hyman et al., 1984; Lippa et al., 1992). In the rat, lesions of the entorhinal cortex induce various transcellular changes in the hippocampus including sprouting of remaining afferents, retraction of dendrites, induction of early genes, and activation of glial cells (Rose et al., 1976; Nitsch and Frotscher, 1991, 1992; see Nitsch, 1993). In man, changes in hippocampal dendrites (Flood et al., 1987), axonal sprouting (Masliah et al., 1992) and also an early gene activation (Zhang et al., 1992) have been reported in Alzheimer's disease.

Axonal sprouting has been regarded as a restorative capacity of the brain to establish new connections following lesioning (e.g. Cotman and Nieto-Sampedro, 1985), whereas dendritic changes are presumably persisting alterations of the hippocampal organization (Nitsch, 1993). Although thoroughly studied (West et al., 1975; Reeves and Steward, 1988), the functional implications of the remodeling occurring in the hippocampus following deafferentation of its entorhinal input is still not well understood. However, recent physiological data suggest a permanent alteration of hippocampal signal propagation upon this deafferentation (Nitsch et al., 1992). Both, experimental entorhinal lesions in rats as well as Alzheimer's disease-related neurofibrillary changes in entorhinal cortex neurons in man are associated with disturbances in the cAMP-mediated signal transduction (Ohm et al., 1992, 1993; see Ohm, 1993).

The main purpose of our ongoing project is to establish an experimental model which allows for an analysis of the entorhinal-hippocampal interaction *in vitro*. The minimum demands for such a model should be (I) the possibility to gain access to maturated neurons in order to circumvent for the most part the interference with the process of cell development, (II) the opportunity to take advantage of this accessibility for longer periods of time and (III) the preservation of the main morphological and neurochemical features of the system, i.e., conservation of cell types, expression of neuroactive substances and main afferent connections. The controlled access to identified cell types may eventually allow the study of the selective vulnerability and peculiar reaction modi of these cells within an organotypic environment. The organotypic slice culture as first described by Gähwiler (1981) using the roller tube technique and further developed by Stoppini et al. (1991) employing the interface technique is a promising approach to fulfill the above mentioned criteria.

Materials and methods

Culture preparation and maintenance

Twelve-thirty days old female and male Sprague-Dawley rats were used in this study. Following decapitation, the brains were rapidly removed from the skull under aseptic conditions and immediately placed in oxygenated cooled medium (6°C). After dissecting the adjacent brain structures, the hippocampus and entorhinal cortex were cut into 250 μm-thick slices on a tissue chopper. The slices were placed on a sterile, porous (0.4 μm) membrane (Millicell-CM, Millipore). Two slices each consisting of the hippocampus and the adjacent entorhinal cortex were placed on one membrane. The membranes were transferred into a tissue culture plate with each dish containing 1 ml medium. Incubation was allowed for three weeks at 35°C in a 5%-enriched CO_2 atmosphere. Medium was changed three times a week.

For preparation and culturing, different media were used. The medium for preparation consists of Minimum Essential Medium (MEM), (Gibco, No. 079-01012). Medium for cultivation contained the following mixture: MEM 50%, horse serum 25% (Gibco, No. 034-06055H), Hank's Balanced Salts 25% (Gibco, No. 041-04020H) and Penicillin/Streptomycin 10.000 U/100,000 μg/ml (Seromed, No. K082). A pH of 7.35 was used for preparation, for cultivation a pH of 7.3 was used for animals younger than 21 days. Slices of rats older than 21 days were kept at pH 7.4.

Biocytin labeling

Biocytin (5% in Tris buffer 0.05 M, pH 7.6; Sigma Chemicals) was applied iontophoretically (7 μA, 7 sec on and 7 sec off) or manually by a Hamilton syringe into the area of interest in slices 14 days in vitro. After further incubation for 24–30 hours, slices were immersed in a fixative containing 4% paraformaldehyde and 1% glutaraldehyde in phosphate buffer (PB) for 1 hour. Following several washes in PB and pretreatment with Triton 0.3% in 0.1 M PB (pH 7.4) for 2 hours to allow better tissue penetration, slices were incubated with an avidin-biotin complex (ABC, Vector Labs) for 24 hours. For ABC detection, slices were placed in 0.1% diaminobenzidine (DAB) and 0.03% H_2O_2 for about 10 minutes. After intensification with osmium tetroxide (0.5% for 1–2 minutes), slices were dehydrated, mounted and coverslipped with Eukitt (Merck).

Calbindin immunocytochemistry

Slices maintained 21 days *in vitro* were immersed in a fixative containing 4% paraformaldehyde and 0.1% glutaraldehyde in PB for 15 minutes and postfixed in 4% paraformaldehyde in PB for 1 hour. The material was incubated for 6 days in a polyclonal rabbit anti-calbindin antibody diluted 1:5.000 in PB containing 1% normal goat serum, 0.5% Triton, and 0.1% NaN_3. After extensive washing in PB, sections were incubated in a biotinylated anti-rabbit IgG (Vector Labs., diluted 1:250 in PB) at room temperature for 2 hours. Following washing in PB, sections were incubated for 2 hours in ABC-reagent, diluted 1:250 in PB. For visualization of the tissue-bound peroxidase, sections were placed 10–15 minutes in a solution containing 0.1% DAB and 0.03% H_2O_2. After several rinses in PB, the DAB-reaction product was intensified with osmium tetroxide (0.5% for 1–2 minutes) and slices were dehydrated, mounted and coverslipped.

Results

Complex slice cultivation

As a general rule, cultivation was regulary possible with slices of animals up to postnatal day 15 (p15). Culturing of slices from rats at the age of p16–p21 was possible but higher failure rates were observed when compared to younger animals (p12–p15). Obtained from older animals (p22–p30), only few if any slices could be cultivated despite several modifications have been tested such as adding various amounts of glucose, different slice thicknesses or use of a different media.

Morphology

The overall structure of the hippocampus proper and the entorhinal cortex was generally well preserved after three weeks in vitro (Fig. 1b). The entorhinal cortex showed the typical laminar arrangement including the lamina dissecans. In the hippocampus, the dentate granule cell layer and the pyramidal cell layer in the Ammon's horn could be easily detected suggesting a high level of preservation of the morphological features in this complex slice culture (Fig. 1b). Biocytin labeling revealed typical stellate neurons in layer II (Fig. 2a) and pyramidal cells in layer III of the entorhinal cortex (Fig. 2b) with morphological characteristics similar to those described by Lorente de Nó (1934). Stellate neurons sent several dendrites arising from their polymorph cell body in all directions, however, mostly towards the pial surface (Fig. 2a). Pyramidal cell bodies gave rise to one apical dendrite which branched on its way to the pial surface and regularly two basal dendrites extending into the deep layers of the entorhinal cortex (Fig. 2b). In most cases, axons of these entorhinal neurons could only be followed for short distances, however, in some instances, axons arising from the cell body were observed traversing the deep layers of the entorhinal cortex heading towards the white matter where they formed branches (Fig. 2b).

Originating from these fibers, we observed axons with numerous varicosities and a typical growth cone heading through the white matter towards the hippocampal formation (Fig. 2c). Granule cells in the dentate gyrus labeled with biocytin were concentrated in a cell layer but in comparison to the granule cells *in vivo* (Lübbers and Frotscher, 1987), they were more loosely distributed (Fig. 2d,e).

Nevertheless, labeled granule cells gave rise to axons emerging from the basal pole of their soma, i.e., the mossy fibers, which traveled through the hilus towards the Ammon's horn (Fig. 2d,e) and formed a characteristic plexus in the stratum lucidum of CA3 (Fig. 2d). Dendrites of granule cells almost exclusively emerged from the apical pole of the soma and extended throughout the molecular layer up to the hippocampal fissure (Fig. 2e). These dendrites displayed numerous typical spines (Fig. 2e).

Immunocytochemistry

Calbindin-immunoreactive cells in the entorhinal cortex were observed mostly in layer II (Fig. 3a). In the dentate gyrus, granule cells were immu-

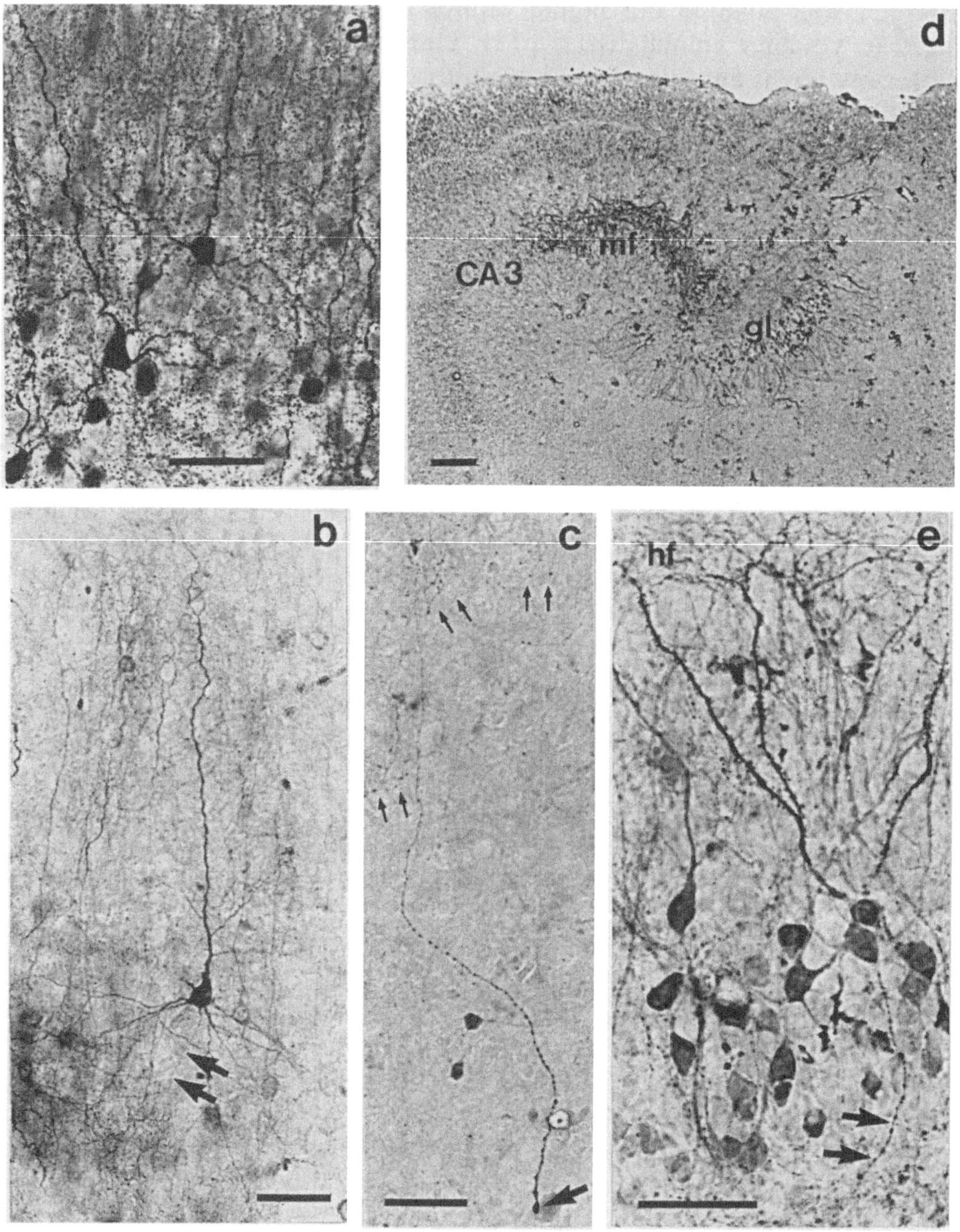

Fig. 2. Biocytin-labeled entorhinal-hippocampal complex slice culture obtained from 15 day old rats maintained for two weeks in culture. **a** Stellate neurons in layer II of the entorhinal cortex exhibit a polymorph cell body and dendrites with a predominant orientation towards the pial surface. **b** Pyramidal cell in layer III of the entorhinal cortex with an apical dendrite heading towards the pial surface. Arrows pointing to an axon emerging from the basal pole of the cell body traversing the deep layers of the entorhinal cortex. **c** Axon in the deep layers of the entorhinal cortex heading towards the hippocampus with a typical growth cone (large arrow) entering the white matter. Note plexus in the deep layers of the entorhinal cortex (small arrows) from where this axon emerges. **d** Biocytin-labeled dentate granule cells with their axons traversing the hilus towards the CA3 region and terminating in the stratum lucidum. Note that

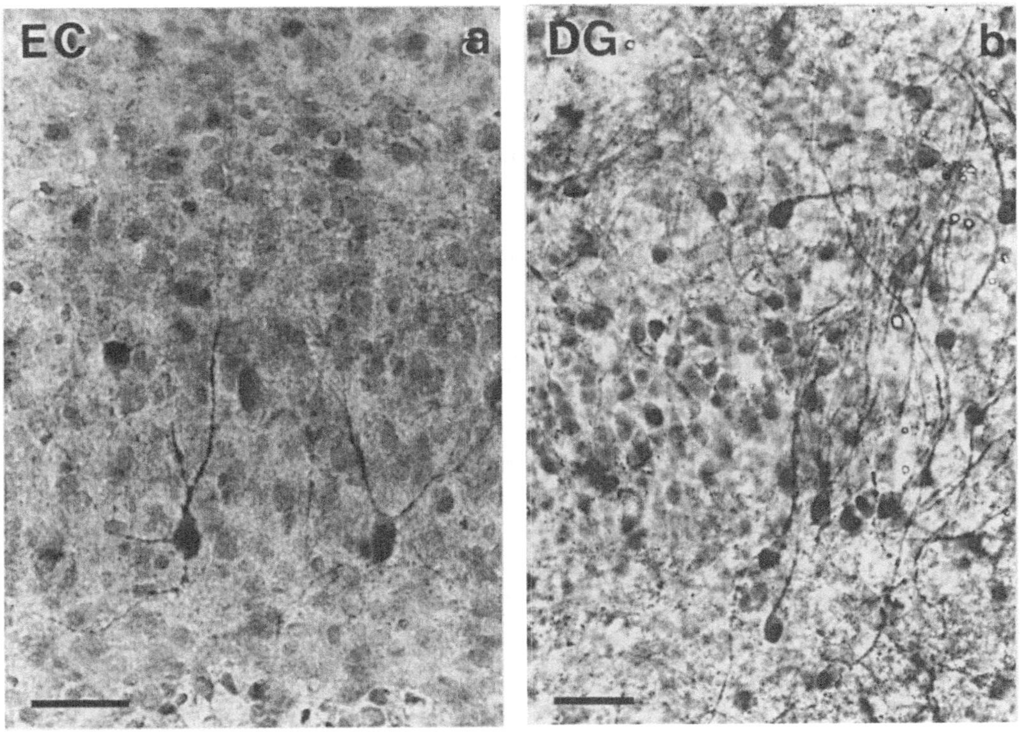

Fig. 3. Entorhinal-hippocampal complex slice culture obtained from 15 day old rats maintained for three weeks in culture immunostained for calbindin. **a** Neurons in layer II of the entorhinal cortex (*EC*). Occasionally, staining of dendritic processses allows for their identification as stellate neurons. **b** Granule cells of the dentate gyrus (*DG*) loosely arranged in their layer containing calbindin. In some instances, also their dendrites extending up to the hippocampal fissure are immunolabeled. Note that some neurons located in the hilus also contain this neuroactive substance. Scale bars: 50 μm

noreactive for calbindin (Fig. 3b). In addition, in the hilar region, non-principal neurons exhibited calbindin immunostaining as known for *in vivo* preparations (Sloviter, 1989) (Fig. 3b).

Discussion

To study the entorhinal-hippocampal interaction, different approaches are possible each afflicted with advantages and disadvantages. Investigations using autopsy material have been performed but face the typical restrictions

---

biocytin labeling is restricted to a part of the granule cell layer (*gl*) and that consequently only a fraction of the mossy fibers contains this dye. **e** Dentate granule cells loosely arranged in their layer with typical spinous dendrites extending up to the hippocampal fissure (*hf*). Arrows pointing to an axon emerging from the basal pole of a granule cell heading towards the hilus. Note the absence of basal dendrites emerging from granule cell bodies. *mf* mossy fibers Scale bars in **a–c**, **e** 50 μm; Scale bar in **d** 200 μm

of human studies (see Ohm, 1993). Studies employing lesions of the entorhinal cortex in living animals (see Nitsch, 1993; Ohm et al., 1993) include the complete reaction of an intact organism but at the same time a lot of uncontrolled factors may influence the experimental outcome. Furthermore, this approach does not allow the direct observation of dynamic changes nor a defined application of substances in order to induce and/or prevent changes in the system. Co-culturing of dissociated cells from the entorhinal cortex and the hippocampus does not lead to an organotypic organization and network interaction. Thus, the rationale of the present work was to establish a model to enable highly controlled studies of the entorhinal-hippocampal interaction *in vitro* by means of morphological, physiological and, particularly, pharmacological approaches. Among the available methods we have choosen the interface technique (Stoppini et al., 1991) which was shown to result in a better preservation of morphological features when compared to the roller tube technique (Li et al., 1993).

The present work focused on basic features of the first element of the various entorhinal-hippocampal–circuits (see Introduction for details), i.e., (I) the preservation of layer II and III projection neurons of the entorhinal cortex which give rise of the perforant path, and (II) the hippocampal granule cells, i.e., the main receptive elements of the perforant path. Our results suggest, that complex–slice cultures of the entorhinal cortex and hippocampus yield an organotypic preparation which retains basic cellular aspects comparable to *in vivo* conditions. Considering morphological aspects, a characteristic hippocampal and entorhinal cytoarchitecture remained conserved in our preparations suggesting the preservation of this pathway in our complex slice cultures. Both entorhinal and dentate neurons showed typical morphological characteristics (Lorente de Nó, 1934; Lübbers and Frotscher, 1987) and exhibited typical neurochemical features (Sloviter, 1989). In contrast to hippocampal cultures obtained from younger animals (p5–8) and cultured with the roller tube technique, we did not observe major architectural or cytological changes such as the lack of the lower blade of the granule cells or the appearance of basal dendrites emerging from these neurons (Heimrich and Frotscher, 1991). It is important to note that we have cultured slices from adolescent animals (p15), a postnatal stage, when, for the most part, developmental processes are completed (e.g., Loy et al., 1977; Bayer, 1980). For example, the postnatal development of the granule cells and the expression of neuroactive substances in non-principal neurons (Bayer, 1980; Nitsch et al., 1990; Seto-Ohshima et al., 1990) are almost at a mature level at p15 which is not the case at earlier postnatal stages, e.g., at p2–8 as usually used for slice culture experiments (Gähwiler, 1984; Heimrich and Frotscher, 1991). Furthermore, the biocytin labeling suggests that also the projection of the granule cells (mossy fibers) to the CA3 pyramidal cells is maintained since a dense suprapyramidal axon plexus was seen after small injections of solubilized biocytin in the granule cell layer of the dentate gyrus. Regarding the perforant path, however, ongoing experiments with other tracers like PHA-L and electronmicroscopical demonstration of typical synapses are required although biocytin-labeled axons were observed originating in the entorhinal cortex running to the hippocampus. In addition, axons with typical growth cones emerging from a plexus in the entorhinal cortex invade the white matter and head towards the hippocampal formation. This finding indicates, that apart from fibers which already form the entorhinal-hippocampal pathway at the day of explantation (Loy et al., 1977; Bayer and Altman, 1987), an additional sprouting of entorhinal axons during the time of culturing takes place which is directed towards the hippocampal formation. It remains open, whether these sprouting axons are a response to tissue preparation of the slices or an ongoing maturational effect.

In conclusion, complex–slice cultures of the entorhinal cortex and hippocampus of adolescent rats appear to be a promising model for studying transneuronal and transcellular changes of the entorhinal-hippocampal interaction *in vitro*. The analysis of changes occurring in this interaction induced by the application of pathological substances including Alzheimer's disease-typical proteins, induction of Alzheimer-related changes (Sautière et al., 1992; Mattson and Barger, 1993) and testing of putatively preventive or therapeutic agents are only some examples for a possible usage of this model.

Acknowledgements

The authors wish to thank Dr. M.R. Celio for providing the calbindin antibody. We are grateful to E. Thielen, R. Zarski and S. Kroeber for excellent technical assistance. This study was supported by grants from the Hirnliga, the Hertie-Stiftung and the Tropon-Werke AG to R. Nitsch and T.G. Ohm.

References

Bayer SA (1980) Development of the hippocampal region in the rat. I. Neurogenesis examined with 3H-thymidine autoradiography. J Comp Neurol 190: 87–114

Bayer BA, Altman J (1987) Directions in neurogenetic gradients and patterns of anatomical connections in the telencephalon. Progr Neurobiol 29: 57–106

Bachevalier J, Beauregard M (1993) Maturation of medial temporal lobe memory functions in rodents, monkeys, and humans. Hippocampus 3: 191–202 (special issue)

Braak H (1980) Architectonics in the human telencephalic cortex. In: Braitenberg V, Barlow HB, Florey E, Grüsser OJ, van der Loos H (eds) Studies of brain function, vol 4. Springer, Berlin Heidelberg New York, pp 1–147

Braak H, Braak E (1991) Neuropathological stageing of Alzheimer-related changes. Acta Neuropathol 82: 239–259

Braak H, Braak E (1993) The entorhinal-hippocampal interaction in mnestic disorders. Hippocampus 3: 239–246 (special issue)

Cotman CW, Nieto-Sampedro M (1985) Progress in facilitating the recovery of function after central nervous system trauma. Ann N Y Acad Sci 457: 83–92

Flood DG, Buell SJ, Horwitz GJ, Coleman PD (1987) Dendritic extent in human dentate gyrus in normal aging and senile dementia. Brain Res 402: 205–216

Frotscher M (1988) Neuronal elements in the hippocampus and their synaptic connections. In: Frotscher M, Kugler T, Misgeld U, Zilles K (eds) Neurotransmission in the hippocampus. Springer, Berlin Heidelberg New York Tokyo, pp 2–19 (Adv Anat Embryol Cell Biol, vol 11)
Gähwiler BH (1981) Organotypic monolayer cultures of nervous tissue. J Neurosci Meth 4: 329–342
Gähwiler BH (1984) Slice cultures of cerebellar, hippocampal and hypothalamic tissue. Experientia 40: 235–308
Grandes P, Streit P (1991) Effect of perforant path lesion on pattern of glutamate-like immunoreactivity in the rat dentate gyrus. Neuroscience 41: 391–400
Heimrich B, Frotscher M (1991) Differentiation of dentate granule cell in slice cultures of rat hippocampus: a Golgi/electron microscopic study. Brain Res 538: 263–268
Hyman BT, Van Hoesen GW, Damasio AR, Barnes CL (1984) Alzheimer's disease: cell-specific pathology isolates the hippocampal formation. Science 225: 1168–1170
Jaffard R, Meunier M (1993) Role of the hippocampal formation in learning and memory. Hippocampus 3: 203–218 (special issue)
Jones RSG, Heinemann U (1988) Synaptic and intrinsic responses of the medial entorhinal cortical cells in normal and magnesium-free medium in vitro. J Neurophysiol 59: 1476–1497
Li D, Field PM, Starega U, Li Y, Raisman G (1993) Entorhinal axons project to dentate gyrus in organotypic slice co-culture. Neuroscience 52: 799–813
Lippa CF, Hamos JE, Pulaski-Salo D, Degenaro LJ, Drachman DA (1992) Alzheimer's disease and aging: effects on perforant pathway perikarya and synapses. Neurobiol Aging 13: 405–411
Lorente de Nó R (1934) Studies on the structure of the cerebral cortex I. The area entorhinalis. J Psychol Neurol 45
Loy R, Lynch G, Cotman CW (1977) Developement of afferent lamination in the fascia dentata of the rat. Brain Res 121: 229–243
Lübbers K, Frotscher M (1987) Fine structure and synaptic connections of identified neurons in the rat fascia. Anat Embryol 177: 1–14
Masliah E, Mallory M, Hansen L, Alford M, DeTeresa R, Terry R, Baudier J, Saitoh T (1992) Localisation of amyloid precursor protein in GAP 43-immunoreactive abberant sprouting neurites in Alzheimer's disease. Brain Res 574: 312–316
Mattews DA, Cotman CW, Lynch G (1976) An electron microscopic study of lesion induced synaptogenesis in the dentate gyrus of the adult rat. I. Magnitude and time course of degeneration. Brain Res 115: 1–21
Mattson MP, Barger SW (1993) Roles for calcium signaling in structural plasticity and pathology in the hippocampal system. Hippocampus 3: 73–88 (special issue)
Nafstad PHJ (1967) An electron microscope study on the termination of the perforant path fibers in the hippocampus and the fascia dentata. Z Zellforsch 76: 532–542
Nitsch R, Soriano E, Frotscher M (1990) The parvalbumin-containing nonpyramidal neurons in the rat hippocampus. Anat Embryol 181: 413–425
Nitsch R, Frotscher M (1991) Maintenance of peripheral dendrites of GABAergic neurons requires specific input. Brain Res 554: 304–307
Nitsch R, Frotscher M (1992) Reduction of posttraumatic transneuronal "early gene" activation and dendritic atrophy by the NMDA receptor antagonist MK-801. Proc Natl Acad Sci USA 89: 5197–5200
Nitsch R, Clussmann H, Heinemann U (1992) A current source density analysis of stimulus induced field potentials in the rat dentate gyrus following entorhinal lesions. Soc Neurosci Abstr 18: 321
Nitsch R (1993) Transneuronal changes in the lesioned entorhinal-hippocampal system. Hippocampus 3: 247–256 (special issue)
Ohm TG (1993) Alterations of signal transduction in the lesioned entorhinal hippocampal system. Hippocampus 3: 131–138 (special issue)

Ohm TG, Schmitt M, Lemmer B, Bohl J (1992) Significant decrease in adenylate cyclase activity precedes the full development of Alzheimer-related neurofibrillary changes. Soc Neurosci Abstr 18/1: 205

Ohm TG, v Dewitz G, Witte K, Nitsch R, Lemmer B (1993) Hippocampal adenylate cyclase activity after unilateral electrolytic lesion of the entorhinal cortex in young adult rats. Abstr, 16th Annual Meeting of the European Neuroscience Association, Madrid. Eur J Neurosci [Suppl] 6: 235

Reeves TM, Steward O (1988) Changes in the firing properties of neurons in the dentate gyrus with denervation and reinnervation: implications for behavioral recovery. Exp Neurol 102: 37–49

Rose G, Lynch G, Cotman CW (1976) Hypertrophy and redistribution of astrocytes in the deafferentiated dentate gyrus. Brain Res Bull 1: 87–92

Sautiere PE, Sindou P, Couratier P, Hugon J, Wattez A, Delacourte A (1992) Tau antigenic changes induced by glutamate in the rat primary culture model: a biochemical approach. Neurosci Lett 140: 206–210

Seto-Ohshima A, Aoki E, Semba R, Emson PC, Heizmann CW (1990) Appearence of parvalbumin-specific immunoreactivity in the cerebral cortex and hippocampus of the developing rat and gerbil brain. Histochemistry 94: 579–589

Sloviter RS (1989) Calcium-binding protein (calbindin D28k) and parvalbumin immunocytochemistry: localisation in the rat hippocampus with specific reference to the selective vulnerability of hippocampal neurons to seizure activity. J Comp Neurol 280: 183–196

Steward O (1976) Topographic organisation of the entorhinal area to the hippocampal formation. J Comp Neurol 167: 285–314

Stoppini L, Buchs PA, Muller D (1991) A simple method for organotypic cultures of nervous tissue. J Neurosci Meth 37: 173–182

Storm-Mathisen J, Ottersen OP (1984) Neurotransmitters in the hippocampal formation. In: Reinoso-Suarez F, Afmone-Marsan C (eds) Cortical integration. Raven Press, New York, pp 105–130

West JR, Deadwyler S, Cotman CW, Lynch G (1975) Time-dependent changes in commisural field potentials in the dentate gyrus following lesions of the entorhinal cortex in adult rats. Brain Res 97: 215–233

Witter MP, Amaral DG, Van Hoesen GW (1989) The entorhinal-dentate projection in the macaque monkey: topographical organisation along the longitudinal axis of the hippocampus. J Neurosci 9: 216–228

Witter MP (1993) Organisation of the entorhinal-hippocampal system: a review of the current anatomical data. Hippocampus 3: 33–44 (special issue)

Zhang P, Hirsch EC, Damier P, Duckaerts C, Javoy-Agid F (1992) c-fos protein-like immunoreactivity: distribution in the human brain and over-expression in the hippocampus of patients with Alzheimer's disease. Neuroscience 46: 9–21

Zipp F, Nitsch R, Soriano E, Frotscher M (1989) Entorhinal fibers form synaptic contacts on parvalbumin — immunoreactive neurons in the rat fascia dentata. Brain Res 495: 161–166

Authors' address: Dr. S. Diekmann, Zentrum der Morphologie, Universitätsklinik Frankfurt, Theodor-Stern-Kai 7, D-60590 Frankfurt am Main, Federal Republic of Germany.

J Neural Transm (1994) [Suppl] 44: 73–85

The use of ion-sensitive electrodes and fluorescence imaging in hippocampal slices for studying pathological changes of intracellular Ca^{2+} regulation

P. Schubert, F. Keller, Y. Nakamura*, and **K. Rudolphi****

Department of Neuromorphology, Max Planck Institute for Psychiatry, Munich, Federal Republic of Germany

Summary. The physiological regulation of the intracellular Ca^{2+} homeostasis and its pathological alteration has been studied in rat and gerbil hippocampal slices using ion-sensitive electrodes and the fluorescence imaging technique. The ischemia-induced intracellular Ca^{2+} rise, accentuated in the synaptic/dendritic layer of the vulnerable CA1 neurons was observed in vivo and could be replicated at an accellerated time course in the "ischemic" hippocampal slice superfused with unoxygenated, glucose-free medium. The intracellular Ca^{2+} loading, thought to be instrumental for the generation of postischemic nerve cell damage, seems to result from an increased Ca^{2+} release out of intracellular stores as well as from an enhanced synaptic Ca^{2+} influx. The latter is attributed to a depolarization-induced opening of the voltage-dependent Ca^{2+} channels and to an uncontrolled influx through "upregulated" NMDA receptor-operated channels. Such an ischemia-induced upregulation which is reported to occur physiologically by the activation of PKC, is reflected by the selective loss of the depressive control of the synaptic NMDA Ca^{2+} influx by adenosine. Ischemia also leads to a hypertrophy of astrocytes which may go along with an impairment of their physiological function to take up glutamate adding to the extracellular rise of the excitotoxic amino acids. A pathological activation of microglial cells and their transformation into macrophages, known to release oxygen radicals, may further add to neuronal damage. The observed neuroprotection by adenosine can be primarily ascribed to its limiting effect on a pathological membrane depolarization and its deleterious consequences. The more powerful neuroprotection by propentofylline, thought to act analogue to adenosine, seems to be achieved by additional mechanisms. This pharmacon depresses the ischemia-induced neuronal Ca^{2+} loading in vivo and in vitro, prevents the activation of astrocytes and interferes with the transformation as well as with the free radical formation of microglia-derived

*On leave from Ehime University, Department of Physiology, Japan
**Hoechst Ag, Frankfurt a.M., Federal Republic of Germany

macrophages as demonstrated in complementary studies with fluorescence techniques on cell cultures.

Cognitive functions are related to the integrity of the entorhino-hippocampal system (see Kadar, this volume) which is particularly vulnerable in neurodegenerative disorders (Ohm and Nitsch, this volume). This makes the hippocampus an adequate cortical structure for the experimental evaluation of the basic mechanisms underlying the age-dependent impairment of brain functions. Since the pathogenetic process occurring in vivo is highly complex, in vitro models have to be designed in parallel to in vivo models allowing the investigations of defined aspects under simplified but largely controlled experimental conditions. The use of cultures from brain tissue or selectively of neurons, glial cells, fibroblasts and transfected cell lines has been shown in a number of elegant studies presented at this symposium. We want to propose the hippocampal slice preparation as a complementary in vitro model which allows to study the ionic mechanisms of pathological nerve cell damage and possibilities of neuroprotection in an integrated cortical system where the natural neuronal circuitry and the interconnections to glial cells are still largely preserved.

A pathological alteration of the intracellular Ca^{2+} homeostasis and of the Ca^{2+}-dependent metabolic regulation including that related to the activity state of protein kinase C (PKC) is thought to be instrumental for the generation of nerve cell damage. There is evidence that an excessive intracellular Ca^{2+} loading and PKC activation plays a role for the pathogenesis of aging deseases. For example, an activation of cholinergic receptors leading via phospholipase C to an activation of the phosphatidylinositol (PI)-turnover and PKC is apparently involved in the secretion and processing of the Alzheimer amyloid precursor (Nitsch, this volume). The toxic β-amyloid peptide was found to cause an increased glutamate toxicity and pathological Ca^{2+} loading which mediates neurofibrillary degeneration (Mattson and Smith-Swintowsky, this volume). Similar ionic mechanisms seem to be responsible for the ischemia-induced functional impairment of the aging arteriosclerotic brain.

The main topic of our group is to study the regulation of the intracellular Ca^{2+} homeostasis and the pathological alteration of particularly those cellular control systems which are relevant for the generation of postischemic nerve cell damage and which are possible targets for endogenous and pharmacological neuroprotection. The hippocampal slice preparation seems to be particularly suitable as experimental model for the following reasons: 1. brain ischemia -hypoxia or -hypoglycemia can be mimicked by superfusing the slice with a medium lacking oxygen and/or glucose. Changes in the subcellular Ca^{2+} content and the degree of structural damage in the presence and absence of neuroprotective agents can be visualized morphologically by electronmicroscopic histochemistry. 2. Changes in nerve cell activity can be monitored by intra- and extracellular recordings. The use of ionselective electrodes allows to monitor the stimulus train-evoked neuronal Ca^{2+} influx under various experimental conditions. 3. The fluorescence imaging tech-

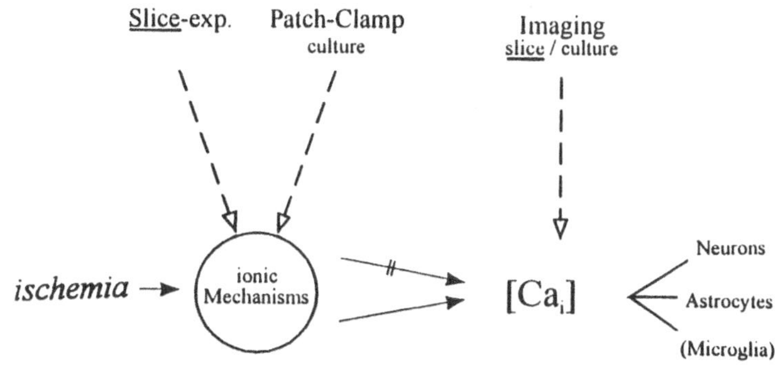

Fig. 1. Strategy for the in vitro study of pathological changes of intracellular Ca^{2+} regulation (see text)

nique can be applied allowing the dynamic measurement of a wide range of intracellular parameters relevant for the ischemia-induced pathogenetic process including changes in the intracellular concentration of Ca^{2+} as well as of PKC activity. The information obtained from the slice expriments can be complemented and extended by investigations on selective neuron or glial cell cultures by analogue measurements with fluorescence imaging-techniques or by patch clamping (Fig. 1). Thus, the hippocampal slice preparation may be used as an experimental tool which may help to bridge the detailed information about the pathologically altered functions of individual cells in culture with the results from in vivo experiments.

Similar as in gerbils after transient forebrain ischemia in vivo, we found in rat hippocampal slices after 5 min superfusion with "ischemic" (unoxygenated, glucose-free) medium a histochemically demonstrable Ca^{2+} loading and structural damage of the vulnerable CA1 pyramidal neurons (DeLeo et al., 1987; Dux et al., 1992). However, the time courses of the ischemia-induced events were quite different. Whereas several days were required to observe manifest changes in vivo, 30 min reperfusion with normal oxygenated glucose-containing medium was sufficient to produce in vitro a heavy Ca^{2+} accumulation in the mitochondria and ultrastructural damage. Both were almost prevented by adding 30 μM adenosine to the ischemic medium and aggravated in the presence of the adenosine receptor antagonist theophylline. In contrast to the less affected presynaptic axon terminals, the dendrites of the pyramidal cell neurons revealed large edematous swellings, densely packed with Ca^{2+} deposits. The postsynaptic accentuation of the ischemia-induced Ca^{2+} loading points to a synaptically evoked increased Ca^{2+} influx which is controlled, to some extent, by adenosine. This may be the result of an enlarged membrane depolarization by excessive extracellular glutamate leading to the opening of voltage-operated Ca^{2+} channels or could be due to the additional opening of NMDA receptor-operated ion channels which are present in the CA1 pyramidal cell dendrites at particularly high amounts (Monaghan et al., 1983).

Neuronal Ca^{2+} influx measured by ion sensitive electrode

We investigated the regulation of the synaptically evoked neuronal Ca^{2+} influx in rat and gerbil hippocampal slices with the aid of ion-sensitive electrodes (for prepartion of the electrodes see Heinemann et al., 1977). Here, we focussed on the parameters which control the NMDA receptor-mediated Ca^{2+} influx including its modulation by adenosine and on the ischemia-induced neuronal Ca^{2+} loading leading to nerve cell damage (for a summarizing review see Rudolphi et al., 1992). Two combined recording/Ca^{2+} sensitive electrodes were used to measure the evoked decreases of the extracellular Ca^{2+} concentration (Ca^{2+} transients) in response to repetitive synaptic activation in the dendritic layer (Srad) and in the soma layer (Spyr) of the CA1 neurons. We have evidence that the amplitudes of the Ca^{2+} transients reflect the amount of the stimulus train-evoked neuronal Ca^{2+} influx. In response to afferent 20 Hz/10 sec synaptic activation, a Ca^{2+} transient is seen in Srad as well as in Spyr. On the other hand, in response to direct activation of the pyramidal neurons by antidromic stimulation of their efferent axons, a Ca^{2+} transient is only seen in the soma layer where antidromic population spikes are produced (Fig. 2). This indicates that the generation of the Ca^{2+} transients in Srad and Spyr are independent events

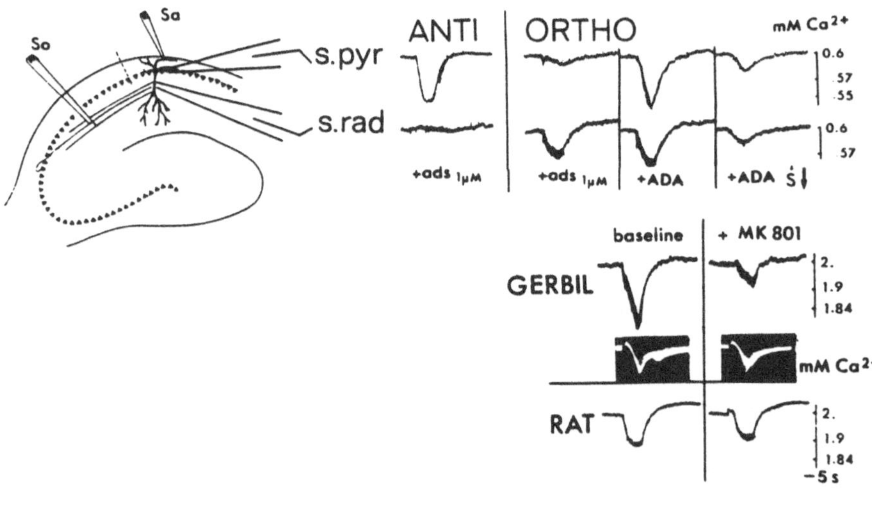

Fig. 2. Stimulus train-evoked Ca^{2+} influx measured by two Ca^{2+}-sensitive electrodes in the CA1 soma layer (s.pyr) and dendritic layer (s.rad) of hippocampal slices. *Left:* diagram showing the positions of the stimulating (So, Sa) and recording electrodes. *Right:* recordings of 20 Hz/10 sec stimulus train-evoked decreases of the extracellular Ca^{2+} concentration and of field potentials (black inserts), elicited by synaptic activation (ORTHO) or by antidromic activation (ANTI) of the CA1 pyramidal neurons. *Top:* changes of the Ca^{2+} transients in s.rad and s.pyr by enzymatic degradation of 1 μM adenosine with deaminase (ADA) are shown at the same (middle) and at a reduced (right) stimulus intensity. *Bottom:* Recordings of Ca^{2+} transients in s.rad from a gerbil and from a rat hippocampal slice; differential effect of MK801

reflecting the synaptic and somatic Ca^{2+} influx respectively. Accordingly, this arrangement allows to differentiate between experimentally induced changes of the activity-evoked neuronal Ca^{2+} influx at the synaptic and at the somatic sites.

Adenosine acts at both sites. It depresses the activity-evoked synaptic Ca^{2+} influx and also the somatic Ca^{2+} influx (Table 1; Fig. 2). We have evidence that the latter is achieved by an additional local mechanism, i.e. by an adenosine-mediated maintenance of the feed back GABA inhibition during repetitive activation. Patch clamp experiments on cultured hippocampal neurons revealed that adenosine primarily acts at the postsynaptic site by increasing the membrane K^+ and Cl^- conductance (Trussel and Jackson, 1985; Mager et al., 1990), thus limiting the amount of the activity evoked depolarization of the postsynaptic membrane. This, in turn, reduces the opening for the voltage-dependent Ca^{2+} channels and also adds to the control of the NMDA receptor-operated ion channels by the membrane potential.

The control of the dangerous neuronal NMDA Ca^{2+} influx resembles that by a high security lock. It requires two keys: the interaction of the agonist, glutamate, with the NMDA receptor which triggers the opening of the connected channel and in addition, a sufficient membrane depolarization which removes the voltage-dependent Mg^{2+} block of the channel (Nowak et al., 1984). Because of this voltage sensitivity, a critical input frequency is reguired to generate a sufficient postsynaptic depolarization able to open the NMDA Ca^{2+} channels. Under physiological conditions, the additional generation of a postsynaptic NMDA Ca^{2+} influx acts as a "synaptic amplifier" and contributes to the phenomenon known as frequency potentiation (Herron et al., 1986). Another physiological mechanism which is linked to neuronal NMDA Ca^{2+} influx is the generation of synaptic long term potentiation (LTP) which can easily be achieved in a hippocampal slice by high frequency (ca. 100 Hz) synaptic activation (Dunwiddie and Lynch, 1978). Under pathological conditions, the generation of NMDA Ca^{2+} influx seems to run out of control contributing to nerve cell death. Several factors could account for a pathologically increased neuronal NMDA Ca^{2+} influx: 1. an enlarged receptor activation — there is evidence that this occurs in the aging brain (Müller, this volume); 2. an excessive membrane depolarization, as occuring during brain ischemia and 3. a reduction of the Mg^{2+}-dependent voltage sensitivity of the NMDA Ca^{2+} channel.

The amount of NMDA Ca^{2+} influx upon repetitive synaptic activation was determined by measuring the propertion of the Ca^{2+} transients which could be blocked by antagonists of the NMDA receptor/channel complex. Here, we used the competitive antagonist, 2-amino-5-phosphonovaleric acid (APV) which acts at the receptor site and the non-competitive antagonist MK801 which acts at a channel site as Mg^{2+} and glycine. Blocking the contribution NMDA Ca^{2+} influx to the total stimulus train-evoked synaptic Ca^{2+} influx with MK801 revealed marked species differences. In contrast to gerbil slices in which a significant amount of NMDA Ca^{2+} fluxes was found to be generated in normal medium by 20 Hz afferent stimulation, only a

Table 1. Modulation of the activity evoked synaptic Ca^{2+}-influx

I. Change of Ca^{2+}-transients by removal of 1 μM adenosine action or of Mg^{2+}			
Ads/Mg	20 Hz	100%	(n = 30)
~~Ads~~/Mg (ADA)	20 Hz	+130* ± 72%	(n = 7)
~~Ads~~/Mg (DPCPX)	20 Hz	+140** ± 78%	(n = 23)
Ads/~~Mg~~	20 Hz	+322** ± 233%	(n = 32)
20 μM Ads/~~Mg~~	20 Hz	not significant	(n = 33)
II. APV-sensitive proportion of Ca^{2+}-transients (NMDA Ca^{2+} influx) in:			
Ads/Mg	20 Hz	not significant	(n = 17)
~~Ads~~/Mg (DPCPX)	20 Hz	30** ± 14%	(n = 8)
Ads/Mg	10 Hz	not significant	(n = 6)
~~Ads~~/Mg (DPCPX)	10 Hz	38** ± 15%	(n = 6)
Ads/~~Mg~~	20 Hz	46** ± 16%	(n = 40)
20 μM Ads/~~Mg~~	20 Hz	48* ± 17%	(n = 6)

The experiments were performed on rat hippocampal slices. Recordings of the evoked decreases of the extracellular Ca^{2+} concentration (Ca^{2+} transients), elicited by afferent 10 sec stimulation at the indicated frequency, were taken from the CA1 stratum radiatum. The "baseline" superfusion medium (2 ml/min at 32°C) contained in mM: NaCl 124, $NaHCO_3$ 25.7, $CaCl_2$ 2.4, KH_2PO_4 1.25, $MgSO_4$ 2.4, KCl 3.3, glucose 10 and 1 μM adenosine. I. *Ads/Mg* Ca^{2+} transients generated in the baseline medium = 100%; *~~Ads~~/Mg* change of Ca^{2+} transients by adding either adenosine deaminase (ADA, 5 μg/ml) or the A1 receptor antagonist DPCPX (50 nM) to the baseline medium; *Ads/~~Mg~~*: change of Ca^{2+} transients by removing Mg^{2+} ions from the baseline medium; *20 μM Ads/~~Mg~~*: change of Ca^{2+} transients by adding 20 μM adenosine to Mg^{2+}-free medium. II. The proportion of stimulus train-evoked NMDA Ca^{2+} influx was determined in the indicated media by measuring the % depression of the Ca^{2+} transients by APV. $^{*}p < 0.05$, $^{**}p < 0.01$ singnificance

small or even no generation of NMDA Ca^{2+} influx was found in rat hippocampal slices (Fig. 2). This suggests that the activation of the synaptic NMDA Ca^{2+} amplifier is facilitated in gerbils compared to rats.

The physiological control via the Mg^{2+}-dependent voltage sensitivity of the NMDA receptor-operated channel was investigated in rat hippocampal slices. Removal of the Mg^{2+} ions from the superfusion medium led to a threefold increase of the 20 Hz/10 sec stimulus train evoked Ca^{2+} transient. 46% of this Ca^{2+} transient could be blocked by APV and represents NMDA Ca^{2+} influx which was minimal in the presence of Mg^{2+} (Table 1). This indicates the band width of the control exerted by the voltage-sensitivity of the NMDA Ca^{2+} channel. Accordingly, all the mechanisms which influence the postsynaptic membrane potential should be powerful instruments to regulate the amount of NMDA Ca^{2+} influx at the synapse elicited by a given input. As stated before, adenosine is such an endogenous modulator which stabilizes the membrane potential limiting membrane depolarization. Therefore, adenosine should be expected to control the activity-evoked synaptic NMDA Ca^{2+} influx.

We tested this by measuring the APV-sensitive proportion of the stimulus train evoked Ca^{2+} transients in the presence and during blockade of adenosine action (Schubert and Mager, 1991; Schubert et al., 1993). 1 μM adenosine completely prevented the synaptic NMDA Ca^{2+} influx by a 20 Hz input, whereas a blockade of the A_1 receptor mediated adenosine action by 8-cyclopentyl-1,3-dipropylxanthine (DPCPX) allowed a large NMDA Ca^{2+} influx in response to 20 Hz and even 10 Hz afferent activation (Table 1). However, if the NMDA Ca^{2+} channels had lost their voltage sensitivity in Mg^{2+}-free medium, even enlarged adenosine did no longer reduce the synaptic NMDA Ca^{2+} influx (Table 1). We conclude from these findings that adenosine (in the physiological extracellular concentration range) determines the input frequency required to generate synaptic NMDA Ca^{2+} influx and that this action relies on the voltage sensitivity of the channel.

However, the latter may be changed under physiological and pathological conditions. It has been recently reported by Ben-Ari's group that an activation of PKC reduces the Mg-dependent voltage sensitivity of the NMDA Ca^{2+} channel and that this may contribute to the initiation of LTP (Ben-Ari et al., 1992). PKC activation may be also elicited pathologically, e.g. in conjunction with the Alzheimer desease (see above) or with some delay during the postischemic phase as experimentally shown in the rat hippocampus (Onodera et al., 1989). Such a PKC-induced "upregulation" of the NMDA receptor-operated channel implies that synaptic NMDA Ca^{2+} influx can be already generated at low input frequencies and can be no longer influenced by modulators of the membrane potential like adenosine. We, therefore, used the adenosine-mediated modulation of the NMDA Ca^{2+} influx as a test paradigm to evaluate whether transient brain ischemia in vivo may cause a delayed upregulation of the synaptic NMDA Ca^{2+} influx during the postischemic phase.

Hippocampal slices from gerbils which had survived 5 min bilateral carotid occlusion for 18–24 hours and slices from normal gerbils were investigated simultaneously in the same experiment (see also Schubert et al., 1993). In the absence of adenosine action, the proportion of the synaptic NMDA Ca^{2+} influx, generated by 2 Hz synaptic activation, was not significantly different in normal and postischemic slices. Also, the modulatory power of 1 μM adenosine to depress the non-NMDA receptor-mediated synaptic Ca^{2+} influx (tested in the presence of APV) was not changed by preceeding in vivo ischemia. In contrast, if the modulatory power of adenosine was tested under conditions when NMDA Ca^{2+} influx was allowed to be generated (in the absence of APV), marked differences were observed between normal and postischemic slices. Whereas the proportion of the adenosine-depressed synaptic Ca^{2+} influx increased by more than 100% in normal slices, the depression of the Ca^{2+} transients by 1 μM adenosine remained almost unchanged in the postischemic slices. This indicates that transient brain ischemia causes a selective loss of the adenosine-mediated control of the synaptic NMDA Ca^{2+} influx which relies on the voltage sensitivity of the NMDA Ca^{2+} channel. This is consistent with the interpretation that a pathological PKC activation during the postischemic phase reduces the voltage sensitivity of the NMDA Ca^{2+} channel which does no longer allow its control by adenosine. This is supported by the

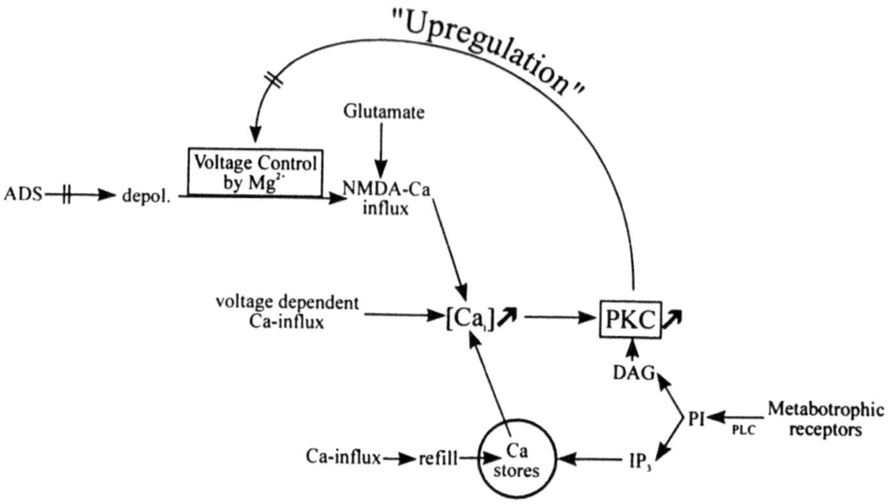

Fig. 3. A possible chain of the PKC-dependent metabolic events contributing to nerve cell damage

finding that pretreatment of hippocampal slices (n = 35) with a PKC activator, phorboldibutyrate, for 1 hour led to a similar loss of the 1 μM adenosine action on the activity-evoked synaptic Ca^{2+} influx.

The proposed chain of the PKC-dependent metabolic events which may contribute to pathological nerve cell damage is summarized in Fig. 3. A pathological activation of metabotropic receptors and of the PI turnover (see above) could be a primary trigger for an uncontrolled PKC activation. This, in turn, may reduce the Mg^{2+} -dependent voltage sensitivity of the NMDA Ca^{2+} channel complex and allow synaptic NMDA Ca^{2+} influx at reduced, physiologically ineffective input frequences. Such an upregulation of the NMDA Ca^{2+} influx during the postischemic phase could be a major factor which causes an increased vulnerability of the CA1 pyramidal neurons and contributes to their delayed death.

Another consequence of a pathological activation of the PI turnover in conjunction with an increased neuronal Ca^{2+} influx can be an increased Ca^{2+} mobilization from the IP_3-and/or Ca^{2+}-sensitive intracellular stores. The possible contribution of such an increased intracellular Ca^{2+} mobilization to the ischemia-induced excessive Ca^{2+} loading of the nerve cells can be investigated in hippocampal slices by means of the fluorescence imaging technique.

Fluorescence imaging technique for studying changes in intracellular Ca^{2+} concentration

In our group a device has been developed and installed by R. Uhl, which allows the selection of any small bands of light (13 nm) within a range of wavelengths from 270 to 700 nm and to change between different wavelengths of excitation within a few milliseconds. This allows us to make use of many different fluorescent indicators which may be imaged quasi simultaneously. Primarily concentrations of various kinds of ions can be measured, but recently distribution of protein kinase C (Khalil and Morgan, 1991) concentration of cAMP (Adams et al., 1991), or distribution and activity of calmodulin (Gough and Taylor, 1993) have been added to the list. The ions which can be measured are calcium (Grynkiewicz, 1985; Minta et al., 1989), sodium (Deri and Adam-Vizi, 1993), chloride (Verkman, 1990), and protons (Ou-yang et al., 1993). Other characteristics that can be assessed include, for example, intracellular free radical formation, survival of cells under noxious or toxic conditions and the distribution of actin filaments by rhodamine-phalloidin (Van Resandt et al., 1985).

The device uses a xenon lamp, which provides an almost even spectrum ranging from UV-light to low infra-red. A parabolic mirror sends the light to a grating which splits it into its spectrum. Rotation of the grating around its vertical axis determines which part of the spectrum is focussed into a glass fiber coupled to the excitation path for epifluorescence of a Zeiss Axiovert 100. The dichroic mirrors and cutoff filters for fluorescence can be chosen as appropriate for the dye or combination of dyes used. The fluorescence of the stained sample is picked up by a CCD-camera which is operated in a so called slow scan mode. In this mode, the light is gathered on the camera target for a preset period of time ranging from ten milliseconds to several seconds. Cooling of the camera by a Peltier-element minimizes the thermal noise. The camera as well as the monochromator are under control of a personal computer with a software designed for this system (Fig. 4).

The dynamics of the ischemia-induced intracellular Ca^{2+} loading and the preferential accentuation of the pathological process in the dendritic layer of the highly vulnerable CA1 neurons could be evidently demonstrated by applying the fluorescence imaging technique to the hippocampal slice preparation. As stated before, the time course of the pathological process is largely accelerated in vitro. Already 6 min after starting the superfusion of unoxygenated, glucose-free medium, a strictly localized increase of the intracellar Ca^{2+} was seen in the CA1 stratum radiatum sparing the soma layer and the CA3 region (Fig. 4). The ischemia-induced intracellular Ca^{2+} rise also occurs, somewhat reduced, in the absence of extracellular Ca^{2+} ions pointing to a major contribution of a Ca^{2+} release from the intracellular stores. Since the Ca^{2+} loading was sensitive to dantrolene, those stores which are operated by cytosolic Ca^{2+} seem to be responsible for the intracellular Ca^{2+} mobilization (Mitani et al., 1993). The xanthine derivative, propentofylline, delayed the onset of the ischemia-induced Ca^{2+} rise (Kadoya et al., 1992). This is in agreement with our earlier observation that a powerful neuroprotection against postischemic Ca^{2+} loading and nerve cell damage could be achieved in vivo by pre- and posttreatment of gerbils with propentofylline (DeLeo et al., 1987). Studies are in progress to identify the metabolic sites at which this pharmacon, thought to act analogue to adenosine, interferes with the pathological chain of events.

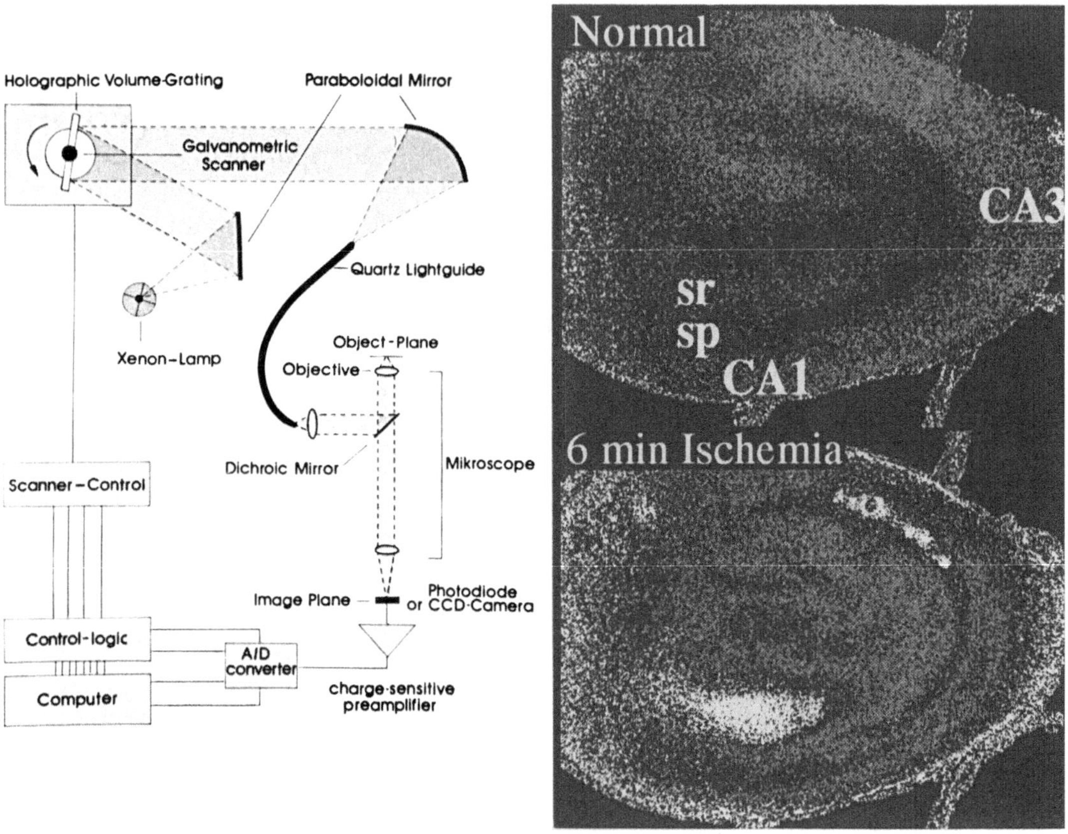

Fig. 4. *Left:* schematic illustration of the fluorescence imaging system developed by Prof. R. Uhl (see text). *Right:* ischemia induced intracellular Ca^{2+} elevation in a rat hippocampal slice. The slice was stained with Fluo3-AM and kept in the perfusion chamber installed on the stage of an epi-fluorescence microscope (5× objective lens). Temperature-controlled (35°C) perfusion solution was changed from a normal oxygenated medium to a nitrogen-bubbled glucose-free medium. Images of the fluorescence signal of Fluo-3, reflecting intracellular Ca^{2+}, were taken for 20 min with 15 sec interval. Two images before (*top*) and 6 min after changing medium (*bottom*) are shown. Marked intracellular Ca^{2+} elevation induced by the ischemic condition was observed in the synaptic reagion (sr) but not in soma layer (sp) of CA1 pyramidal cells

Possible contribution of glial cells to brain damage

It also remains to be clarified whether an ischemia-induced alteration of glial cell function is involved in the process leading to nerve cell death. Thus, the physiological action of astrocytes to take up glutamate from the extracellular space can be reversed by excessive depolarization and by an extracellular increase of K^+ (Barbour et al., 1988). We know from in vivo experiments on gerbils that transient brain ischemia leads to an early hyperplasia of astrocytes which can be detected in the hippocampal CA1 region before significant nerve cell damage (DeLeo et al., 1987). Treatment with propentofylline prevents this astrocyte reaction suggesting that this pharmacon may also interfere with the presumed ischemia-induced change of the astrocyte function. Studies performed with the fluorescence imaging technique on the hippocampal slice model as well as on astrocyte cultures are in progress focussing on the role of the intracellular Cl^- regulation for the physiological and pathological astrocyte function.

Another recently emphasized aspect is the possible contribution of microglial cell activation, found during the course of many brain deseases, to pathological nerve cell damage. Activated microglial cells and/or microglia-derived macrophages release a number of potentially toxic agents including cytokines and free oxygen intermediates (Banati et al., 1993). The finding of Beyreuther (see his chapter in this volume, pp. 219–230) that free oxygen radicals and the related formation of H_2O_2 are involved in the processing of the amyloid precursor protein seems to be particularly interesting in this context and may point to a pathogenetic role of activated and transformed microglial cells in the Alzheimer desease. We used fluorescence techniques to test possibilities of pharmacological interference with activated microglial cells in cultures. For example, we determined the intracellular formation of free oxygen radicals respectively H_2O_2 by measuring the oxidation of dihydrorhodamine into the fluorescent rhodamine and found a marked inhibition by propentofylline (Schubert et al., 1992; Banati et al., 1994). Furthermore, by using the uptake of fluorescent microbeads as a test parameter, we obtained evidence that propentofylline inhibits the transformation of cultivated microglial cells into macrophages (Nakamura and Schubert, in preparation).

The described experiments may illustrate the advantages of a combined use of Ca^{2+}-sensitive electrodes and the fluorescence imaging technique to investigate Ca^{2+}-dependent mechanisms of nerve cell damage and neuro-

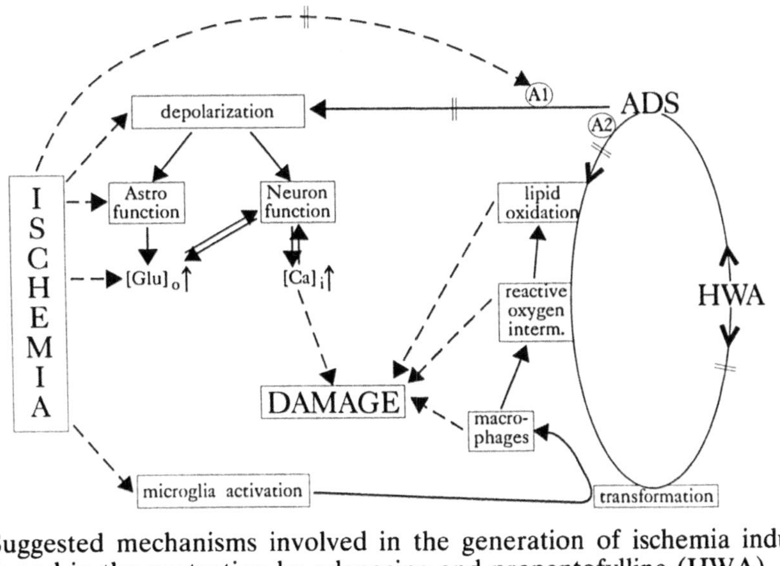

Fig. 5. Suggested mechanisms involved in the generation of ischemia induced nerve cell death and in the protection by adenosine and propentofylline (HWA). For details see Summary

protection in hippocampal slices (in conjunction with studies on individual cell cultures). The results, in particular concerning the ischemia-induced impairment of the interconnected functions of neurons and glial cells are diagramatically shown in Fig. 5 and interpreted in the Summary section.

Acknowledgements

We thank R. Gleich for her engaged cooperation. This study was supported by SFB220 and by the Hoechst AG, Werk Albert, Wiesbaden.

References

Adams S, Harootunian A, Buechler Y, Taylor S, Tsien R (1991) Fluorescence ratio imaging of cyclic AMP in single cells. Nature 349: 694–697

Banati R, Gehrmann J, Schubert P, Kreutzberg G (1993) Cytotoxicity of microglia. Glia 7: 111–118

Banati R, Schubert P, Rothe G, Gehrmann J, Rudolphi K, Valet G, Kreutzberg G (1994) Modulation of intracellular formation of reactive oxygen intermediates in peritoneal macrophages and microglia/brain macrophages by propentophylline. J Cereb Blood Flow Metab 14: 145–149

Barbour B, Brew H, Attwell D (1988) Electrogenic glutamate uptake in glial cells is activated by intracellular potassium. Nature 335: 433–435

Ben-Ari Y, Aniksztejn L, Bregestovski P (1992) Protein kinase C modulation of NMDA currents: an important link for LTP induction. TINS 15: 333–339

DeLeo J, Tóth L, Schubert P, Rudolphi K, Kreutzberg G (1987) Ischemia-induced neuronal cell death, calcium accumulation and glial response in the hippocampus of the Mongolian gerbil and protection by the xanthine derivative HWA 285. J Cereb Blood Flow Metab 7: 745–752

Deri Z, Adam-Vizi V (1993) Detection of intracellular free Na^+ concentration of synaptosomes by a fluorescent indicator. Na^+-binding benzofuran isophthalate: the effect of veratridine, ouabain, and α-latrotoxin. J Neurochem 61: 818–825

Dunwiddie T, Lynch G (1978) Long-term potentiation and depression of synaptic responses in the hippocampus: localization and frequency dependency. J Physiol 276: 353–367

Dux E, Schubert P, Kreutzberg GW (1992) Ultrastructural localization of calcium in ischemic hippocampal slices: the influence of adenosine and theophylline. J Cereb Blood Flow Cell Metab 12: 520–524

Gough AH, Taylor DL (1993) Fluorescence anisotropy imaging microscopy maps calmodulin binding during cellular contraction and locomotion. J Cell Biol 121: 1095–1107

Grynkiewicz G, Poenie M, Tsien RY (1985) A new generation of Ca^{2+} indicators with greatly improved fluorescence properties. J Biol Chem 260: 3440–3450

Heinemann U, Lux H, Gutnick M (1977) Free extracellular calcium and potassium during paroxysmal activity in cerebral cortex of the cat. Exp Brain Res 27: 237–243

Herron C, Lester R, Coan E, Collingridge G (1986) Frequency-dependent involvement of NMDA receptors in the hippocampus: a novel synaptic mechanism. Nature 322: 265–268

Kadoya F, Mitani A, Arai T, Kataoka K (1992) Effects of propentofylline on hypoxia-hypoglycemia induced calcium accumulation in gerbil hippocampal slices. J Cereb Blood Flow Metab 12: 301–305

Khalil RA, Morgan KG (1991) Imaging of protein kinase C distribution and translocation in living vascular smooth muscle cells. Circ Res 69: 1626–1631

Mager R, Ferroni S, Schubert P (1990) Adensosine modulates a voltage-dependent chloride conductance in cultured hippocampal neurons. Brain Res 532: 58–62

Minta A, Kao J, Tsien R (1989) Fluorescent indicators for cytosolic calcium based on rhodamine and fluorescein chromophores. J Biol Chem 264: 8171–8176

Mitani A, Yanase H, Sakai K, Wake Y, Kataoka K (1993) Origin of intracellular Ca^{2+} elevation induced by in vitro ischemia-like condition of hippocampal slices. Brain Res 601: 103–110

Monaghan T, Holets V, Toy D, Cotman C (1983) Anatomical distribution of four pharmacologically distinct ^{3}H-L-glutamate binding sites. Nature 306: 176–178

Nowak L, Bregestovski P, Ascher P, Herbert A, Prochianz A (1984) Magnesium gates glutamate-activated channels in mouse central neurones. Nature 307: 462–465

Onodera H, Araki T, Kogure K (1989) Protein kinase C activity in the rat hippocampus after forebrain ischemia: autoradiographic analysis by ^{3}H-phorbol 12,13-dibutyrate. Brain Res 481: 1–7

Ou-yang Y, Merllergard P, Siesjö BK (1993) Regulation of intracellular pH in single rat cortical neurons in vitro: a microspectrofluorometric study. J Cereb Blood Flow Metab 13: 827–840

Rudolphi K, Schubert P, Parkinson F, Fredholm B (1992) Neuroprotective role of adenosine in cerebral ischemia. TIPS 13: 439–445

Schubert P, Banati R, Dux E, Gehrmann J, Rudolphi K, Kreutzberg G (1992) Reactive oxygen intermediates in microglial cells differentiating into brain macrophages: depression by propentophylline are relation to adenosine action. In: Krieglstein J, Oberpichler H (eds) Pharmacology of cerebral ischemia 1992. Wissenschaftliche Verlagsgesellschaft, Stuttgart, pp 461–467

Schubert P, Keller F, Rudolphi K (1993) Depression of synaptic transmission and evoked NMDA Ca^{2+} influx in hippocampal neurons by adenosine and its blockade by LTP or ischemia. Drug Dev Res 28: 399–405

Schubert P, Mager R (1991) The critical input frequency for NMDA receptor-mediated neuronal Ca^{2+} influx depends on endogenous adenosine. Int J Purine Pyrimidine Res 2: 11–16

Trussel LO, Jackson MB (1985) Adenosine-activated potassium conductance in cultured striatal neurons. Proc Natl Acad Sci 82: 4857–4861

Van Resandt R, Marsman H, Kaplan R, Davoust J, Stelzer H, Stricker R (1985) Optical fluorescence microscopy in three dimensions: microtomoscopy. J Microsc 138: 29–34

Verkman A (1990) Development and biological applications of chloride-sensitive fluorescent indicators. Am J Physiol 259: C375

Authors' address: Dr. P. Schubert, Department of Neuromorphology, Max-Planck-Institute for Psychiatry, Am Klopferspitz 18a, D-82152 Martinsried, Federal Republic of Germany

J Neural Transm (1994) [Suppl] 44: 87–95

The cultured fibroblast model

J. P. Blass

Altschul Laboratory for Dementia Research, Burke Medical Research Institute, Cornell University Medical College, White Plains, New York, U.S.A.

Summary. Cultured skin fibroblasts have been successfully used to elucidate the molecular and biochemical bases of a large number of inborn errors of metabolism which cause neurological disease. The fibroblast model is now being extended to studies of genetic diseases of the nervous system with later clinical onset, including Alzheimer's Disease (AD). Persistence in the cultured cells of an abnormality found in AD brain provides evidence that the abnormality in brain is not simply a reflection of anatomic or chemical damage in the clinically affected tissue. Biochemical and cell biological studies of AD fibroblasts are succeeding in implicating molecules which warrant further study at the molecular genetic level. In neurological diseases with an important genetic component, including AD, studies of fibroblasts can be a useful adjunct to studies of brain.

Introduction

Cultured skin fibroblasts have, over the last 30 years, proven a useful adjunct to direct studies of brain in the elucidation of molecular and biochemical aspects of genetic diseases of the nervous system. Genomic DNA is, at least at birth, the same in all nucleated cells from a single individual, including brain cells and cultured fibroblasts. Depending on the extent of transcription and translation of a gene in fibroblasts, the products of that gene can also be studied conveniently in these cultured cells at the mRNA/cDNA and protein as well as DNA levels. Changes which persist in fibroblasts in serial culture are, however, not necessarily linked directly to an abnormal gene. Abnormalities in genomic DNA can affect the transcription of other genes. Abnormalities in the structure or the amount of a protein often alter the function of the organelles of which that protein is a part or of metabolic pathways in which that protein participates, leading to secondary, "downstream" effects. Such secondary effects, which can sometimes be detected by changes in cell function and therefore on a cell biological level, may provide clues from which one can try to trace back to a primary, genetic abnormality.

This article focuses on ways in which fibroblasts have been used successfully to study hereditary diseases of the nervous system. Illustrative examples are provided from classic studies of inborn errors and from recent studies of Alzheimer's disease. Other abnormalities in AD fibroblasts have been listed elsewhere (Blass, 1993a).

Inborn errors of the nervous system

Inborn errors of the nervous system provide numerous examples of the utility of cultured fibroblasts in studying diseases of the nervous system whose etiology involves a major genetic component, including for instance the gangliosidoses and other disorders of lysosomes and Lesch-Nyhan syndrome, an inborn error associated with profound and stereotyped psychiatric manifestations (Stanbury et al., 1983).

Refsum's disease provides a particularly clear illustration of the potential contribution of the cultured fibroblast model for elucidating the biochemical and genetic basis of an inborn error affecting the nervous system (Steinberg, 1983). Clinically, Refsum's disease (hederopathia atactica polyneuritiformis) is a form of spinocerebellar ataxia with prominent neuropathy and without clinically significant symptoms and signs outside the nervous system. The time of clinical onset can vary from childhood to later adult life. Although clinically Refsum's disease is a disease of the nervous system, biochemically it is a disease of the liver and the kidney. They are the only two tissues in the body which normally contain the perioxosomal phytanic acid oxidase which is the genetically defective component in this condition. Deficiency of this liver-kidney enzyme leads to the failure to metabolize phytanic acid, a branched-fatty acid derived from the normal diet. Phytanic acid eventually accumulates in the bodies of affected patients in quantities which become toxic to the nervous system. Fortunately for investigators, the gene of interest is normally transcribed in cultured skin fibroblasts and the enzyme is normally present. Characterization of the enzyme defect in fibroblasts from Refsum's Disease patients and their framilies played a critical role in elucidating the biochemical and genetic basis of the disorder (Herndon et al., 1969).

The studies of Refsum's disease illustrate well the value of studying cultured fibroblasts *in addition to* studying pathologically affected tissues. By the time patients with Refsum's disease are diagnosed clinically, their tissues contain high levels of phytanic acid. There is *no direct, experimental* technique to exclude the possibility that the enzymatic defect demonstrated in the clinically affected tissues or in the whole patients is in some way secondary to the accumulation of phytanic acid. This problem was circumvented experimentally by studying cultured skin fibroblasts (Herndon et al., 1969). After several passages in serial culture, fibroblasts from the Refsum patients like those from controls contained no detectable levels of phytanic acid. However, the defect in phytanic acid oxidase could still be demonstrated in Refsum fibroblasts, in studies using tracer quantities of radioactively highly-labelled phytanic acid of nearly negligible mass (Herndon et al., 1969). This result showed clearly that the defect was not due to the inhibition of phytanic acid oxidase by the excess amounts of phytanic acid stored in Refsum tissues in vivo.

If Refsum's disease were being studied for the first time in the 1990s, molecular studies would probably be done relatively early rather than after most of the biochemical studies were completed. However, demonstration of a defect in phytanic acid oxidase would still be necessary to justify the effort to clone the gene, and the fibroblast studies would still be useful to demonstrate by *direct, unequivocal experiment* the significance of the molecular abnormality.

Enzymatic and genetic studies in energy metabolism in *AD*

A defect in energy metabolism is one of the best-documented and most consistent abnormalities in AD. As reviewed elsewhere (Blass and Gibson, 1991; Blass, 1993), evidence for this abnormality comes from studies of brain in vivo, of brain biopsies ("ex vivo"), and of brain obtained at autopsy. The discussion below concentrates on enzymatic and genetic data.

Defective activity of three metabolic enzymes which use thiamin has been documented by our group and others in AD brain (Sheu et al., 1985, 1988, 1993; Gibson et al., 1988; Paoletti et al., 1990; Yates et al., 1990; Mastrogiacomo et al., 1993). There are no contravening reports. Two of these enzymes are mitochondrial multi-enzyme complexes which catalyze steps in the quantitatively major pathway of oxidative metabolism in the brain, namely the pyruvate dehydrogenase complex (PDHC) and the α-ketoglutarate dehydrogenase complex (KGDHC). The third is transketolase (TK), a cytoplasmic enzyme which catalyzes the rearrangement of phosphorylated sugars in the pentose phosphate pathways. Measurements of these enzymes in brain have at least two severe limitations. Firstly, there is no direct experimental procedure to rule out the possibility that the defects in activity are not secondary to brain damage in this disorder, even though the defects are found even in regions such as caudate nucleus which are usually largely spared of histological damage in AD (Sheu et al., 1985). Secondly, PDHC and TK are stable post-mortem but KGDHC is not (Gibson et al., 1988; Mastrogiacomo et al., 1993; Sheu et al., 1993). Despite attempts to match AD and control brains for post-mortem time and agonal state as well as for age and sex, the possibility that the decreased activity of KGDHC in AD brain results from agonal or post-mortem change can not be definitively ruled out by studies limited to brain obtained at autopsy.

Studies of cultured skin fibroblasts from AD and control patients were therefore undertaken in fibroblasts from patients with familial AD (FAD), modelled on studies of inborn errors discussed above. Two laboratories (Peterson and Goldman, 1986; Sims et al., 1987) found that subtle but statistically significant abnormalities of oxidative metabolism persisted in cultured AD fibroblasts. In the course of those studies, it became clear that reproducible and interpretable results with the AD fibroblast model required extraordinarily meticulous attention to the replicability of culture conditions, including but not limited to matching AD and control cells not only for chronological age and sex of donor but also for biological age in culture, ie cumulative population doubling levels (cPDL) and percentage life span completed (Tofel-Grehl et al., 1992). In meticulously cultured fibroblasts, reductions were found to persist in cultured AD fibroblasts in the activities of TK (Paoletti et al., 1990; Jung et al., 1993) and of KGDHC but not of PDHC (Sheu et al., 1993). Immunoblotting studies were therefore undertaken of TK and KGDHC but not of PDHC in the histologically normal AD cells, to test for abnormalities at the protein level. Immunoblotting studies of fibroblast TK after isoelectric focusing were consistent with either abnormal proteolysis of TK or of a form of TK unusually susceptible to hydrolysis (Paoletti et al., 1990; Jung et al., 1993). Immunoblotting studies of KGDHC were also suggestive (Sheu et al., 1993). KGDHC consists of a structured array of three proteins: E1k, E2k, and E3. Immunoblots in both AD fibroblasts and AD brain provided no evidence for an abnormality in either E1k or E3, but suggested the presence in AD fibroblasts of a cross-reacting immune material (CRIM) to E2k. This CRIM has not been further characterized at the protein level, in part because the sequence of human E2k was not available at the time these studies were done. Instead, we proceeded to the molecular level.

The cDNA was cloned and the chromosomal localization determined for both human TK (Jung and Sheu, in preparation) and human E2k (Ali et al., 1993). A 14q24.3 locus is linked to AD in approximately half of FAD kindreds with disease onset before age 65, including a large Canadian kindred which was included in our studies (Pericek-Vance et al., 1992; Sheu et al., 1993; Blass et al., 1993b). E2k was localized both to chromosome 1p31 and to chromosome 14q24.3. PCR and direct sequencing indicated that the 1p31 site is a pseudogene (specifically, a retroposon). The 14q24.3 site appears to be the true gene. Studies to date, using both cosmids and yeast artificial chromosomes (YACs) constructed using DNA from the FAD region of 14q24.3 are consistent with the presence of the human E2k gene within the FAD region. Neither TK (Jung et al., in preparation) nor E1k (Koike et al., 1992; Cai et al., in preparation) localize to chromosome 14. Previous studies by others have shown that E3 is on chromosome 7 (Pons et al., 1988). On biochemical as well as genetic grounds, E2k is an attractive candidate for a functionally significant dominantly inherited mutation, since insertion of an aberrant E2k into the ordered KGDHC complex might reasonably be expected to disrupt the structure and therefore the function of the KGDHC complex. We are therefore concentrating on molecular genetic studies of E2k in AD.

These studies of AD have used a classic "inborn error" approach: identification of an abnormality in the whole patient; studies of the basis of the abnormality in clinically affected tissues; more precise specification of the abnormality utilizing the cultured fibroblast model; and subsequent molecular studies at the cDNA and gene levels, including utilization of data from linkage studies. At the present time, E2k appears to be a reasonable candidate gene for some forms of early onset FAD; whether it is or is not in fact one of the genes for FAD can only be determined by molecular genetic studies now underway.

Functional studies of AD fibroblasts

As noted in the first section of this paper, fibroblasts can express abnormalities associated with an inherited disease not only at the gene and gene product level but also at a functional level. If the inborn error leads, directly or indirectly, to abnormalities in a housekeeping function or other function common to brain cells and to fibroblasts, then the fibroblast model can provide a useful tool to elucidate the mechanisms involved, without the inherent complexities of interpreting results obtained with anatomically damaged tissues from sick people. However, the significance of cell biological abnormalities detected in the fibroblast model must always be tested by direct studies of the brain or other clinically important tissue in the disease being studied.

An example of the successful use of fibroblasts to examine a cell-biological abnormality in AD is recent work by Huang and Gibson (1993) demonstrating abnormal β-adrenergic receptor-stimulated cAMP generation in AD cells. Extensive studies using pharmacological probes on meticulously cultured fibroblasts allowed them to implicate an abnormality in the coupling of G proteins to the β-adrenergic receptor in the cultured AD cells. Others have found abnormalities of G-protein function in AD brain (Smith et al., 1987; Danielson et al., 1988; DeKeyser et al., 1990). Thus, abnormalities of G-protein function exist in the clinically significant AD tissue as well as in the cultured fibroblast model. Which of the large number of G proteins is or are specifically involved has not, however, been determined as yet.

Studies in AD of signal transduction in general (Huang and Gibson, 1993) and of G protein function in particular illustrate again the value of using the cultured fibroblast model as well as the clinically damaged tissue to elucidate mechanisms. Quantitative and perhaps even qualitative changes in signalling mechanisms could be expected to accompany the loss of neurons and synapses and other neuropathological alterations characteristic of AD brain. *No direct experimental* means is available to rule out the possibility that the changes in G proteins and other signal transduction mechanisms in AD brain are not simply consequences of synaptic alterations in the anatomically altered tissue. However, the demonstration that abnormalities in signal transduction persist in the anatomically normal AD fibroblasts in serial culture indicates clearly that these changes are somehow linked to the fundamental cellular aberration in this disorder. If the combination of studies in AD brain and studies in AD fibroblasts implicate a specific G protein, then the possibility of a genetic defect in that protein would be worth examining by molecular genetic techniques.

Fibroblasts cultured from neural sources

In studying diseases of the nervous system including AD, the cultured fibroblast model would clearly be more useful if it could be used to study not only housekeeping enzymes but also genes and proteins found primarily if not uniquely in brain cells (such as tau proteins). Our group has taken two approaches to this problem. In the first, skin fibroblasts were cultured in the presence of very high ("industrial") levels of the growth factors which favor the expression of neuronal properties in cultures of embryonic brain. The skin fibroblasts cultured under these conditions contained detectable levels of neurofilament proteins (NFs) and of neuron-specific enolase (SNE), which are characteristic constituents of adult neurons (Blass et al., 1992). However, the quantities of neurofilaments were too low for convenient experimentation.

The second approach was to culture fibroblast-like cells from brain. Two ideas prompted these studies. One was the assumption that fibroblast-like cells could be cultured from brain, as from other tissues which contain connective tissue components within their substance. The second was that the fraction of the genome expressed in fibroblast-like cells cultured from brain might differ from that expressed in fibroblasts cultured from skin, since cells in brain would have been exposed to higher concentrations of brain-specific growth and differentiation factors than cells in skin. In the event, cells which express experimentally convenient amounts of NSE and of NFs have proven easy to culture from young adult rat brain (Black et al., 1993), young adult human brain obtained as a consequence of therapeutic neurosurgery (Black et al., 1992), adult and elderly human brain obtained at routine autopsy (Blass et al., 1993), and more recently from leptomeninges obtained from elderly subjects at routine autopsy (Blass et al., 1993a; DeGiorgio et al., in preparation). The precise morphological and biochemical properties of these cells vary slightly with the tissue of origin and more markedly with the culture conditions used. However, under all conditions studied cells from all these sources contain convincing, easily studied amounts of NSE and of NF-M and usually of NF-H. Preliminary results suggest that these cells may also contain readily detectable amounts of adult tau proteins.

The provenance of these cells is not clear. The cultures are pleomorphic. They contain spindle-shaped "fibroblastoid" cells, polygonal "epithelioid" cells, and occasional cells with long processes. The cells from autopsy brain react with an antibody to the B component of proline hydroxylase, but this component is not unique to collagen-forming cells (Blass et al., 1993). Positive identification of these cells as "brain fibroblasts" would be premature. Their identification as neurons or "neuronal precursors" is clearly not justified at this time, and they show no properties characteristic of glia including no GFAP staining. Studies including transplantation experiments are underway to try to identify the biological potentials of these cells more completely. At the present time, the appellation "cells cultured from brain" (CCFBs) is convenient and implies no unproven assumptions.

From the viewpoint of biochemists and molecular biologists studying disease, the critical question about these cells is not what they are but whether or not they are of any use. As discussed elsewhere (Blass and Degiorgio, 1992), the "CCFBs" may prove more useful than skin fibroblasts *if* they express some of the genes and proteins and therefore some properties of interest which are normally expressed in neurons but not in human cells which can be easily obtained in serial cultures from disease and control subjects. This possibility is under active investigation in regard to tau proteins from AD and control subjects.

Conclusion

The cultured fibroblast model appears to be useful for answering certain types of molecular and biochemical questions about inherited diseases of the nervous system, but can not be used to answer a variety of other important questions which require a behaving organism, or intact brain, or cells which express a panoply of neuronal or glial properties. In the study of AD and other diseases of the nervous system with an important hereditary component, cultured fibroblasts are an adjunct and not a substitute for studies of brain and other clinically affected tissues.

Acknowledgement

These studies were supported by grants from the Overbrook Foundation, the Will Rogers Institute, the Winifred Masterson Burke Relief Foundation, and the NIA (AG03853, AG09014, AG05119, and AG05144).

References

Ali G, Cai X, Sheu K-FR, Szabo P, Cooper AJL, Blass JP (1993) α-ketoglutarate dehydrogenase: a candidate gene for Alzheimer's disease? J Neurochem 61: 5177

Black RS, Bouvier MM, Sheu K-FR, Darzynkiewicz Z, Blass JP (1992) Presence of typical neuronal markers in serially cultured cells from adult human brain. J Neurol Sci 107: 223–232

Blass JP (1993a) Metabolic alterations common to neural and non-neural cells in Alzheimer's disease. Hippocampus 3: 45–54

Blass JP (1993b) Pathophysiology of the Alzheimer's syndrome. Neurology 43 [Suppl]4: S25–S38

Blass JP, DeGiorgio LA (1992) Tissue culture and studies of human brain disease. J Neurol Sci 113: 1–3

Black RS, DeGiorgio LA, Sheu K-FR, Darzynkiewicz Z, Duffy JT, Blass JP (1993) Expression of neuronal proteins in a cell line derived from adult rat brain. J Neurochem 62: 2132–2140

Blass JP, Gibson GE (1991) The role of oxidative abnormalities in the pathophysiology of Alzheimer's disease. Rev Neurol (Paris) 147: 513–525

Blass JP, Markesbery WR, Ko L, DeGiorgio L, Sheu K-FR, Darzynkiewicz Z (1993) Presence of neuronal proteins in serially cultured cells from autopsy human brain. J Neurol Sci 121: 132–138

Blass JP, Sheu K-FR, Ko L, Bancher CL, Wisniewski HM, Bouvier M, Duffy JT, Baker AC (1992) Presence of low amounts of "neuronal" antigens in cultured human skin fibroblasts. J Neurol Sci 107: 223–232

DeKeyser J, Ebinger G, Vauquelin G (1990) D1-dopamine receptor abnormality in frontal cortex points to a functional alteration of cortical cell membranes in Alzheimer's disease. Arch Neurol 47: 761–763

Gibson GE, Sheu K-FR, Blass JP, Baker A, Carlson KC, Harding B, Blass JP (1988) Reduced activities of thiamine-dependent enzymes in the brains and peripheral tissues of patients with Alzheimer's disease. Arch Neurol 45: 836–840

Herndon JH Jr, Steinberg D, Uhlendorf BW, Fales HM (1969) Refsum's disease: characterization of the enzyme defect in cell culture. J Clin Invest 48: 1017–1024

Huang H-M, Gibson GE (1993) Altered β-adrenergic receptor-stimulated cAMP formation in cultured skin fibroblasts from Alzheimer donors. J Biol Chem 268: 14616–14621

Jung E-H, Sheu K-FR, Blass JP (1993) An enzymatic and immunochemical analysis of transketolase in fibroblasts from Wernicke-Korsakoff syndrome. J Neurol Sci 114: 123–127

Koike K, Urata Y, Goto S (1992) Cloning and nucleotide sequence of the cDNA encoding human 2-oxoglutarate dehydrogenase (lipoamide). Proc Natl Acad Sci USA 89: 1963–1967

Mastrogiacomo F, Bergeron C, Kish SJ (1993) Brain α-ketoglutarate dehydrogenase activity in Alzheimer's disease. J Neurochem 61: 2007–2014

Paoletti F, Mocali A, Marchi S, Sorbi S, Piacentinni S (1990) Occurrence of transketolase abnormality in extracts of foreskin fibroblasts from patients with Alzheimer's disease. Biochem Biophys Res Commun 172: 396–401

Pericek-Vance M, Foncin JF, Montesi F, et al (1992) Genetic evidence for a novel familial Alzheimer's locus on chromosome 14. Nature Genet 2: 330–334

Pons G, Raefshy-Estrin C, Carothers DJ (1988) Cloning and DNA sequence in the dihydrolipoamide dehydrogenase component of human α-ketoacid dehydrogenase complexes. Proc Natl Acad Sci USA 85: 1422–1426

Peterson C, Goldman JE (1986) Alterations in calcium content and biochemical processes in cultured skin fibroblasts from aged and Alzheimer donors. Proc Natl Acad Sci USA 83: 2758–2762

Sheu K-FR, Clarke DD, Kim YT, Blass JP, Harding BJ, DeCicco J (1988) Studies of the transketolase abnormality in Alzheimer's disease. Arch Neurol 45: 841–845

Sheu K-FR, Cooper AJL, Lindsay JG, Blass JP (1993) Abnormality of the α-ketoglutarate dehydrogenase complex in fibroblasts from familial Alzheimer's disease. Ann Neurol 35: 312–318

Sheu K-FR, Kim YY, Blass JP, Weksler M (1985) An immunochemical study on the deficit of pyruvate dehydrogenase in Alzheimer's disease brain. Ann Neurol 17: 444–449

Sims NBR, Finegan JM, Blass JP (1987) Altered metabolic properties of cultured skin fibroblasts in Alzheimer's disease. Ann Neurol 21: 451–457

Smith CJ, Perry EK, Perry RH, Fairbairn AF, Birdsall NJM (1987) Guanine nucleotide modulation of muscarinic cholinergic receptor binding in post-mortem human brain — a preliminary study in Alzheimer's disease. Neurosci Lett 82: 227–232

Stanbury JB, Wyngaarden JB, Fredrickson DS, Goldstein JL, Brown MS (eds) (1983) The metabolic basis of inherited disease, 5th edn. McGraw Hill, New York

Steinberg D (1983) Phytanic acid storage disease (Refsum's disease). In: Stanbury JB, Wyngaarden JB, Fredrickson DS, Goldstein JL, Brown MS (eds) The metabolic basis of inherited disease, 5th edn. McGraw Hill, New York, pp 731–747

Tofel-Grehl B, Blass JP, Gibson GE (1992) Heterogeneity in CPDL to phase-out among fibroblasts from different Alzheimer families. Mol Biol Cell 35: 1898
Westlind-Danielsson A, et al. (1988) VIP-sensitive adenylate cyclase, guanylate cyclase, muscarinic receptors, choline acetyltransferase and acetylcholinesterase, in brain tissue afflicted by Alzheimyer's disease/senile dementia of the Alzheimer type. Neurobiol Aging 9: 153–162
Yates CM, Butterworth J, Tennant MC, Gordon A (1990) Enzyme activities in relation to pH and lactate in postmortem brain in Alzheimer-type and other dementias. J Neurochem 55: 1624–1630

Author's address: J.P. Blass, M.D., Ph.D., Altschul Laboratory for Dementia Research, Burke Medical Research Institute, Cornell University Medical College, White Plains, NY 10605, U.S.A.

J Neural Transm (1994) [Suppl] 44: 97–109

Psychometric testing in rats during normal ageing
Procedures and results

H. Bubna-Littitz[1] and **J. Jahn**[2]

[1] Institut für Physiologie and [2] EDV-Zentrum, Veterinärmedizinische Universität, Wien, Austria

Summary. Examples are given for assessing parameters for motor activity learning and memory. By using the method of the open field the authors could verify that motor activity decreases with age and that it underlies obviously endogenous rhythms which undergo age dependent alterations (e.g. diminution of the amplitude). From the results of an investigation on the influence of ageing and stress on learning ability and memory of the rat the authors draw the following conclusions: Learning ability declines with advancing age. The effect of ageing on memory could not be clearly shown. The effect of the stressors used (crowding and noise) seems to depend on the season of the year (circannual biorhythm).

Introduction

The interpretation of neurochemical, neurohistological, etc. data obtained from the brain often requires the assessment of functional parameters of the brain. One powerful indicator of the functional state of the brain is behaviour. Psychometric parameters describe parts of the behaviour of the experimental animal qualitatively and or quantitatively. In the following, examples are given for parameters concerned with motor activity, learning and memory.

Age related changes of the motor activity of the rat

We can distinguish between reactive and spontaneous motor activity (Hofecker et al., 1978; Shirley, 1929). Spontaneous motor activity occurs without any detectable alteration in the environment. But one should be aware of the fact that putting the animal in an apparatus for measuring motor activity is a stimulus itself which can provoke motor activity. There exists a lot of systems developed to measure motor activity. The most important of them have been described by P. Silverman, in his excellent book on the laboratory rat (Silverman, 1978). One method for examining spontaneous motor activity is illustrated by the following investigation upon the influence of ageing on the behaviour of the rat in the open field.

The motor activity is an important indicator of the functional state of the system and the neuronal and motoric mechanisms involved (Hofecker et al., 1978). A change of this complex system will be followed by a change in motor activity. A lot of experiments showed that the motor activity decreases during ageing (Dawson and Crowne, 1988; Hofecker et al., 1978; Kment et al., 1982; Shirley, 1929; Sloanaker, 1907). These experiments were unable to detect changes in the spatial pattern of the movements. A part of the motor activity is caused by exploratory activity. The amount of exploratory activity is perhaps subjected to age changes, too. One method to investigate the spatial pattern of motor activity and exploratory activity is the open field (OF; Hall, 1934).

Motor activity depends not only on age but also on the time of the day in consequence of circadian biorhythms. Biorhythms are "feed-forward mechanisms" which prepare the organism for rhythmically recurring environmental changes (e.g.: day-night). With increasing age, rhythmicity becomes weaker and weaker (diminution of the amplitude, shift of the phase, (Bubna-Littitz, 1980, 1981; Pav et al., 1981). The old organism depends more and more on the feed-back mechanisms. Therefore the adaption to the time-dependent alterations of the environment becomes weaker and weaker.

Open field experiments, known from literature, did not last longer than 30 minutes and therefore don't give any information about the circadian biorhythm. That is the reason why we developed a system, allowing us to observe the animals for 24 hours with a minimal amount of labour and staff (Zoufal, 1988).

Methods

Animals

36 male Sprague-Dawley rats (OFA 17) from the ageing cohort of our institute were used (Table 1). The rats were housed in groups of six in Makrolon^{R}-IV-cages under standard conditions (25°C, light from 6 a.m. to 6 p.m., Altromin^{R} rat diet and tap water and libitum).

Observation system

Panasonic WV 1850G Videocamera PC AT 286 with a frame grabber, infrared illumination (Viptronic, Vipdula 810 IR), graphic monitor and printer.

Procedure of observation

We used a squared open field with a side length of 1.2 m and walls 0.5 m high. The OF was illuminated by infrared light (Viptronic, Vipdula 810 IR). In the middle of one side

Table 1. Age of the groups

Group A: 5–9 months ($\bar{x}$ = 7 months)
Group B: 13–17 months ($\bar{x}$ = 15 months)
Group C: 24–28 months ($\bar{x}$ = 26 months)

of the wall there was a water drinking device. Just before the dark phase of our light regimen one of the animals was put into the OF and was observed for 24 hours in complete darkness. The centre of gravity was calculated 8 times per second by the computer and the coordinates were stored. The software was developed by the Technische Universität Wien (Zoufal, 1988).

The following parameters were calculated for each one of the 24 hours observatinal time:

a) Entrances: The OF was subdivided into 9 squares equal in size. The number of entrances into each of these subsquares was computed.
b) Duration of stay in each of the subsquares in percent.
c) Duration of stay in the marginal area and the centre of the OF. The marginal area was defined 15 cm inwards from the wall.
d) Traversions: The number of crossings through the OF. This is a measure for the exploratory activity of the animal.

Statistics applied

Between groups: Kruskal-Wallis-Test and Dunn-Test
Within groups: Wilcoxon-Test (comparison between L-phase and D-phase)
Friedman-Test followed by Wilcoxon-Wilcox-Test

Results

Whole time of observation

a) Entrances per hour: The young (A) and adult animals (B) show significantly higher values than the older ones (Median: A: 5.5; B: 3.0; C: 0.8; B/C: $p < 0.05$; A/C: $p < 0.01$).

b) Traversions per hour: Again, the young (A) and adult (B) animals show significantly higher values than the old ones (Median: A: 2.0; B: 1.2; C: 0.0; B/C: $p < 0.05$; A/C: $p < 0.01$).

c) Time spent in the marginal area: The old animals (C) almost stay for the whole observational time in the marginal area (A: 95.9%; B: 97.6%; C: 99.9%; B/C and A/C: $p < 0.01$).

Comparison between the original dark-phase (D) and light-phase (L)

Entrances: During the D-phase the animals of groups A and B show remarkably higher values than during the L-phase (A: 9.3; 1.9; B: 3.8; 1.7).

These values differ significantly ($p < 0.01$). In group C no difference could be detected by statistical analysis.

Traversions per hour: The young and adult rats cross the OF during the L-phase more frequently (Median: 3.0; 1.9) than during the D-phase (Median: 0.8; 0.5). This difference is significant in both groups ($p < 0.01$; $p < 0.05$). Group C showed no traversions at all.

Duration of stay in the marginal area: Only in group A a significant difference between L- and D-phase could be detected ($p > 0.01$; Median: 92.8%; 99.0%).

Comparison of the time-course of the parameters

Entrances per hour (Fig. 1)*:* The curves were drawn using the median-values. The original dark phase is indicated by a bar. Group A and B show three well marked peaks. The curve of group C is always near the O-line.

Traversions (Fig. 2)*:* Group A shows a well marked peak during the D-phase. This peak corresponds with the third peak of the entrances. In group B are two peaks, which correspond with the second and third peak of the entrances.

Duration of stay in the marginal area (Fig. 3)*:* In the space of time in which in group A the highest numbers of traversions occur can the shortest duration of stay in the marginal area be detected. The curves of group A and B slope down and afterwards increase in the last third of the D-phase.

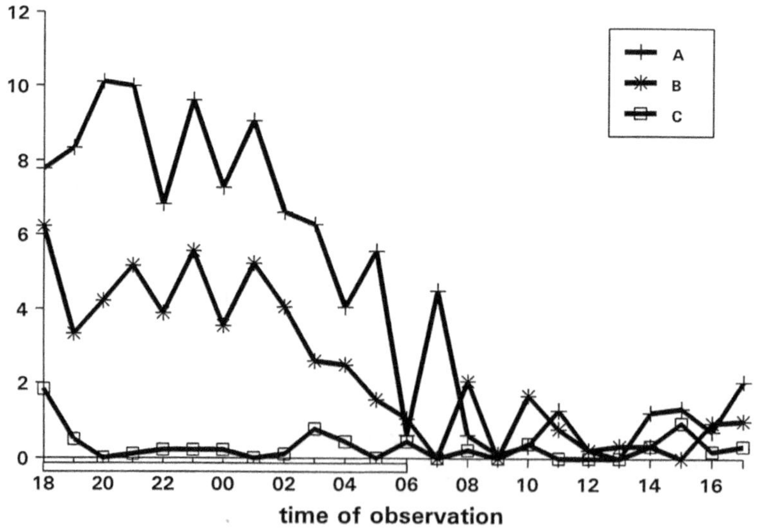

Fig. 1. Time course of the parameter entrances into the subsquares during the whole observational period. For drawing the curves we used the median of the single hours of observation

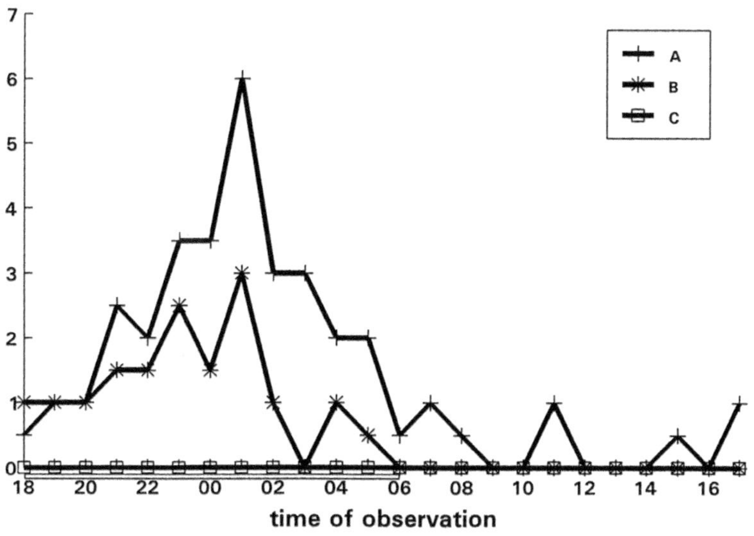

Fig. 2. Time course of the parameter traversion through the OF (compare Fig. 1)

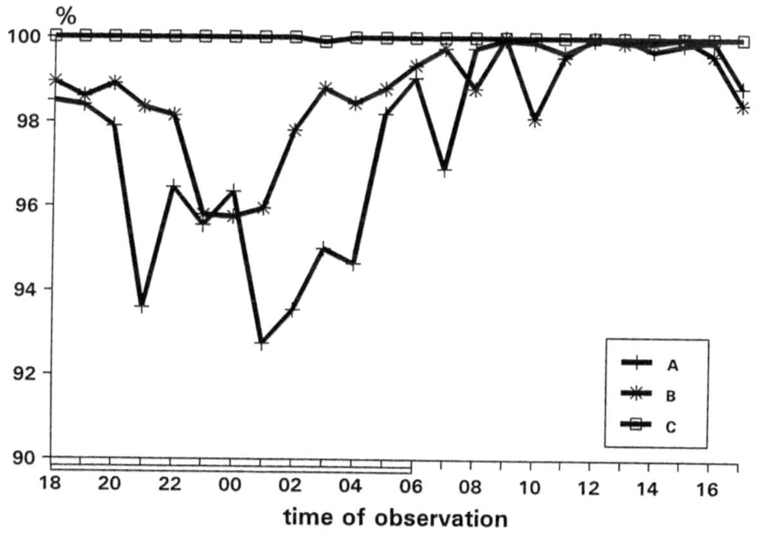

Fig. 3. Time course of the parameter duration of stay in the marginal area (compare Fig. 1)

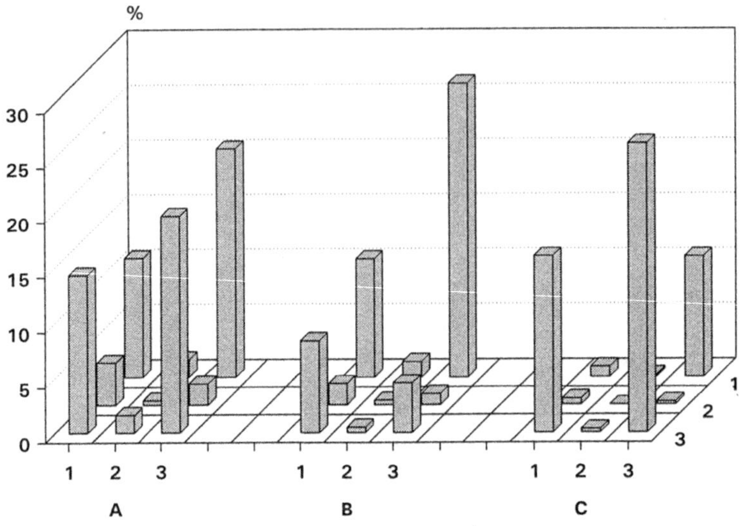

Fig. 4. Duration of stay in the subsquares during the whole observational time in the three age-groups. The height of the bars indicates the median of the percentage of the whole observational time

During the L-phase the curves of all groups are near the 100%-line. The old animals are in the marginal area throughout the whole time of observation.

Duration of stay in the subsquares during the whole observational period (Fig. 4)

The values are the percentage of time of the whole observational period. The water was provided in field 1.2 (top row) and rats were put into the OF in field 3.2 (bottom row).

All animals preferred the corners (square 1.1, 1.3, 3.1 and 3.3). They rarely used the centre (square 2.2). This preference is more pronounced in group B than in group A. The older animals markedly prefer the corners adjacent to the square, where they were put in the OF (subsquare 3.2).

Discussion

The number of entrances indicates that motor activity in the rat decreases with age. This fact has been described by other authors, too (Dawson and Crowne, 1988; Shirley, 1929; Sloanaker, 1907). Exploratory activity seems to decrease faster than motor activity. If the mean of entrances and the mean of traversions of group A is 100%, group B shows 56% of the entrances and 44% of the traversions. In group C the number of entrances is 17% and the number of traversions 5.5%. These data show that exploratory activity is markedly dimished in rats aged 13–17 months and nearly ceased in rats aged 24–28 months. This can also be derived from the values of the parameters "duration of stay in the marginal area". The rats of group C nearly exclusively use the marginal area. The preference of the corners adjacent to the subsquare where the animals were put into the OF also hints at low exploratory activity. The animals of the other groups use all four corners at nearly the same intensity. The early cessation of exploratory activity can also be seen from the fact that the adult rats use the middle marginal subsquare less frequently than the young rats.

The comparison of the L-phase with the D-phase indicates a diminution of the amplitude of the circadian rhythmicity of the motor activity with increasing age. No rhythmicity could be detected in group C. This diminution of the amplitude becomes especially evident when one is looking at time course of the parameters (Figs. 1–3). These biorhythms are free-running for the observation was performed in complete, constant darkness (Halberg and Lee, 1974). Free-running of biorhythms has also been described by other investigators (Albers et al., 1981; Aschoff, 1957; Bünning, 1977). Observation during a normal light-regimen could simulate an endogenous circadian biorhythm.

From the time course curves (Figs. 1–3) of the groups A and B we conclude that the following pattern of behaviour in the OF is reasonable: About two hours after the beginning of the D-phase there is a first peak of motor activity. During this peak the rats use only the marginal area. Therefore the first peak of the curve of the entrances is not accompanied by a peak in the "traversions". When the third peak of the parameter "entrances" occurs there also occurs one in the curve drawn for the parameter 'traversions". This indicates that during the third burst of motor activity the rats explore the central area of the OF. Not only the preference for the marginal area but also the preference for the corners of the OF can be explained by the orientation behaviour of the rat: When there are no visual cues the rat orientates by means of its vibrissae and therefore moves along the wall. Optimal contact with vibrissae and body is provided in the corners and therefore the rats must feel more secure there.

The neglectable exploratory activity observed in the old animals might be interpreted as strategy to compensate the decreased ability to adapt to new environmental conditions in the old organism: The cessation of exploration helps the aged organism to avoid new environmental situations which perhaps could not be met by it. Our results show that the observation of rats in an OF for longer periods of time is a powerful method to detect age dependent changes in the time course of motor activity and exploratory activity.

Learning and memory — theoretical aspects

Firstly we should define the term learning: Learning is the change of the probability for the occurrence of behavioural patterns in certain configurations of stimuli. This change of probability must be due to an earlier confrontation with this or a similar situation. It need not be due to injury of the organism or to spontaneous developmental processes (Hofstätter, 1957). This definition is illustrated by the following example: A rat is put into a box with two levers which can be pressed by the animal. When the right lever is pressed a pellet of food is offered to the animal. No food is offered when the left lever is pressed. At the beginning of the experiment the chances of the animal manipulating the right or the left lever are equal (0.5). During the experiment the probability will shift to the right lever. When the animal has learned the task thoroughly the probability of pressing the right lever will be 1.0 and of pressing the left one 0.0. This shift should not be brought about by injury. Should the motor system of the left paw be disturbed we could also perhaps observe an increase in the probability of pressing the right lever. Spontaneous developmental processes are not relevant in our experimental work on learning. Nevertheless here is an example as completion: In a defined developmental state of the animal the probability for sexual behaviour increases dramatically but it is not learned but on the contrary it is innate. In selecting a task for testing the learning ability of animals one should pay attention to the learning disposition (Eibl-Eibesfeldt, 1987), of the animals used in the experiment. We can differentiate between innate, acquired and actual learning disposition.

Innate learning disposition

Its basis are essentially the proporties of the sensory system and the innate behavioural patterns. For example rats in their natural environment have to orientate in a complex system of tunnels and are very skilled in manipulating objects. Therefore rats learn easily tasks which require orientation or manipulation of objects (e.g.: T-maze, Skinner box).

Acquired learning disposition

It depends on the biography of the animal. Rats housed in cages provided with ladders, wheels, boxes, platforms and facilities for exploratory behaviour show increased cortex weight and higher learning ability compared to animals reared in the usual colony cages or kept in isolation (Bennet et al., 1970). In consequence of these results, especially in gerontological experiments we should ask ourselves whether the observed decrease of learning ability in aged rats is due to ageing per se and/or due to deprivation of external stimuli.

Actual learning disposition

It depends on the animals actual state of health, motivational state, the phase of its biorhythms and so on. The description of the following experiment should give some insight in performing learning experiments and also how to test memory.

Gerontological investigation into the influence of stress on the learning ability (L) and memory (M) of rats

Method

The influence of noise and overcrowding on L and M was investigated by a longitudinal cohort study of 258 male Sprague Dawley rats. The rats were aged 6 months at the beginning and 29 months at the end of the study. From 6 months onwards one group was exposed to noise (*group B*) or *overcrowding (group C)*. Group A served as *control*. As soon as the animals reached the age of 8 months, 2 animals of each group were trained by means of a semi-automatic multiple T-maze with 7 choice points (Bubna-Littitz et al., 1980). From the definition of the term "learning" given by Hofstätter (1957) we derived the parameter "probability" (P) for L and M: We computed the probability of the occurrence of at least 0 to 7 correct decisions at random during a single run through the maze (see Table 2). We related these values to the number of

Table 2. Probability

N	P
0	1.000
1	0.992
2	0.938
3	0.773
4	0.500
5	0.277
6	0.063
7	0.008

N Number of correct decisions. *P* Probability for the occurence of at least. N correct decisions at random during a single run

Table 3. Age of the four age groups

Age group	Age range (months)	X of the age	Number of animals in the age groups A	B	C
I	8–11	10	24	24	24
II	11–17	14	20	20	19
III	20–23	21	23	24	24
IV	25–29	27	19	16	21

correct decisions observed during the single maze run. The rats were trained for 5 consecutive days 3 times per day. 16 days after the last training M was tested by 3 runs. For the statistical evaluation we computed the mean values of P of the 15 maze runs for each animal (L) and the mean values of P of the 3 "memory runs" (M). Each group (A, B, C) had been divided into 4 age groups (I, II, III, IV; see Table 3).

Experimental conditions

Noise: 5 days per week, 2 hours per day, pulse sound, 100 dB, sweeps 10.000–150 Hz. After the first training the animals were not exposed to noise again.
Overcrowding: 12 instead of 6 rats per Makrolon IV cage
Motivation: 60 hours deprivation from water
Reinforcement: 1 min. water drinking in the goal box

Results (L)

Comparing the 4 age groups within the experimental groups (Fig. 5b) one can see a continues decline of L with age in group A. In group B and C age groups I and III show better L than age groups II and IV. In group C these differences are statistically significant. The comparison of the 3 experimental groups within the 4 age groups (Fig. 5a) shows the following trends: In age groups I and II overcrowding and noise seem to improve L. In age group IV

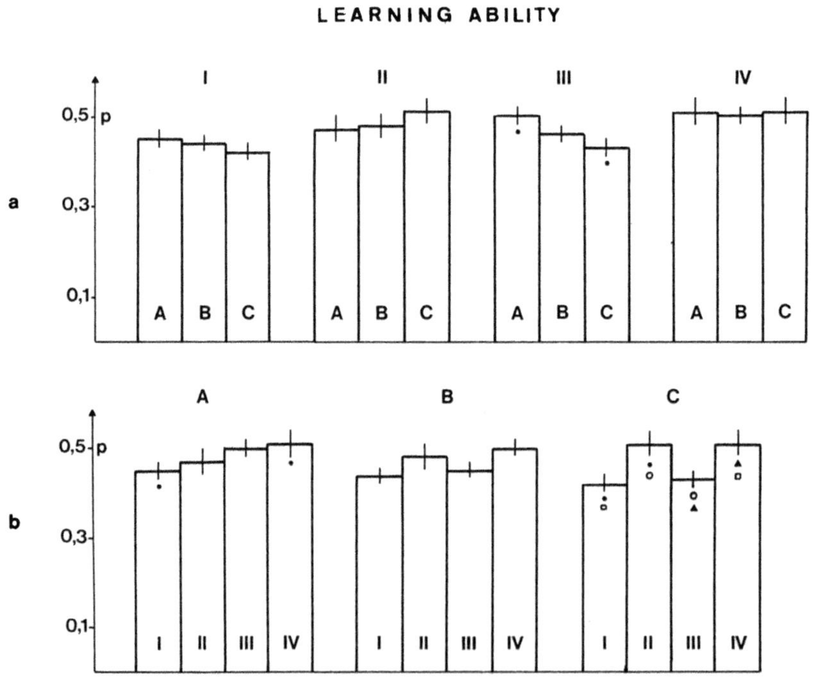

Fig. 5. a Comparison of L within the age-groups. **b** Comparison of L within the experimental groups. Significant differences are indicated (e.g.: *). The hight of the bars indicates x ± s_X

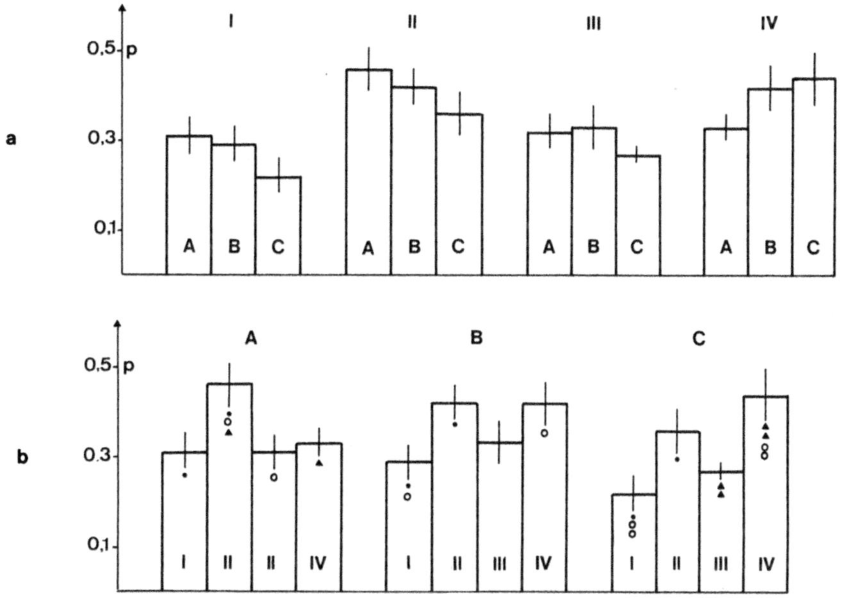

Fig. 6. a Comparison of M within the age groups. **b** Comparison of M within the experimental groups (compare Fig. 5)

an effect of these influences on L cannot be detected. In age group II groups B and C revealed — insignificantly — weaker L than the control group.

Results (M)

Within age group I, II and III group C reveals always better M than group A (although not significant; see Fig. 6a). In age groups I and II the values of group B are situated between those of the groups A and C. In age group IV this trend is reversed: B and C show weaker M than group A. Comparing the age groups within the experimental groups (Fig. 6b) group B and C provide a picture similar to that of L: age groups I and III show better M than age groups II and IV. In group A age group II shows weaker M than the age groups I, III and IV.

Discussion (L)

We believe that the pattern of histograms in Fig. 5b which can be observed in group B and C is caused by seasonal changes of the effects of the experimental influences exerted by us on group B and C: Animals trained at similar seasons of the year (I and III, II and IV; see Fig. 7) reveal similar L.

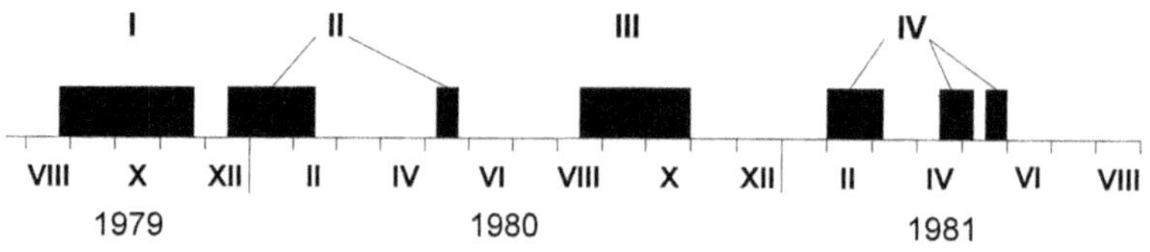

Fig. 7. Time table of the learning and memory tests. From the table you can see in which season of the year the different age-groups had been trained

Moreover in age group I and III especially overcrowding improves L but diminishes it in age group II (see Fig. 5a). In age group IV L is neither affected by noise nor by overcrowding (Fig. 5a).

Discussion (M)

In contrast to L M showed no detectable decrease with age in group A (see Fig. 6b). The weaker M in age group II in all 3 experimental groups compared with the age groups I and III might be due to a circannual rhythm (see Fig. 6b). The diminution of M in age group IV in the groups B and C might be caused by a seasonal change of the effects of noise and overcrowding. Another explanation for this observation is Selye's "tear and wear hypothesis" ("the length of life depends inversely on the rate of living"). In the age groups I–III overcrowding (I, II and III) and noise (I and III) improve M compared with the control group (see Fig. 6a).

We believe that the inclusion of the effect of circannual rhythms on L, M and on the susceptibility of the rats to our experimental influences into our interpretation is justified by the seasonal changes of other parameter values examined in these animals.

General conclusions

In planning studies involving the assessment of psychometric parameters in the rat one should take into account the influence of circadian and circannual biorhythms. In studies not presented here we could show that circadian rhythms also influence learning ability. Especially in longitudinal gerontologic studies circannual rhythms can exert an influence on the results.

References

Albers HE, Gerall AA, Axelson JF (1981) Circadian rhythm dissociation in the rat: effects of longterm constant illumination. Neurosci Lett 25: 89–94

Aschoff J (1957) Aktivitätsmuster der Tagesperiodik. Naturwissenschaften 44: 361–367

Bennet LE, Rosenzweig MR, Diamond MC (1970) Time course of effects of differential experience on brain measures and behaviour of rats. In: Byrne W (ed)

Molecular approaches to learning and memory. Academic Press, New York, pp 55–89
Bubna-Littitz H (1980) Biologische Rhythmenforschung und ihre Beziehung zur Klinik. Wien Tierärztl Mschr 67: 127–133
Bubna-Littitz H, Pav E, Skalicky M (1981) Untersuchungen zu den Tagesschwankungen der Konzentration von Blutzellen bei jungen und alten Ratten. Wien Tierärztl Mschr 68: 383–387
Bubna-Littitz H, Windischbauer G, Lechner P (1980) Durchführung von Lern- und Merkversuchen an Ratten mittels eines halbautomatischen Mehrfach-T-Labyrinthes sowie deren Auswertung. In: Windischbauer G, Keck G (eds) Bildverarbeitung und Musterverarbeitung in der Medizin. Österr Ges für Biomedizinische Technik, Graz, S 146–150
Bünning E (1977) Die physiologische Uhr. Circadiane Biorhythmik und Biochronometrie. Springer, Berlin Heidelberg New York, S 1–79
Dawson KA, Crowne DP (1988) Longitudinal development of activity in long evans rats. J Gerontol 43: 85–86
Eibl-Eibesfeldt I (1987) Grundriß der vergleichenden Verhaltensforschung. Piper, München, S 376–386
Halberg F, Lee JK (1974) Glossary of selected chronobiologic terms. In: Scheving LE, Halberg F, Pauly JE (eds) Chronobiology. Thieme, Stuttgart, pp 37–50
Hall CS (1934) Citation from Silverman P (1978) Animal behavior in the labratory. Chapman & Hall, London, p 380
Hofecker G, Kment A, Niedermüller H (1978) Die motorische Aktivität als Altersparameter der Ratte. Akt Gerontol 8: 271–279
Hofstätter PR (1957) Psychologie. Fischer, Frankfurt, S 195
Holmgren HJ, Swenson A (1953) Der Einfluß des Lichtes auf den 24-Stunden-Rhythmus der Aktivität, des Leberglykogens und der Körpertemperatur. Acta Med Scand 278: 71–76
Kment A, Hofecker G, Skalicky M, Niedermüller H (1982) Der Einfluß anhaltend erhöhter Populationsdichte auf die Spontanaktivität der alternden Ratte. Akt Gerontol 12: 128–133
Pav E, Bubna-Littitz H, Skalicky M, Hofecker G (1981) Zirkadiane Studien an jungen und alten Ratten. Akt Gerontol 11: 12–16
Shirley M (1929) Spontaneous activity. Psychol Bull 26: 341–364
Silverman P (1978) Animal behaviour in the laboratory. Chapman & Hall, London
Sloanaker JR (1907) The normal activity of the white rat at different ages. J Comp Neurol Psychol 17: 342–359
Zoufal A (1988) Videoauswertungssystem für Tierbeobachtungen. Diplomarbeit, TU Wien

Authors' address: Doz. Dr. med. vet. H. Bubna-Littitz, Institut für Physiologie, Veterinärmedizinische Universität, Linke Bahngasse 11, A-1030 Wien, Austria.

J Neural Transm (1994) [Suppl] 44: 111–132

Ultrastructural changes in brain parenchyma during normal aging and in animal models of aging*

C. Struys-Ponsar[1], **A. Florence**[2], **A. Gauthier**[2], **R. R. Crichton**[2], and **Ph. van den Bosch de Aguilar**[1]

Laboratoires de [1] Biologie cellulaire et de [2] Biochimie, Université Catholique de Louvain, Louvain-la-Neuve, Belgique

Summary. During aging, the brain parenchyma of animals and humans share many similarities, both in the gray and the white matter. Unfortunately, until now, neither aged animals nor animal models reproduce the two hallmarks of aging of the human brain: senile plaques and tangles. Therefore, observations performed on animals are limited to some aspects of the involutive process which affects brain parenchyma during aging and their appropriateness to the human situation. One striking aspect concerns the occurence of vacuolated necrotic cells whose number increases with advancing age. These cells can constitute markers of the brain involutive process and they characterize, both in animal and human, the more vulnerable areas of the brain affected by the neuronal rarefaction. Experimental animal models can be used to study the various conditions which sustain the cell survival and to determine, at the cellular level, the factors leading the brain parenchyma to an irreversible state of degradation.

Introduction

During aging, the histological aspect of both grey and white matter is profoundly modified throughout the nervous system. These histological manifestations have been recognized since a long time and the observations done in aged animal and human brain have accumulated in the literature over many years. These studies raise two main problems. First, the type and the extent of the alterations concerned in both normal aging and pathological manifestations of Alzheimer's disease (AD) in the first publications. Only recently, the majority of the studies, supported by various diagnostic tests, give more accurate information with regard to the pathological status

* This work was supported by grants from the "Fonds National de la Recherche Scientifique" of Belgium. The following text presents research results of the Belgian Incentive Program "Health hazards" initiated by the Belgian State — Prime Minister's Service — Science Policy Office. The scientific responsibility is assumed by its authors

and, due to the challenge represented by AD, most of the histological studies concentrate on this pathology leaving normal aging relatively neglected. The second problem is the difficulty to parallel the alterations observed in aged animals to those manifested in human, either healthy or suffering from AD. On the basis of morphological and/or behavioral studies, many similarities have been established between animals and humans, but until now no aged animal or animal model mimics the whole involutive process which affects human brain during aging and dementia pathogenesis. Therefore, morphological observations performed in animals are limited to some aspects, more or less relevant, of the human phenomenom.

1. A general involution of the brain parenchyma is found both in human and animals. There is a strong agreement about the weight loss in the aging human brain, associated with a decrease in brain volume. In the gray matter, the weight loss is correlated with cell loss, mainly neurones, since astrocytes are less affected or their number can even increase. Both small and large neurones are affected but the level of neuronal loss depends on the area, some structures being more vulnerable while others seem to be spared, suggesting that each region of the brain may age differently (Haug et al., 1983). The process is exacerbated in AD, frontal and temporal cortical areas and basal nuclei being particularly affected. The neuronal loss and the progressive decrease in the synaptic connections by restriction of the dendritic fields weaken the neuronal network, thereby enhancing the involutive process (Coleman and Flood, 1987). In the white matter, there is a reduction in the fibre population associated with a decrease in the amount of myelin. The greater loss of large neurones and the subsequent degeneration of their large myelinated axons can enhance this process during the last period of life. The progressive corticocortical disconnection leads to the isolation of cortical functions disrupting then the cohesive manner in which the cortex normally processes information (Morrisson and Hof, 1992). In the most impaired area, nerve cell necrosis and loss as well as degeneration of the processes can give to the cerebral parenchyma a spongy aspect. The same involutive trends are observed in animals but similarities of age-related changes in human and animal brains are limited to sharply defined brain regions, e.g. primary visual cortex and CA1 of the hippocampus (Flood and Coleman, 1988).

2. Deposition of senile plaques in the brain parenchyma is an obvious alteration in aged brain and as they are found in greater number and extent in AD brains, they are considered as a typical marker of this pathology (Brion and Duyckaerts, 1991; Braak and Braak, 1991). However, the way they are formed and the factors which cause their induction still remain unclear. Of particular interest is the deposit of aluminium in the brain parenchyma, which is found in the elderly and patients suffering from AD. The deposits are associated with plaques and, although its action remains enigmatic, aluminium is suspected to be an environmental factor leading to the AD pathological status (Candy et al., 1986; Martyn et al., 1989; Perl and Good, 1992). Structures similar to human senile plaques, clusters of

distended argentophilic neurites around amyloid deposits, have been described in aged animals such as rhesus monkeys (Wisniewski et al., 1973; Struble et al., 1985; Selkoe et al., 1987), Sprague-Dawley rats (Vaughan and Peters, 1981), a polar bear (Cork et al., 1988) and a lemurian primate (Bons et al., 1991).

3. Neurofilament impairments are responsible for the appearance of the tangles, which deposit in the neurones. Like senile plaques, tangles are typical of aged and AD brains although both alterations, either alone or together, can be found in other cerebral pathologies. Tangles have also been extensively investigated (Brion and Duyckaerts, 1991), but as in the case of senile plaques, the factors which induce them are unknown. Tangles identical to those seen in AD have not been demonstrated in nonhuman species, but neurofibrillary degeneration with abnormal filamentous profiles have been observed in aged animals such as rhesus monkeys (Wisniewski et al., 1973), a rat of the Kyoto strain (Knox et al., 1980), an asiatic brown bear (Cork et al., 1988), and a lemurian primate (Bons et al., 1991). Recently, accumulated neurofilaments showing helical profiles have been identified in the sensory part of the trigeminal nucleus in aged Wistar rats (Klosen and van den Bosch de Aguilar, 1993). They correspond to those already described in the spinal ganglia in this strain (van den Bosch de Aguilar et al., 1992) but, as the trigeminal nucleus can be considered as an sensory ganglion secondarily integrated in the brain, these abnormal neurofilaments are not really located in the brain parenchyma itself.

From these recognized general morphological features, it is clear that the histological studies aimed at establishing a comparison between animal and human aging processes in brain are limited to the problem of brain cell survival during aging. This problem has been explored according to two main approaches:

A chronological approach by comparing the course of cerebral aging in animals and in humans in order to identify the similarities shared by the studied species. This approach can lead to the understanding of the factors which induce nerve cell necrosis, the reactions of the brain parenchyma, the particular vulnerability of some brain regions. At this cellular level, one can suppose that the aging phenomenom is basically quite similar in all mammals and therefore the observations done in animals can shed light on the aging process in human.

An invasive approach by setting up experimental animal models in order to mimic the human alterations. This means the use of various attempts, such as administration of tissue or homogenates of human brain to test the action of putative necrotic factors, injection of substances (e.g. bromopyruvate) which interfere with cellular metabolism, overload of toxic elements (e.g. aluminium) which are suspected of involvement in the pathogenesis of AD. These approaches can provide fruitful results as to the reaction of the nervous tissue to a specific intervention (e.g. cell necrosis inducing factors) but the conclusions can be obscured by the lack of a real similarity between the animal and the human alterations.

Material and methods

Wistar male rats were used for all the studies, the young animals were three months old and the old animals were at least two years old. For the histological studies, deeply anaesthetized rats were sacrificed by an intraaortic perfusion of 4% formaldehyde in phosphate buffer (0.1 M, pH 7.4) for light microscopic (LM) studies or the Karnowski fixative mixture (glutaraldehyde-formaldehyde-phosphate buffer, 0.1 M, pH 7.4) for the electron microscopic (EM) studies. After fixation, the brain was removed from the skull and selected areas were dissected out and immersed in the same fixative during one hour. For LM, the pieces were routinely processed for paraffin sections. Sections were stained by the Kluver and Barrera method for general screening or processed for immunocytochemistry with different antibodies (anti-GFAP, anti-MAP2, antiphosphorylated NFH). For EM, the samples were post-fixed in 2% buffered osmium during one hour, alcohol dehydrated and embedded in Epon. Semi-thin sections were stained with toluidine blue and ultra-thin sections contrasted with Reynold's uranyl acetate and lead citrate solutions. Cellular quantifications were established on paraffin or semi-thin sections.

Human brain tissues have been administered in the brain of young rats, either by implantation or by intracerebroventricular (ICV) injection. The tissue was removed post-mortem from temporal cortex of young and elderly humans who died without any apparent neurological symptoms, and patients suffering from AD. The clinical diagnostics were confirmed by brain histopathological studies. Senile plaques and tangles were evident in the aged and AD brains, more abundant in the latter case. For tissue administration, the rat was anaesthetized by an i.m. injection of Hypnorm (Duphar, 0.2 ml per 100 g body weight) and a hole was bored in the right part of the skull (2 mm anterior to the lambda, 2 mm laterally). For implantation, a suction lesion was done in the cortex using a glass microprobe attached to a vacuum apparatus and small human brain fragments were directly pushed into the rat cortex cavity. For ICV injection, brain fragments were homogenized and sonicated for 15 seconds. The homogenates were centrifuged (2 min at 13,800 g) and 30 µl of the pellet were slowly injected into the ventricle through a stereotaxically fixed Hamilton syringe. Tissues of rat temporal cortex, either young or old, were also prepared and injected according to the same protocol. Bromopyruvate was also injected (0.2 µmol/10 µl) in the ventricle. Mock CSF was used as control for the injection experiments.

An experimental aluminium overload was produced by feeding rats with a diet supplemented with aluminium citrate, prepared from aluminium chloride hexahydrate and tri-calcium dicitrate tetrahydrate (Merck) with a stoechiometry 1:1. The total aluminium intake equals 1 g of aluminium element/kg of food. At the end of the overloading period (6 months) the rats were sacrificed and tissues dissected out either for histology or for aluminium determination, according to the method of D'Haese et al. (1985). For this analysis, samples were deposed in extensively washed (1% hydrochloric and nitric acids, each during one week) propylene tubes and weighed. The tissues were digested in 1 ml nitric acid 65% Suprapur (Merck) for five hours at a temperature of 70°C. The sample was recovered in a solution of nitric acid 1% and some of this solution was used as reference. Aluminium level in the tissues was estimated by graphite furnace atomic absorption spectometry (Varian AA 1275 Series). Each sample was measured at least four times and a reproducibility better than five units of absorbance was assumed correct. In addition to its overload by the diet, the action of aluminium was also tested by direct deposition of aluminium chloride (0.343 mg of aluminium citrate powder) on the rat cortex.

Results

Chronological aging

During the course of our studies of aging rat nervous system, histological analysis was performed on several tens of animals aged from 3 to 34 months. Gross morphological aspects of the brain did not greatly vary with advancing age and even the oldest animals showed no conspicuous atrophy of the brain. The weight of the brain remained constant (1.86 g ± 0.22) and no decrease was observed in the old animals. Only in some rare cases, an enlargement of the ventricles was noted and generally the cavities were almost entirely limited to narrow spaces.

The thickness of the cortex was estimated at the occipital level on parasagittal brain sections in young and old animals and showed no atrophy with advancing age (Table 1). An small increased thickness was observed in the layer VI in the old animal but the difference is not statistically significant.

The number of neuronal cells has also been estimated in these layers (Table 2). In all the layers, the layer V excepted, the number of cells diminished from 3 to 10% but the differences are not statistically significant.

From these results, obtained at the light microscopic level, the brain parenchyma appears quite well-preserved during aging and does not reflect profound structural changes as it has been already shown in previous studies

Table 1. Thickness (m ± sd, in μm) of the occipital cortical layers in 3 (n = 3) and 30 (n = 3) months old rats

Cortical layer	Young animals	Old animals
I	163 ± 17	140 ± 20
II–III	360 ± 30	354 ± 19
IV	159 ± 25	157 ± 24
V	456 ± 25	422 ± 32
VI	260 ± 15	312 ± 30
Total	1,368 ± 140	1,385 ± 127

Table 2. Number of neuronal cells (m ± sd per mm^2) in the occipital cortical layers of 3 (n = 3) and 30 (n = 3) months old rats

Cortical layer	Young animals	Old animals	Δ (%)
II–III	985 ± 57	914 ± 53	−7
IV	1,218 ± 45	1,178 ± 35	−3
V	578 ± 64	592 ± 72	+2
VI	1,905 ± 103	1,716 ± 88	−10
Total	4,686 ± 162	4,400 ± 95	−6

Table 3. Number of vacuolated cells (mean percent ± sd) in the gyrus dentatus (GD) and the CA1 and CA3 sectors of the hippocampus in 4 (n = 4), 24 (n = 4) and 29 (n = 4) months old rats

	4 months	24 months	29 months
GD	1.29 ± 0.42	19.97 ± 17*	11.74 ± 6.34*
CA1	1.06 ± 0.83	6.63 ± 4.75*	4.17 ± 2.95*
CA3	3.73 ± 1.34	6.34 ± 3.65*	9.73 ± 3.33*

Newmans-Keuls's test, 24 and 29 m vs 4 m: * $p < 0.05$

(Peters et al., 1983). However, its aspect changes considerably when the tissue is examined on semi-thin sections or at the electron microscope level. Numerous cells, both neurones and astrocytes, typically showed a vacuolated aspect. These cells are characterized by an extensive vacuolation of their cytoplasm and remnants of cell organelles are deposited between the vacuoles (Figs. 1–6). The vacuoles likely originate from distended cisterns of the reticulum (Fig. 8) or from altered mitochondria. They can be limited by membranes and in some places multilamellated membranes are formed by numerous parallel laminae of various lengths (Figs. 7 and 8). The nucleus is apparent in the centre of the cell and generally presents a dense granular content. The nuclear membrane remains continuous but the cell membrane can be disrupted at several places, opening windows between the cell and the extracellular space. In the astrocytes, whose processes can be hypertrophied, large amounts of aggregated filaments and glycogen bodies deposit (Figs. 5 and 9). In the surrounding neuropile, empty cell processes, whose profiles are ill-defined, are intermingled with well-preserved processes (Figs. 10 and 11). These vacuolated cells give to the parenchyma a spongy texture which clearly appears at 24 months as discrete patches and progressively extends afterwards. These cells could represent the population of cells which are undergoing the necrotic process accompanying chronological aging. They are scattered in the cortex, where their number increases from 1–2% at 3 months to 6–8% at 24–30 months, depending on the region, and they reach their full extension in the hippocampus, which is the more affected region by cellular necrosis, particularly the gyrus dentatus (Table 3).

Fig. 1. Semi-thin section in the gyrus dentatus of a 3-month-old rat (×720)

Fig. 2. Semi-thin section in the gyrus dentatus of a 24-month-old rat showing vacuolated cells (arrows) (×720)

Fig. 3. Semi-thin section in the CA1 sector of the hippocampus of a 24-month-old rat showing vacuolated cells (×900)

Fig. 4. Electron micrograph of the CA1 sector of the hippocampus of a 3-month-old rat showing well-preserved nervous tissue (×4,300)

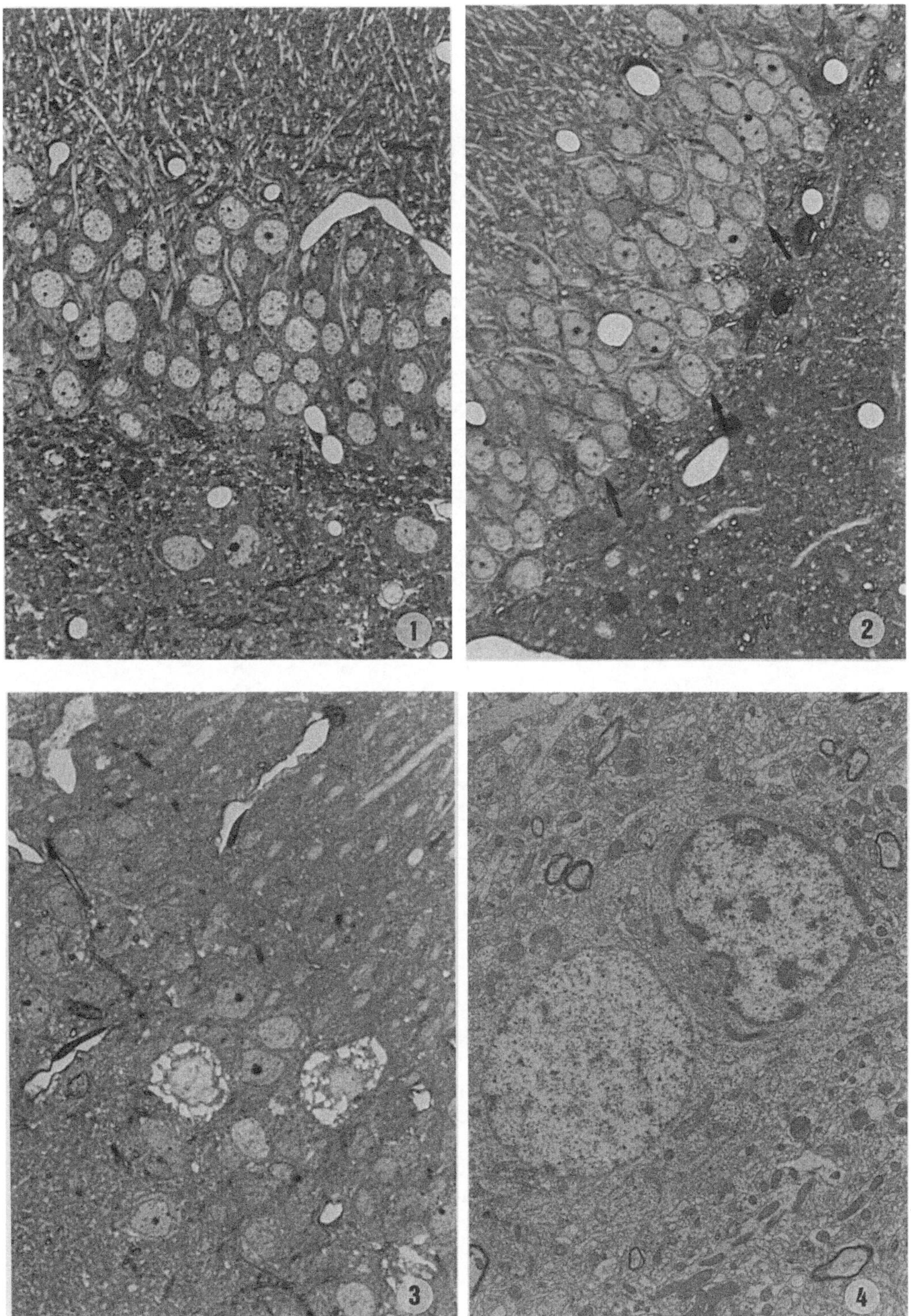
1
2
3
4

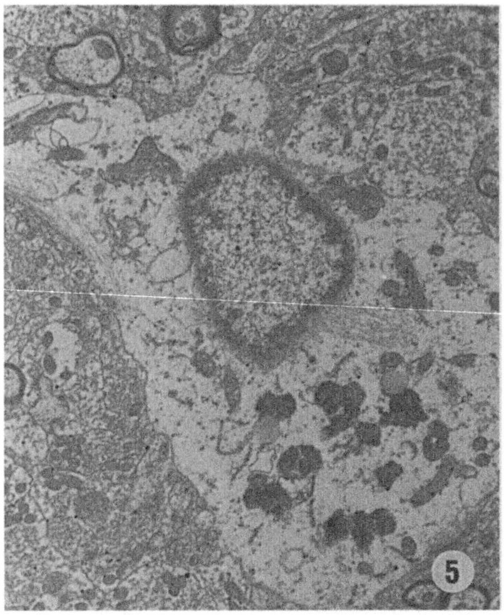

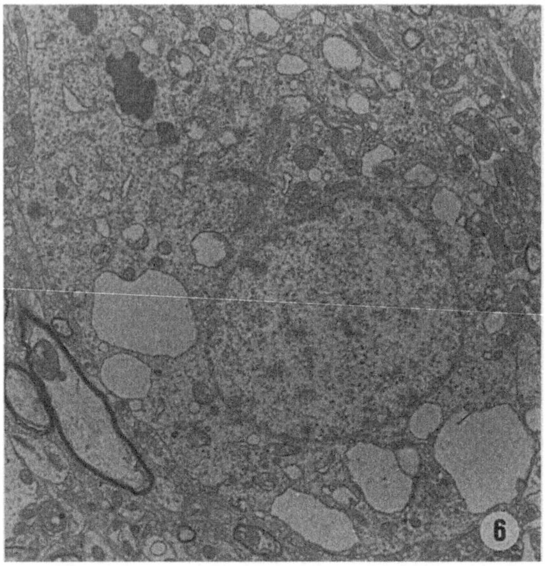

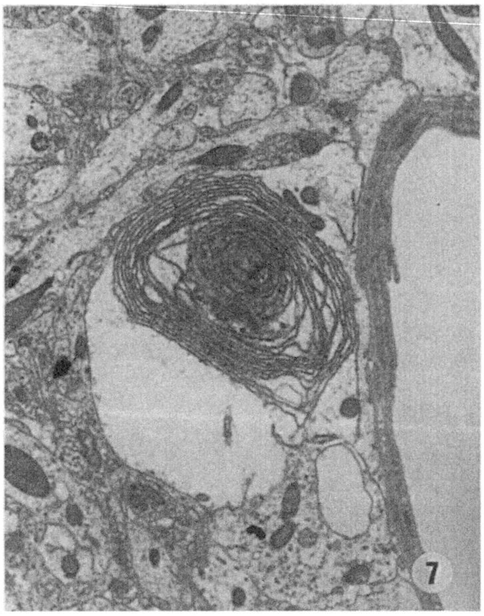

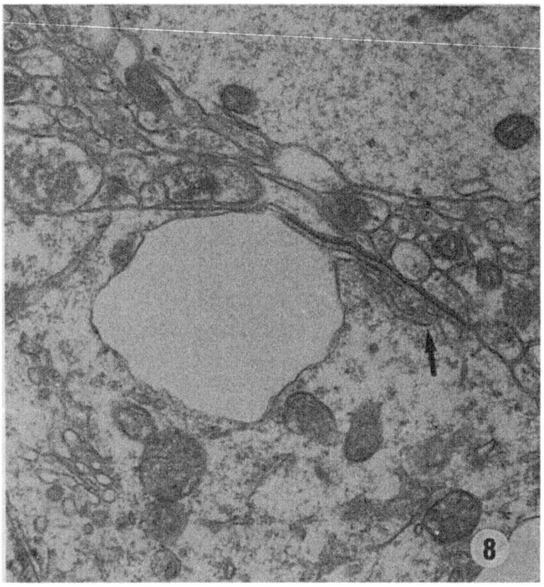

Fig. 5. Necrotic astrocyte in the hippocampus of a 32-month-old rat, showing filament aggregates, lipofuscins and vacuoles (×4,600)

Fig. 6. Necrotic neurone in the parietal cortex of a 32-month-old rat, showing lipofuscins and vacuoles (×4,500)

Fig. 7. Multilamellated whorl in an end foot astrocyte process; parietal cortex of a 32-month-old rat (×7,300)

Fig. 8. Multilamellated membranes and cytoplasmic vacuole emerging from submembrane cytoplasmic cistern (arrow); parietal cortex of a 32-month-old rat (×17,600)

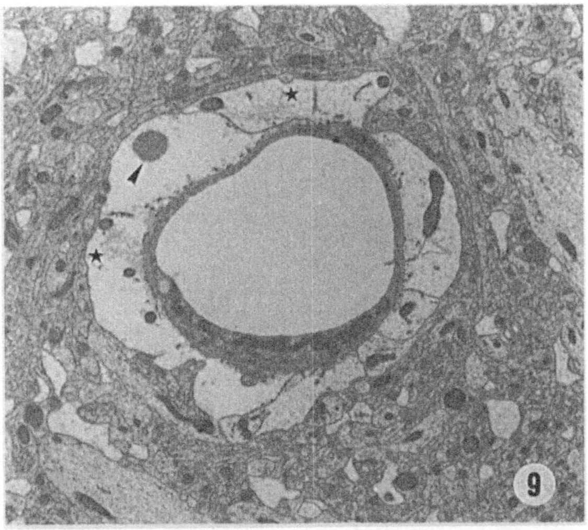

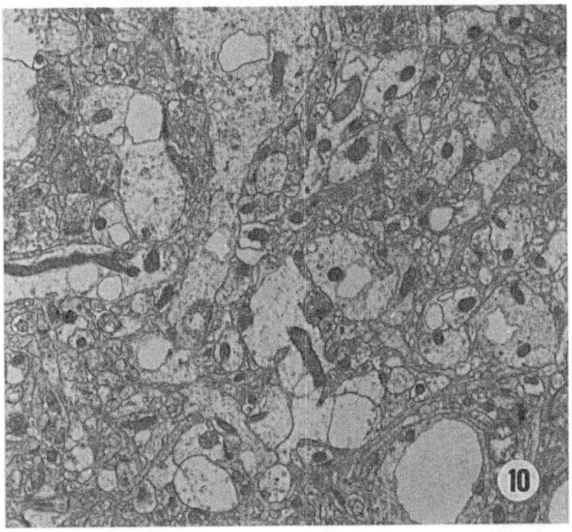

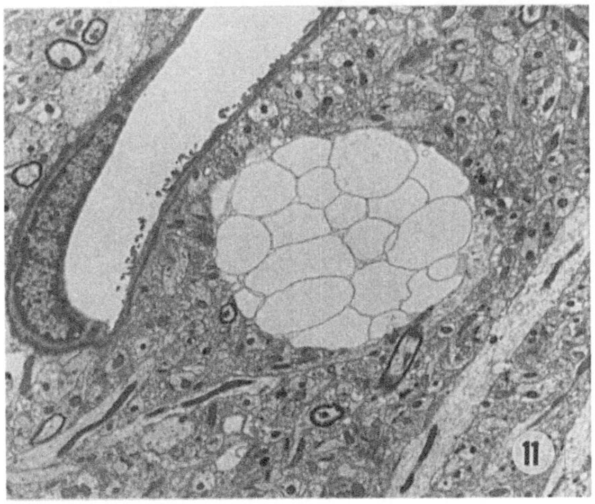

Fig. 9. Astrocytic processes around a vessel showing glycogen body (arrow) and filaments aggregates (asterisks); parietal cortex of a 32 month-old rat (×5,100)

Fig. 10. Empty and vacuolated processes in the hippocampal neuropile of a 32-month-old rat (×4,600)

Fig. 11. Hypertrophied dendrite in the parietal cortex of a 23-month-old rat (×4,000)

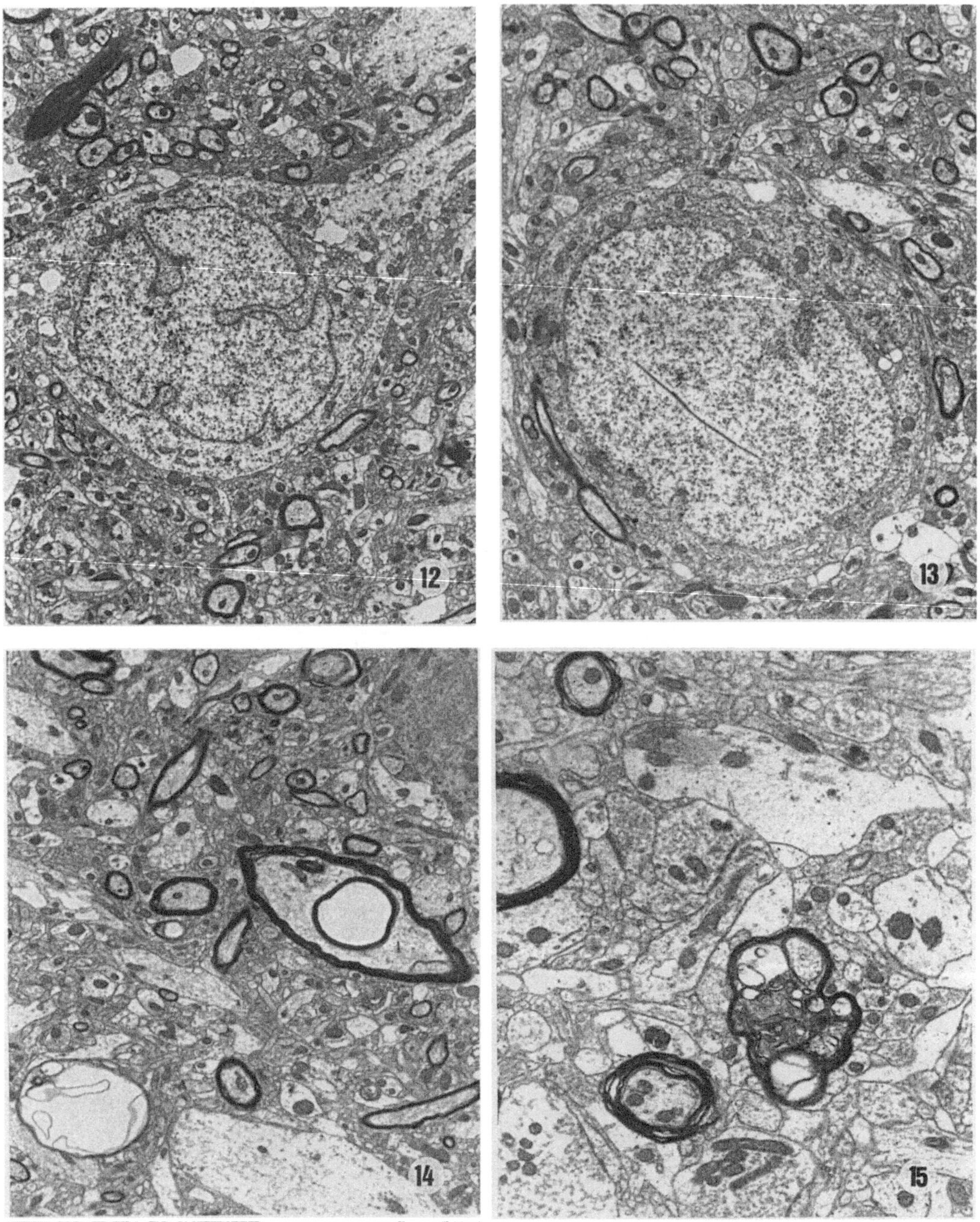

Fig. 12. Infolded nucleus in a parietal cortex neurone of a 32-month-old rat (×5,000)

Fig. 13. Cristalloid inclusion in the nucleus of a parietal cortex neurone of a 32-month-old rat (×5,300)

Figs. 14 and **15.** Various aspects of alterations in the myelin sheaths in the parietal cortex of a 32-month-old rat (Fig. 14 ×5,100; Fig. 15 ×9,400)

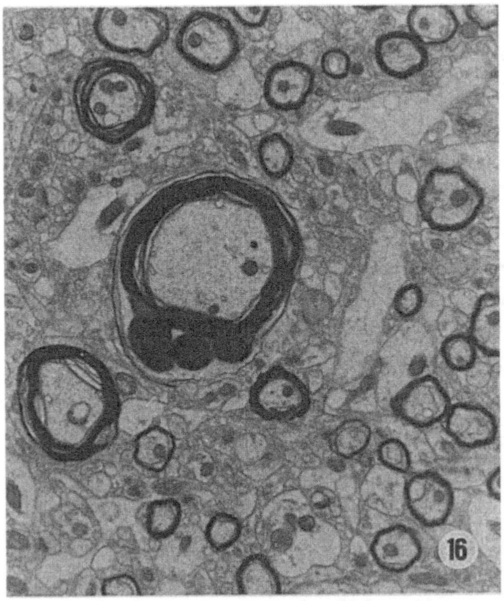

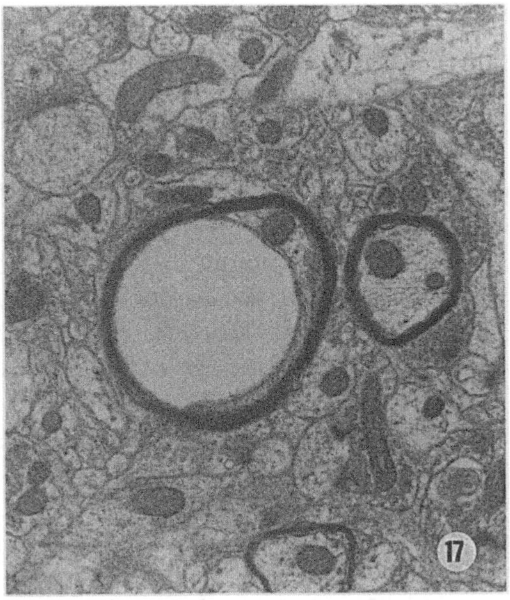

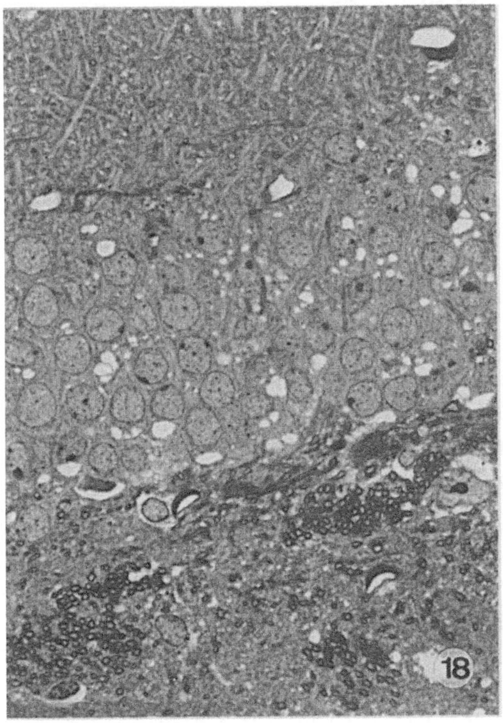

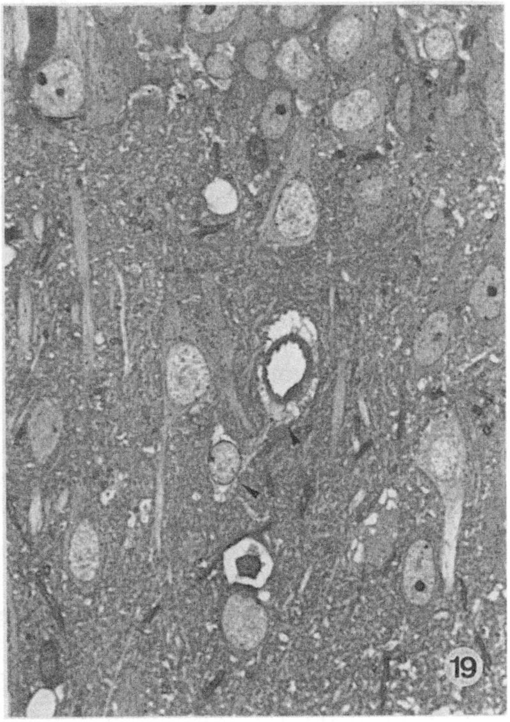

Fig. 16. Myelin budding in the parietal cortex of a 32-month-old rat (×7,000)

Fig. 17. Dystrophic axon in the parietal cortex of a 32-month-old rat (×13,200)

Fig. 18. Semi-thin section in the gyrus dentatus of a 6 months aluminium overloaded rat (×720)

Fig. 19. Necrotic astrocyte and degenerated end-feet processes (arrow); semi-thin section in the stratum molecularis of the hippocampus of a 6 months aluminium overloaded rat (×920)

Compared to these vacuolated cells, the other modifications observed at the ultrastructural level in the aged rat brain appear quite discrete, except the lipofuscins which invade the neurones, astrocytes and microglial cells. The nucleus of the neurones can be folded and sometimes presents cristalloid or filamentous inclusions (Figs. 12 and 13). The oligodendrocytes show no major change except the occurence of inclusions such as lamellar bodies and lipid vacuoles.

In the white matter, the most apparent changes concern the myelinated processes which are deeply modified by the demyelinating-remyelinating and degenerating mechanisms. The structure of the myelin sheath can be disturbed, its thickness can decreased and the regular disposition of the lamellae perturbed (Figs. 14 and 15). Separated lamellae of myelin forming isolated buds can evaginate into neighboring astrocytic processes or invaginate into the axoplasm (Fig. 16). Typical degenerating and dystrophic axons are also encountered (Fig. 17). The alterations which mark the white matter during aging are closely similar to those observed in the peripheral nerve at the same age. These alterations have been recently reviewed extensively (van den Bosch de Aguilar et al., 1992).

Administration of human brain tissue

When human brain tissue fragments were implanted in the cavity bored in the young rat brain cortex, a scar reaction developed in the brain. Seven days after the lesion, the cavity was isolated from the neighbouring cortex by a layer similar to the pia mater and astrocytes joined the capillaries to restore the blood-brain barrier. Fifteen days after the lesion, all the cellular debris, originating from the implanted human tissue or released during the lesion, have been cleaned by macrophagic cells. A particular reaction of the astrocytes occured in the rat cortex only after implantation of AD tissue. In this case, some twisted filaments can appear among the normal straight filaments at the tip of the astrocytic processes, close to the cavity. They can represent phagocyted fragments of human tangles present in the implant (van den Bosch de Aguilar et al., 1984).

Injection of human brain tissue homogenates in the ventricle of the young rat brain induced the appearance of vacuolated cells in the cerebral parenchyma (van den Bosch de Aguilar et al., 1986). In order to evaluate the level of impairment, these cells were counted two weeks after the injection in the parieto-occipital cortex and the underlying hippocampus in the contralateral brain at the same level as the site of injection. Quantification in the contralateral brain was chosen to avoid artefacts due to the injection procedure. Cell vacuolation was induced by human brain homogenates regardless of the age or the pathological status of the donor. A higher level was observed when elderly or AD homogenates were injected. The same difference occured, to a lesser extent, when young and old rat brain homogenates were used (Table 4). Injection of elderly and AD brain homogenates produced learning deficits in rat behavior (Hoyer et al., 1986).

Table 4. Number of vacuolated cells (mean percent ± sd) in the cortex and the hippocampus of the young rat brains following ICV injections of mock CSF (n = 3) and homogenates from young (n = 4), elderly (n = 4) and AD (n = 3) human brain tissue or young (n = 3) and old (n = 3) rat brain tissue

	Rat cortex	Rat hippocampus
Mock CSF	0.2 ± 0.3	3.3 ± 1.7
Human young	1.75 ± 2.2	4.5 ± 2.3
Human elderly	5 ± 1.8*	17.7 ± 3.4**
Human AD	11.7 ± 6.1*	32.7 ± 11.5**
Rat young	1.2 ± 1.5	1.5 ± 0.6
Rat old	2.3 ± 2.2	6 ± 1.4*

Student's t test, AD vs old and old vs young: ** $p < 0.01$, * $p < 0.05$

Injection of bromopyruvate

Bromopyruvate has been tested with regard to one major finding in AD, the severe decrease of brain glucose utilization along with reduced activities of enzymes acting in glucose breakdown. The pyruvate dehydrogenase complex is crucially involved in this process and can be inhibited by bromopyruvate which functions as a suicide inhibitor of the complex and therefore can mimic the diminution of brain glucose utilization observed in AD. Injection of bromopyruvate induced an extensive necrosis in the rat brain parenchyma which presented a spongiform aspect. The necrosis was associated with various glucose metabolic changes in the hippocampus but not in the cortex. Bromopyruvate-injected rats show severe deficits in learning and memory (Hoyer et al., 1986). These deficits can be linked to the massive impairment observed in the hippocampus, which has not been quantified at the cellular level due to the wide spreading of the alterations and their apparent lack of specificity for a defined cell population.

Administration of aluminium (Figs. 18–27)

Administration of aluminium chloride directly on the rat cortex induced a selective destruction of the astrocytes in the underlying parenchyma, shown by the disappearance of the reactivity for anti-GFAP, whereas the neuronal cells remained relatively unaltered and their reactivity for the anti-MAP2 and anti-phosphorylated NFH was not abolished (Figs. 20–22).

Administration of aluminium citrate in the diet leads to an overload in the brain (30 fold increase after 6 months) not obtained until now by other methods (Table 5).

Table 5. Aluminium content in the liver and in different brain regions (μg/g wet tissue ± sd) after overloading by aluminium citrate in the diet

	0 month	1 month	3 months	6 months
Liver	0.9 ± 0.04	0.9 ± 0.08	2.5 ± 0.14	16.6 ± 0.29
Spinal cord	1.6 ± 0.32	2.2 ± 0.21	2.5 ± 0.75	71.1 ± 0.25
Frontal cortex	3.3 ± 0.09	4.4 ± 0.06	8.2 ± 0.94	75.0 ± 0.42
Temporal cortex	3.2 ± 0.70	4.2 ± 0.15	6.3 ± 0.75	53.0 ± 0.51
Parietal cortex	1.4 ± 0.05	3.2 ± 0.19	4.3 ± 0.47	65.0 ± 0.31
Hippocampus	2.2 ± 0.08	2.2 ± 0.11	3.9 ± 0.28	†
Cerebellum	1.1 ± 0.30	†	1.9 ± 0.37	†

Number of rats in each group = 8
Student's t test, 6 vs 1 and 3 months: $p < 0.01$. † data not included because of too high variability in the samples (>3 ppb)

In the brain of the overloaded rats, vacuolated cells scattered in the whole parenchyma gave to the tissue a spongy aspect (Figs. 18 and 19). The astrocyte cellular body and the processes were both affected by vacuolation and this was particularly evident for the vessel abutting astrocyte end feet which were swollen and devoid of any organised structure (Figs. 23 and 24). Neurones were also affected but their alterations appeared more discrete than those present in astrocytes (Fig. 25). Vacuoles can also invade the neuronal body, increasing in this way the spongy aspect of the tissue. Vacuoles can occur in the cell body and the processes, leading to focal swelling of the neuronal outlines. Three modifications were observed in the neuronal nucleus, separately or together, or sporadically in the astrocytes. The nuclear membrane can form numerous shallow indentations (Fig. 25) whereas the nucleoplasma can contain membrane-limited vacuoles (Fig. 26), generally empty, or filamentous or crystalline inclusions (Fig. 27). The nuclear inclusions were mainly observed in large neurones whose rough endoplasmic reticulum is well developed.

Discussion

It is clear that actually, no aging animal nor animal experimental model adequately reflects the entire constellation of alterations that characterize human aging and AD. Except some structures, sporadically described in aged animals and already mentioned in the introduction, which show some similarities with the senile plaques and the tangles, these two hallmarks of the human aging and dementia have never been currently found in aged animals nor induced in animals by experimentation. In this perspective, systematic studies of aged animal brain give quite disappointing results. In contrast, taking precautions in the interpretation of the results, the phenomenom of cellular rarefaction, which affects both animal and human

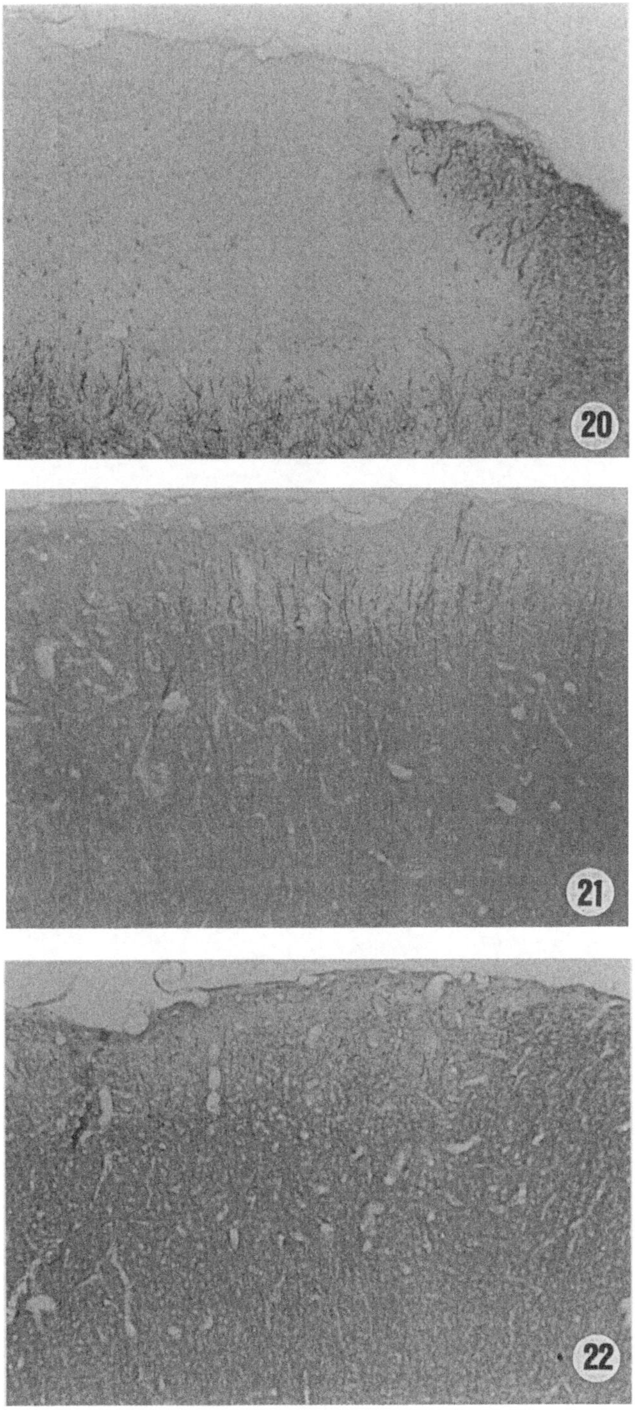

Fig. 20–22. Sections in the parietal cortex of 3-month-old rats after deposition of aluminium citrate on the cortex (×190).

Fig. 20. anti-GFAP

Fig. 21. anti-MAP2

Fig. 22. anti-phosphorylated NFH

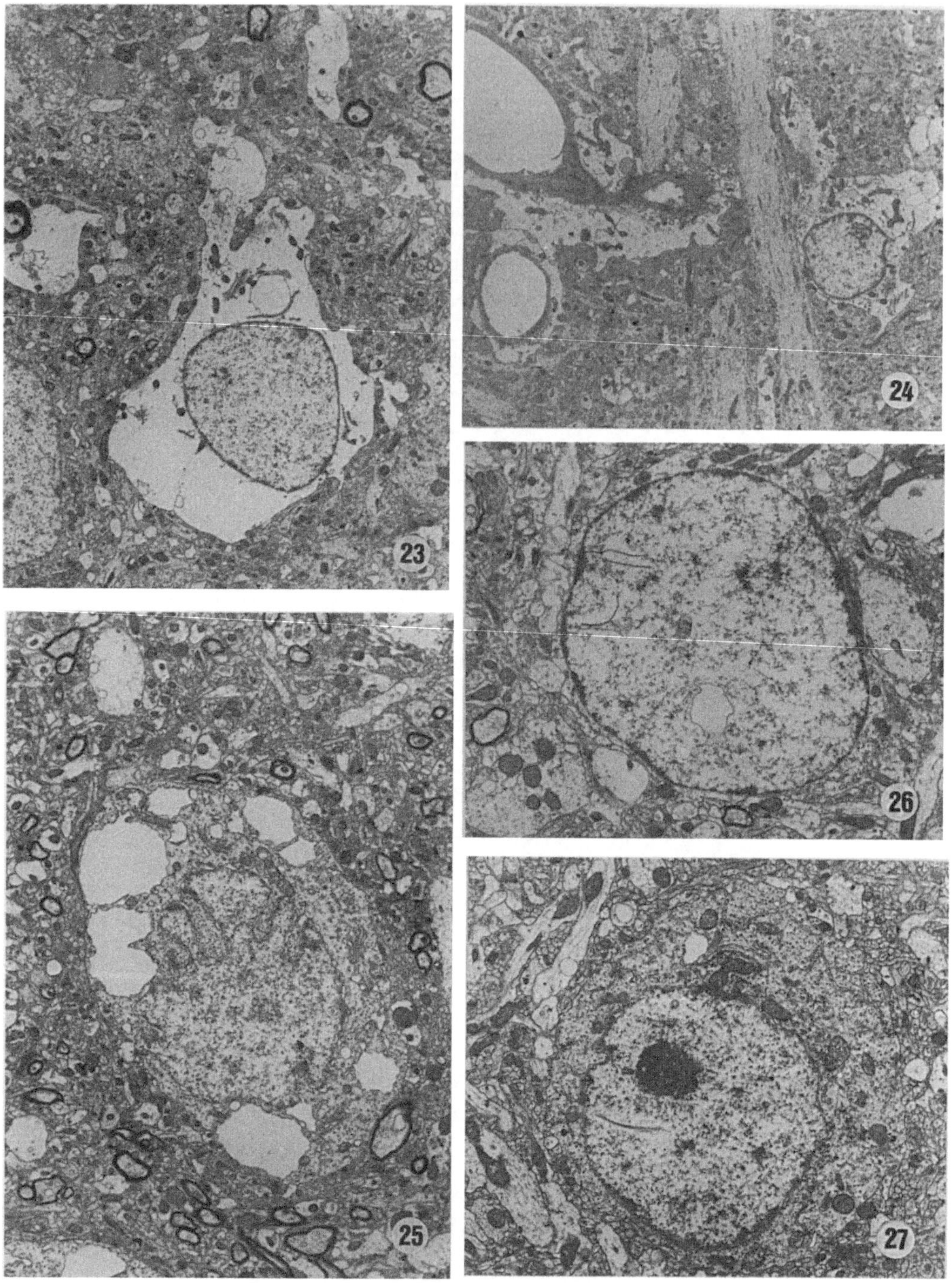

Fig. 23–27. Morphological aspects of the parietal cortex of rats overloaded with aluminum during 6 months

Fig. 23. Necrotic astrocyte (×3,800)

Fig. 24. Degenerated astrocytic processes (×1,950)

Fig. 25. Vacuolated neurone showing an infolded nucleus (×5,000)

Fig. 26. Intranuclear vacuoles and filamentous inclusions in an astrocyte (×5,700)

Fig. 27. Intranuclear cristalloid inclusions in a neurone (×5,700)

during aging, can be fruitfully studied in animals, particularly in the rodents for the facility of housing and experimentation.

There is little doubt that decreased density in neurones occurs in several regions of aged rat brain. However, the extent of this decrease and its relevance for human aging still remain debated due to the variety of the histological procedures, the differences in the methods of quantification, the discrepancy between the ages of the animals studied, the pathological status of the human brain.... All these parameters can cloud the analysis and bias the results (for review, see Coleman and Flood, 1987). On the other hand, neuronal rarefaction is accompanied by a reaction of the astrocytes which occupy, at least in part, the brain territory deserted by the necrosed neurones. These two linked processes can explain how a decreased neuronal density is not reflected by an obvious atrophy of the brain. Moreover, in rodents, unlike in humans, the brain, like the body, can gain weight with advancing age.

Cell necrosis during aging, which is a process widely recognized in all mammals, and its underlying mechanisms are difficult to analyse. Normally, this process is slow and progressively develops in the brain, affecting isolated cells or small groups of cells. Therefore it leaves few if any traces in the brain tissue. The cell remnants can be cleaned by the neighbouring astrocytes or, eventually, by some wandering macrophagic cells. With advancing age, the process of necrosis considerably increases invading the whole brain but it preferentially affects, both in human and rodents, the cortex and more extensively the hippocampus. The fundamental mechanisms which lead to neuronal rarefaction and astrocyte reaction and their increased manifestations with advancing age appear to be shared by human and rodents and, at the cellular level, show similarities enough to validate investigations in animals.

The vacuolated cells which are one of the striking features of the aging brain parenchyma likely represent a cell population undergoing necrosis. This type of cell has already been described during aging in healthy humans (Ball, 1977) and animals (De Estable Puig and Estable-Puig, 1975; Kadar et al., 1990; Lolova, 1991) as well as in AD, both in biopsies (Flament-Durand and Couck, 1979) and in post-mortem fixed tissues (Okamoto et al., 1991). Their increased number in the aged brain probably reflects the shifting of the brain capacities to compensate cellular alterations towards an unreactive state in which alterations accumulate. This involutive process starts in the more vulnerable area of the brain, the hippocampus, both in animal and human, in this latter species the cellular degenerating processes being accompanied by heavy deposits of plaques and tangles (Braak and Braak, 1991). Although the animal models fail to produce tangles and plaques in the rat brain, they induce in a short time a cellular degeneration in the same area as those affected during chronological aging and with the same morphological symptoms. Therefore, they can help to elucidate the mechanisms underlying the cell necrosis process and the particular susceptibility of the vulnerable areas.

Administration of rat or human brain tissue homogenates induces cell necrosis in the receiver rat cortex and hippocampus, the latter being more extensively affected. The level of necrosis is statistically related to the age of the donor and, in the case of human tissue, to the pathological status. This would signify that the homogenates contain some cell necrosis inducing factor whose level increases with advancing age and pathological status (van den Bosch de Aguilar et al., 1986). Due to the wide spectrum of substances contained in crude brain homogenates, these experiences were not easily interpretable. Recent data on the possible action of amyloid, which is contained in the human homogenates, can provide new views on these results. The neurotoxic effect of β amyloid protein is still questionable. Its neurotoxicity, particularly on the hippocampal cells, has been showed either in vivo after intracerebral injection in the rat (Kowall et al., 1991; Frautschy et al., 1991; Emre et al., 1992) or in vitro (Cotman et al., 1992). Other studies, using prolonged delivery of amyloid in the brain either by an osmotic pump (Games et al., 1992) or embedded in a cocoa butter matrix (Clemens and Stephenson, 1992), fail to show any toxicity. The results obtained with the crude homogenates, whose amount in β amyloid increases with age and the AD status, are in favour of its neurotoxicity. This action can be enhanced by the sonication of the homogenates, which could disrupt the aggregates present in the human brain, as only injection of homogenates and not implantation of tissues leads to cellular necrosis in the receiver brain. However, the necrotic action can be also due to other substances than amyloid, as homogenates of aged rat brain, which does not contain amyloid deposits, also induce cellular necrosis, although to a lesser extent than the human brain. Moreover, as astrocytes also undergo necrosis, these substances could be directed not only against the neurones but also against astrocytes.

The role of astrocytes in the involution of the brain during aging has been relatively neglected until now. Injury or neurodegeneration induces an astrocytic response known as reactive gliosis, where astrocytes increase in size and perhaps in number. At the same time, their processes become larger and more numerous. The astrocytic response to neuronal injury is increased in aged animals (Ross and Morgan, 1993; Bronson et al., 1993). The astrocytes have a beneficial function for the neurone by stabilizing its environment and by releasing survival factors but they can also have a deleterious effect on neuronal regeneration by blocking axonal elongation, thereby impeding the traumatized neurones to be sustained by the trophic sustances released from their targets. It can be proposed that during aging a synergistic process occurs implying both neurones and astrocytes. Paralell to the neurodegeneration, injury of the astrocytes, which can be induced under various conditions, could initiate neurotoxic processes, e.g. release of excitatory amino acids (LoPachin and Aschner, 1993). Morphologically such affected astrocytes indeed appear swollen and vacuolated. Alteration of the astrocytes could therefore exacerbate the naturally occuring neuronal degeneration and, moreover, they could participate in the deposit of β amyloid (Siman et al., 1989).

The factors which shift the astrocyte function from its beneficial role to a deleterious one remains unknown.

One of these factors could be a dysregulation in glucose metabolism, of which the major part is contributed by the astrocytes in the brain, as shown by the action of bromopyruvate (Hoyer et al., 1986; Hoyer, 1988) or streptozotocin (Nitsch et al., 1989; Nitsch and Hoyer, 1991) when injected in rat brain. The molecular actions of these substances on the nervous parenchyma remain to be elucidated but the imbalance in the brain energy metabolism could lead to an increased neuronal stress and necrosis.

Another factor could be aluminium which acts both on neurones and astrocytes and whose role is suspected in AD pathogenesis (Martyn et al., 1989; Meiri et al., 1993). The alterations observed in the brain parenchyma could be the manifestations of various deleterious effects of aluminium at the cellular level. In the pericaryon, it has been shown that aluminium can act on energy metabolism by impeding glucose availability in the cell (Lai and Blass, 1984), by favouring the production of hazardous free radicals and enhancing the rate of lipid peroxidation (Gutterridge et al., 1985; Evans et al., 1992). In susceptible animal species, aluminium causes neurofilamentous deformations, mainly in projection neurones. As some of the neuronal populations are the same as those affected in AD, it was suggested that their particular cytoskeletal oragnization may underly a tendency to degenerate under diverse conditions including exposure to aluminium (Kowall et al., 1989). In the nuclei, aluminium increases the stability in linker histone-DNA complexes by packing chromatin into inactive condensed states (Lukiw et al., 1989). It has been suspected to induce polymerisation of actin and tubulin and could be responsible for the assembly of lamins in paracrystalline arrays (Aebi et al., 1986). Similar nuclear inclusions are also observed in aged rats (Brion et al., 1982). Astrocytes are involved in the maintenance of ionic balance, mainly potassium and calcium. Imbalance in calcium homeostasis in brain leads to proteolysis, increased amino acid release and increased calcium influx. The increased calcium content modifies the osmolarity of the cell, inducing the entry of solutes and vacuolation (Nicotera et al., 1990). Glutamine synthetase in the astrocytes transforms glutamate into glutamine which is used by the neurones. This enzyme is particularly susceptible to the action of free radicals and a decrease in its activity leads to an increase in the extracellular glutamate content, toxic for the neurones, by favouring the influx of calcium ions (Pellegrin-Giampietro et al., 1988). Finally, it has been recently proposed that aluminium may accelerate proteolytic processing of the β-amyloid precursor protein by suppression of the inhibitor domain thus leading to accumulation of β-amyloid peptide and initiation of senile plaques (Clauberg and Joshi, 1993).

It is clear that chronological aging does not reflect faithfully in the animal brain parenchyma all the symptoms observed in the human. In fact, the only striking feature shared by animal and human is the progression of the cellular necrosis, which evidently affects the same vulnerable areas. The necrotic phenomenom is manifested in the animal brain by vacuolated cells, both neurones and astrocytes, and as this type of cell is observed in the human brain when neuropathological studies are performed on well-preserved tissue (e.g. biopsies), one can suppose that, at the cellular level, the brain parenchymal involution is common. When based on these morphological manifestations of the neuronal rarefaction during aging and its associated astrocyte reaction, animal models of aging are relevant to the human situation. In this perspective, experimental models — such as administration of homogenates or purified fractions of homogenates of human brain; administration of substances which antagonize metabolic pathways; administration of suspected toxic agents, such as aluminium — can help to elucidate the underlying processes which induce and/or enhance cell necrosis. Of particular interest is that these models underline the pivotal role of the astrocyte in brain aging. They could be, more than the neurones, vulnerable to the agressions and therefore constitute the cell population which determine the capacities of the brain parenchyma to resist against the deleterious actions of aging.

References

Aebi U, Cohn J, Buhle L, Gerace L (1986) The nuclear lamina is a meshwork of intermediate-type filaments. Nature 323: 560–564

Ball MJ (1977) Neuronal loss, neurofibrillary tangles and granulovacuolar degeneration in the hippocampus with ageing and dementia. Acta Neuropathol 75: 111–118

Bons N, Mestre N, Petter A (1991) Senile plaques and neurofibrillary changes in the brain of an aged lemurian primate, *Microcebus murinus*. Neurobiol Aging 13: 99–105

Braak H, Braak E (1991) Neuropathological stageing of Alzheimer-related changes. Acta Neuropathol 82: 239–259

Brion JP, Couck AM, Flament-Durand J (1982) Intranuclear inclusions in the neurons of senescent rats. Acta Neuropathol 58: 107–110

Brion JP, Duyckaerts Ch (1991) Neuropathology. In: Signoret JL, Hauw JJ (eds) Maladie d'Alzheimer et autres démences. Flammarion, Paris, pp 117–138

Bronson RT, Lipman RD, Harrison DE (1993) Age-related gliosis in the white matter of mice. Brain Res 609: 124–128

Candy JM, Klinowski J, Perry RH, Perry EK, Fairbairn A, Oakley AE, Carpenter TA, Atack JR, Blessed G, Edwardson JA (1986). Aluminosilicates and senile plaques formation in Alzheimer's disease. Lancet i: 354–357

Clauberg M, Joshi JG (1993) Regulation of serine protease activity by aluminium: implications for Alzheimer disease. Proc Natl Acad Sci USA 90: 1009–1012

Clemens JA, Stephenson DT (1992) Implants containing β-amyloid protein are not neurotoxic to young and old rat brain. Neurobiol Aging 13: 581–586

Coleman PD, Flood DG (1987) Neuron numbers and dendritic extend in normal aging and Alzheimer's disease. Neurobiol Aging 8: 521–545

Cork LC, Powers RE, Selkoe DJ, Davies P, Geyer JJ, Price DL (1988) Neurofibrillary tangles and senile plaques in aged bears. J Neuropathol Exp Neurol 47: 629–641

Cotman CW, Pike CJ, Copani A (1992) β-Amyloid neurotoxicity: a discussion of in vitro findings. Neurobiol Aging 13: 587–590

D'Haese PC, Van de Vyver FL, de Wolff FA, DeBroe ME (1985) Measurement of aluminium in serum, blood, urine and tissues of chronic hemodialyzed patients by use of electrothermal atomic absorption spectrometry. Clin Chem 31: 24–29

De Estable-Puig RF, Estable-Puig JF (1975) Vacuolar degeneration in aging rats. Virchows Arch 17: 337–346

Emre M, Geula C, Ransil BJ, Mesulam MM (1992) The acute neurotoxicity and effects upon cholinergic axons of intracerebrally injected β-amyloid in the rat brain. Neurobiol Aging 13: 553–559

Evans PH, Peterhans E, Bürge T, Klinowski J (1992) Aluminosilicate-induced free radical generation by murine brain glial cells in vitro: potential significance in the aetiopathogenesis of Alzheimer's dementia. Dementia 3: 1–6

Flament-Durand J, Couck AM (1979) Spongiform alterations in brain biopsies of presenile dementia. Acta Neuropathol 46: 159–162

Flood DG, Coleman PD (1988) Neuron numbers and size in aging brain: comparisons of human, monkey, and rodent data. Neurobiol Aging 6: 453–463

Frautschy SA, Baird A, Cole GM (1991) Effects of injected Alzheimer β-amyloid cores in rat brain. Proc Natl Acad Sci USA 88: 8362–8366

Games D, Khan KM, Soriano FG, Keim PS, Davis DL, Bryant K, Lieberburg I (1992) Lack of Alzheimer pathology after β-amyloid protein injection in rat brain. Neurobiol Aging 13: 569–576

Gutteridge JMC, Quinlan GS, Clark IA, Halliwell B (1985) Aluminium salts accelerate peroxidation of membrane lipids stimulated by iron salts. Biochem Biophys Acta 35: 445–447

Haug H, Barmwater U, Eggers R, Fisher D, Kuhel S, Sass NL (1983) Anatomical changes in aging brain: morphometric analysis of the human prosencephalon. In: Cervos-Navarro J, Sarkander HI (eds) Aging, vol 21. Raven Press, New York, pp 1–12

Hoyer S (1988) Glucose and related brain metabolism in dementia of Alzheimer type and its morphological significance. Age 11: 158–166

Hoyer S, Frölich L, Lauterbach M, van den Bosch de Aguilar Ph (1986) Animal models in dementia research: metabolic changes after intracerebroventricular injections of bromopyruvate, AF 64A, and Alzheimer brain homogenates. In: Modern trends in aging research, vol 147. Colloque INSERM-EURAGE. John Libbey Eurotex, pp 541–549

Kadar T, Silbermann M, Brandeis R, Levy (1990) Age-related structural changes in the rat hippocampus: correlation with working memory deficiency. Brain Res 512: 113–120

Klosen P, van den Bosch de Aguilar Ph (1993) Paired helical filament-like inclusions and Hirano bodies in the mesencephalic nucleus of the trigeminal nerve in the aged rat. Virchows Arch B Cell Pathol 63: 91–97

Knox CA, Yates RD, Chen II (1980) Brain aging in normotensive and hypertensive strains of rats. II. Ultrastructural changes in neurons and glia. Acta Neuropathol 52: 7–15

Kowall NW, Beal MF, Busciglio J, Duffy LK, Yankner BA (1991) An in vivo model for the neurodegenerative effects of β-amyloid and protection by substance P. Proc Natl Acad Sci USA 88: 7247–7251

Kowall NW, Pendelbury WW, Kessler JB, Perl DP, Beal MF (1989) Aluminium induced neurofibrillary degeneration affects a subset of neurons in rabbit cerebral cortex, basal forebrain and upper brainstem. Neuroscience 29: 329–337

Lai JCK, Blass JP (1984) Inhibition of brain glycolysis by aluminium. J Neurochem 42: 438–446

Lolova I (1991) Qualitative and quantitative glial changes in the hippocampus of aged rats. Anat Anz Jena 172: 263–271

LoPachin RM, Aschner M (1993) Glial-neuronal interactions: relevance to neurotoxic mechanisms. Toxicol Appl Pharmacol 118: 141–158

Lukiw WJ, Kruck TPA, Mac Lachan DR (1989) Linker histone-DNA complexes enhanced in the presence of aluminium lactate and implications for Alzheimer's disease. FEBS Lett 253: 59–62

Martyn CN, Osmond C, Edwardson JA, Barker DJP, Harris EC, Lacey FR (1989) Geographical relation between Alzheimer's disease and aluminium in drinking water. Lancet i: 59–62

Meiri H, Banin E, Roll M, Rousseau A (1993) Toxic effects of aluminium on nerve cells and synaptic transmission. Prog Neurobiol 40: 89–121

Morrisson JH, Hof PR (1992) The organization of the cerebral cortex: from molecules to circuits. Discussions Neurosci IX 2: 79

Nicotera P, Bellomo G, Orrenius S (1990) The role of Ca^{2+} in cell killing. Chem Res Toxicol 3: 484–494

Nitsch R, Hoyer S (1991) Local action of the diabetogenic drug, streptozotocin, on glucose and energy metabolism in rat brain cortex. Neurosci Lett 128: 199–202

Nitsch R, Mayer G, Hoyer S (1989) The intracerebroventriculary streptozotocin-treated rat: impairment of cerebral glucose metabolism resembles the alterations of carbohydrate metabolism of the brain in Alzheimer's disease. J Neural Transm 1: 109–110

Okamoto K, Hirai S, Iizuka T, Yanagisawa T, Watanabe M (1991) Reexamination of granulovacuolar degeneration. Acta Neuropathol 82: 340–345

Pellergin-Giampietro DE, Cherici G, Alesiani M, Carla V, Morini F (1988) Excitatory amino acid released from rat hippocampal slices as a consequence of free-radical formation. J Neurochem 51: 1960–1963

Perl DP, Good PF (1992) Aluminium and the neurofibrillary tangle: results of tissue microprobe studies. In: Aluminium in biology and medicine. Wiley, Chichester, pp 217–236 (Ciba Foundation Symposium 169)

Peters A, Feldman ML, Vaughan DW (1983) The effect of aging on the neuronal population within area 17 of adult rat cerebral cortex. Neurobiol Aging 4: 273–282

Ross JR, Morgan DG (1993) The effects of age on glial fibrillary acidic protein RNA induction by fimbria-fornix transection in the mouse brain. Age 16: 15–22

Selkoe DJ, Bell DS, Podlisny MB, Price DL, Cork LC (1987) Conservation of brain amyloid proteins in aged mammals and humans with Alzheimer's disease. Science 235: 873–877

Siman R, Card JP, Nelson RB, Davis LG (1989) Expression of β-amyloid precursor protein in reactive astrocytes following neuronal damage. Neuron 3: 275–285

Strubble RG, Price DL Jr, Cork LC, Price DL (1985) Senile plaques in cortex of aged normal monkeys. Brain Res 361: 267–275

van den Bosch de Aguilar Ph, Goemaere-Vanneste J, Klosen P, Terao E (1992) Ageing changes of spinal ganglion neurons. In: Fujisawa K, Morimatsu J (eds) Development and involution of neurons. Japan Scientific Societies Press, Tokyo, pp 109–150

van den Bosch de Aguilar Ph, Knoops B, Heuschling P, Flament-Durand J, Brion JP, Hoyer S, Frölich L (1986) Morphological aspects of rat brain after intracerebral administration of Alzheimer brain homogenates. In: Modern trends in aging research, vol 147. Colloque INSERM-EURAGE. John Libbey Eurotext, pp 533–540

van den Bosch de Aguilar Ph, Langendries-Weverberg Ch, Goemaere-Vanneste J, Flament-Durand J, Brion JP, Couck AM (1984) Transplantation of human cortex with Alzheimer's disease into rat occipital cortex: a model for the study of Alzheimer's disease. Experientia 40: 402–403

Vaughan DW, Peters A (1981) The structure of neuritic plaques in the cerebral cortex of aged rats. J Neuropathol Exp Neurol 40: 472–487

Vaughan DW, Vincent JM (1979) Ultrastructure of neurons in the auditory cortex of ageing rats: a morphometric study. J Neurocytol 8: 215–228

Wisniewski HM, Ghetti B, Terry RD (1973) Neuritic (senile) plaques and filamentous changes in aged rhesus monkeys. J Neuropathol Exp Neurol 32: 566–584

Authors' address: Dr. Ph. van den Bosch de Aguilar, Université Catholique de Louvain, Laboratoire de Biologie cellulaire, Bâtiment Claude Bernard, Place Croix du Sud, 5, B-1348 Louvain-la-Neuve, Belgique.

J Neural Transm (1994) [Suppl] 44: 133–143

Morphological hippocampal changes during normal aging and their relation to cognitive deterioration

T. Kadar[1], **I. Arbel**[2], **M. Silbermann**[2], and **A. Levy**[1]

[1] Department of Pharmacology, Israel Institute for Biological Research, Ness Ziona, and [2] The Faculty of Medicine, Technion, Haifa, Israel

Summary. Cognitive and memory capacities were assessed in two strains of rats of various age groups prior to histological evaluation of their brains. Male Wistar rats, at the age of 3, 12, 17 and 24 months, were tested using the 8-arm radial maze and male Fischer 344 rats, aged 3, 12 and 16 months, were tested in the Morris water maze. Significant memory impairments were found in both strains already at the age of 12 months in about 50% of the population. Morphological analysis of the brains revealed age-related structural changes in the hippocampal formation starting with the middle-age group. Degenerative CA1 and CA3 pyramidal cells characterized the hippocampus of cognitive-impaired rats, while non-impaired animals exhibited intact hippocampus irrespective of age. This characteristic was supported by quantitative morpho-analysis. The best correlation between the decrease of area or number of cells and working memory impairment was found for CA3 region in both strains. Age-related decline in the density of muscarinic receptors in Wistar rats' brain corresponded with the pattern of cognitive deficit.

The results of the present study support the hypothesis which associates hippocampal integrity with normal memory function. It is concluded that chronological age by itself is not an adequate indicator of age-related brain alterations and individual evaluation of performance, based on behavioral scores, is recommended.

Introduction

The role of the central cholinergic system in memory processes in both normal aging and pathological disorders is now well established (Bartus et al., 1982). Among the various brain regions involved, the hippocampus has received considerable attention as one of the most affected brain regions and it clearly seems to be linked to recent memory processes. Lesions of the hippocampus produce deficits in a wide range of memory tasks including spatial memory (Morris et al., 1978; Olton et al., 1978). It has been shown that aged animals display memory impairments which were related to hippocampal injuries (Landfield et al., 1988). However, variability in learning and memory abilities were shown in many studies, in various species (Rapp et al., 1992), and this was established as a prominent feature of the aging process. Our results, as well as others', suggest that the chronological aging per se does not necessarily reflect biological age and by itself is not an adequate criterion for comparing age-related changes in two populations. Therefore, while studying the mechanism of aging deterioration, senescent animals should be divided into two sub-populations, and individual performance scores must be recorded. Indeed, more recent studies have adopted this concept when investigating various aspects of aging and performance. The relationship of biogenic amines (Gallagher et al., 1990), glucose utilization (Gage et al., 1984; Wree et al., 1991) and hippocampal NGF levels (Hellweg et al., 1990; Henriksson et al., 1992) with the decline in spatial learning ability was shown in rats that were previously divided to sub-groups according to their performance. Moreover, aged rats with learning deficits have impaired hippocampal synaptic plasticity (Geinisman, 1986; Landfield, 1988; Toledo-Morrell, 1988) and discrete neuronal degeneration (Kadar et al., 1990). The purpose of the present study was to correlate between memory performance and hippocampal morphological integrity during normal aging based on individual evaluation. Studies were carried out in two strains of rats in order to confute strain specificity. In addition, the role of muscarinic receptors in these processes was elucidated as it had been suggested that impaired cholinergic transmission might be responsible, at least in part, for the memory decline associated with aging. Examination of structural changes within discrete hippocampal areas, accompanying memory impairments, may provide insight into the cellular markers of aging which are most tightly coupled to senescent cognitive dysfunction.

Materials and methods

Animals

Male Wistar rats (Charles River, England) at the age of 3, 12, 17 and 24 months were tested for their memory performance using the 8-arm radial maze (RAM). In addition, memory functions of male Fischer 344 rats (Charles River, England) aged 3, 12 and 16 months, were monitored utilizing the Morris water maze (MWM).

Behavioral testing

The RAM is a useful tool for monitoring subtle changes in working memory performance. The task involves the successive selection of arms, radiating from the center of the maze in order to obtain a food reward. The paradigm we have used was previously described in detail (Kadar et al., 1990). Briefly, each animal was placed in the center of the maze and was allowed to visit all arms, each of which was baited with one food pellet. Each trial (one per day) lasted until the animal had either visited all 8 arms or until 15 min had elapsed. Rats were trained in the maze for 23 days during which time all responses were recorded. Re-entry into an arm which was visited before was scored as an error. Percentage of total errors expressed working memory performance.

The MWM task enables the testing of rats for their spatial reference memory performance, by following their ability to learn the location of an escape platform, submerged in opaque water. Rats were trained and tested in a white circular metal water maze measuring 140 cm in diameter and 50 cm high, filled to a depth of 25 cm with water (26°C). Milk powder was dissolved in the water in order to make it opaque. A white metal platform was placed 2 cm beneath the water. The pool surface was divided into four imaginary quadrants of equal area and four equally spaced points around the edge of the pool were used as starting positions. Performance in the maze was monitored by a tracking system consisted on an overhead video camera linked to a computerized analysis system (Galai, Israel).

Each rat was given four training trials (120 sec each) per day on four consecutive days and escape latencies (the time to find the platform) were recorded. For each rat, the platform position remained constant throughout the four training days. On day 5 (24 hr after the final training day), the "reversal ability" of the rat was tested. Rats were tested as in the training session, except that the platform position was changed. Reversal testing continued for 4 trials, similarly to each training day.

At the termination of the behavioral studies the animals were sacrificed either by decapitation or by trans-cardiac perfusion (saline followed by 4% buffered formaldehyde), and their brains were removed and processed for histological and histochemical evaluation.

Histology

Brains, submitted for histological evaluation, were fixed in 4% neutral buffered paraformaldehyde and processed routinely for paraffin embedding. Coronal sections, 5 μm thick, were cut serially, and selected sections (mainly from the hippocampal level) were stained with hematoxylin and eosin (H&E) for light microscopy examination. Morphometric analysis was performed using a computerized image analysis system (Galai, Migdal Haemek, Israel). Measurements of cell surface area and number of cells in discrete hippocampal sub-fields were performed as described previously (Kadar et al., 1990). In addition, the number of damaged cells in different hippocampal sub-fields was measured in brains of Fischer rats. Cells having pyknotic shrunken nuclei were counted as damaged cells.

Muscarinic receptor studies

The density of receptors was studied on frozen sections of Wistar rats' brains at various age groups, utilizing the in vitro autoradiography technique. 3H-quinuclidinyl benzylate (3H-QNB) was employed as the ligand for labelling muscarinic receptors. The procedure was described in detail elsewhere (Kadar et al., 1990a). Briefly, sections were incubated in 50 mM sodium-potassium phosphate buffer (pH 7.4), containing 1 nM 3H-QNB (43 Ci/mmole; NEN, USA), for 2 hr at 25°C. Non-specific binding was determined in consecutive sections in the presence of 1 μM atropine.

Data analysis

Two-way analysis of variance (ANOVA) with repeated measures was performed across age groups on the behavioral data. One way ANOVA was carried out for the morphometric data and for the receptor studies. Whenever statistical significant was observed, further comparison between specific groups was carried out, using the Scheffe contrast analysis. The relation between the behavioral data and the structural findings was determined using the Pearson's correlation coefficients.

Results

Behavior

RAM

Figure 1 presents working memory performance of Wistar rats, expressed by the percent of total errors for the various age groups. It is clearly evident that young rats had less errors as compared to older rats and learned the task faster within the first week, until they reached a plateau level of performance. A significant difference was found already at the age of 12 months, as compared to the young 3-months-old group ($P < 0.001$). Thereafter, no further decline was found between the performance of 12 vs. 17 months old rats. However, the older group (24 months) exhibited an additional decrease in the percent of total errors. Evaluation of individual scores revealed memory impairment in 50% of the middle-aged groups (12–17 months) and in 75% of the old rats (24 months). Thus, animals could be divided into sub-populations, according to their performance in the maze.

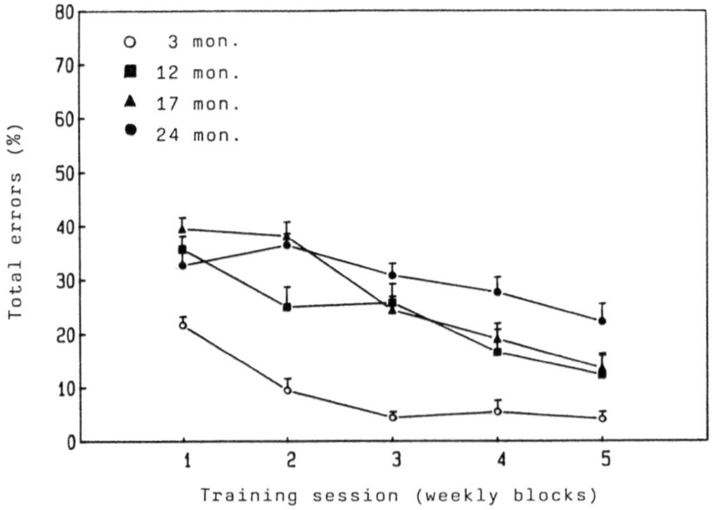

Fig. 1. Working memory performance of Wistar rats at various age groups in the 8-arm radial maze, as expressed by percent of total errors (mean ± SEM). A significant age effect was found between the groups ($F(3,45) = 57.72$; $P < 0.001$), and between the 3 months vs the 12 months groups ($P < 0.005$)

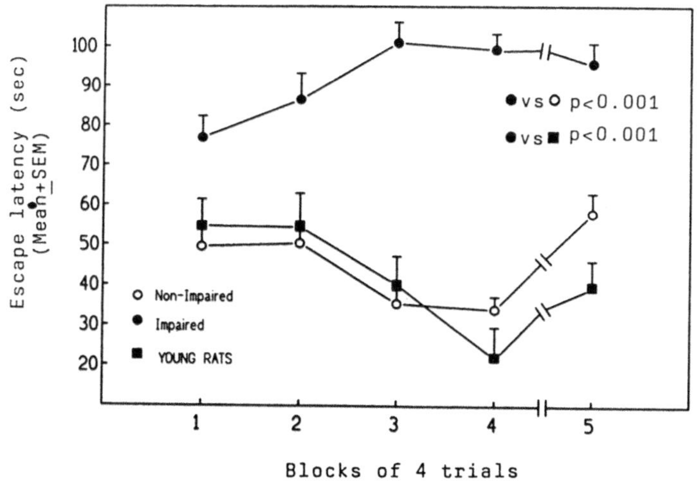

Fig. 2. Escape latency of 3 months old (young) and 12 months old Fischer rats, during training session and reversal test. 12 months old rats were divided into impaired and non-impaired groups, according to their performance on days 3 and 4; data for day 5 represent "reversal test" (see text for further details)

MWM

The escape latencies of the four trials on each day, grouped into daily blocks, are shown in Fig. 2. Scores for young animals decreased significantly over the 4 days of training, from an average of 54 sec on day 1 to 22 sec on the last day of training. In contrast, 12 months old animals had much longer escape latencies throughout the testing period with an average escape latency of 66 sec on the 4th day of training. However, large variability was noticed within the adult group, suggesting different sub-populations. Based on their performance on days 3 and 4, animals were divided into subgroups; rats having escape latency of 45 sec and below were considered as "non-impaired" and those with escape latency of 80 and above were considered as "impaired". An intermediate group, having scores of 46–80 sec included the rest. Figure 2 illustrates the different learning curves of the two sub-groups as compared to young animals. It is apparent that non-impaired animals learned the task similarly to the young group, while impaired rats did not learn the task, and differed significantly from the two other groups during both training and reversal test ($P < 0.001$).

Histology

Distinct morphological changes were noted in the hippocampus of Wistar rats already at the age of 12 months, as compared to the younger group. At this age, degenerative changes were observed in about 50% of the brains mainly in CA1 and CA3 regions. Similar changes were found in the 17

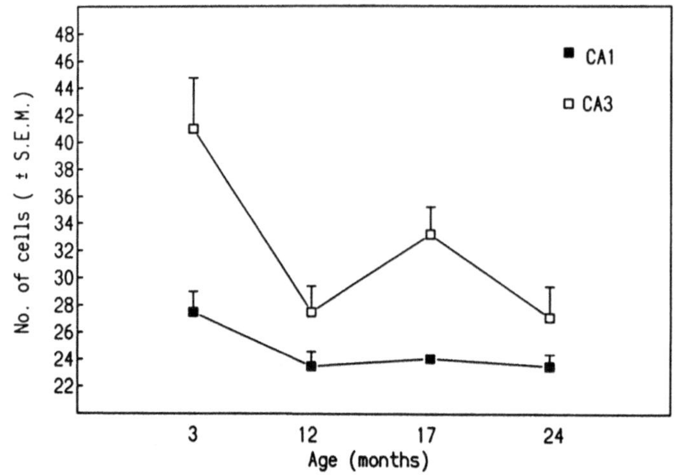

Fig. 3. Morphometric analysis of the number of cells within the hippocampus of Wistar rats at various age groups. Data is given as means ± SEM of cells counted in a frame of 32,000 μ^2, in at least 3 serial sections (n = 5–7). $P < 0.001$ for 3 vs. 12 months group for CA3 region

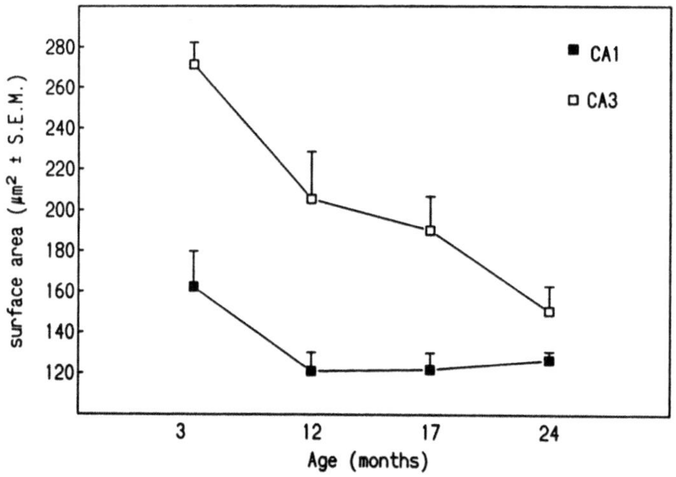

Fig. 4. Morphometric analysis of single cell surface area in CA1 and CA3 hippocampal cells in Wistar rats at various age groups. The mean surface area of single cells was calculated from 120 measurements (20 cells for each animal). $P < 0.001$ for 3 vs. 12 months group for CA3 region

month group, while additional deterioration was found at the age of 24 months. Morphometric analysis of the number and surface area of single cells within CA1 and CA3 sub-fields, shown in Figs. 3 and 4, revealed a general decline in both parameters, starting at the age of 12 months. However, only data for CA3 region were found statistically significant ($P < 0.001$ for 3 vs. 12 mon). When the hippocampal morphology of "impaired" vs "non-impaired" animals was compared, deterioration was most pro-

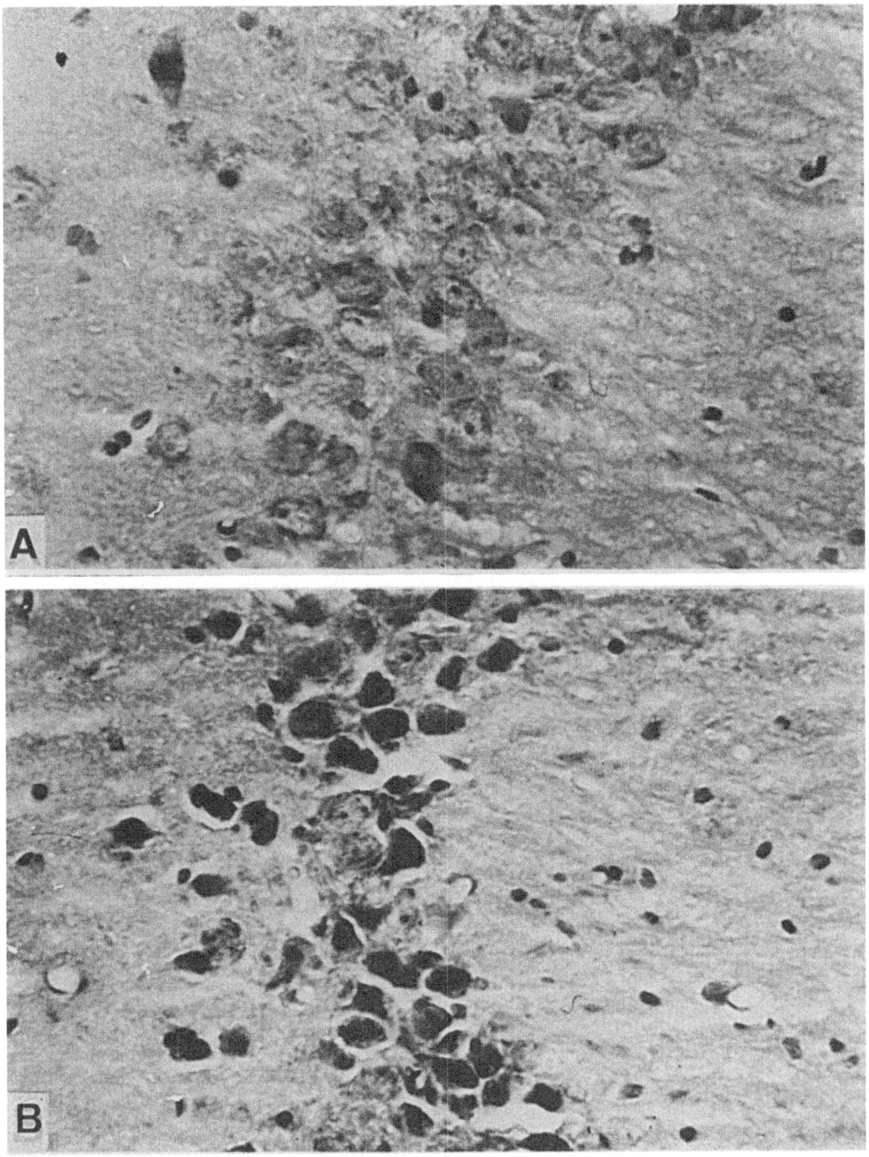

Fig. 5. Sections through CA3 hippocampal region of non-impaired (**A**) and impaired (**B**) 17 month-old Wistar rats. Note degeneration of cells in **B**. H&E staining. Original magnification ×160

nounced in impaired animals. Figure 5 illustrates intact morphology of CA3 region in a non-impaired 17-months-old animal, in contrast to the degeneration found in an impaired rat. Significant correlation was found between the behavioral scores (total errors) and hippocampal degeneration (single cell area) for CA3 cells ($r = -0.544$; $P < 0.01$). Similar pattern of degeneration was found also in cognitive impaired Fischer rats (see Fig. 6).

Morphometric analysis of damaged cells, in CA1 and CA3 sub-fields of Fischer rats, indicated a close resemblance between memory deficits in the two sub-groups, expressed by their escape latencies, and the progressive

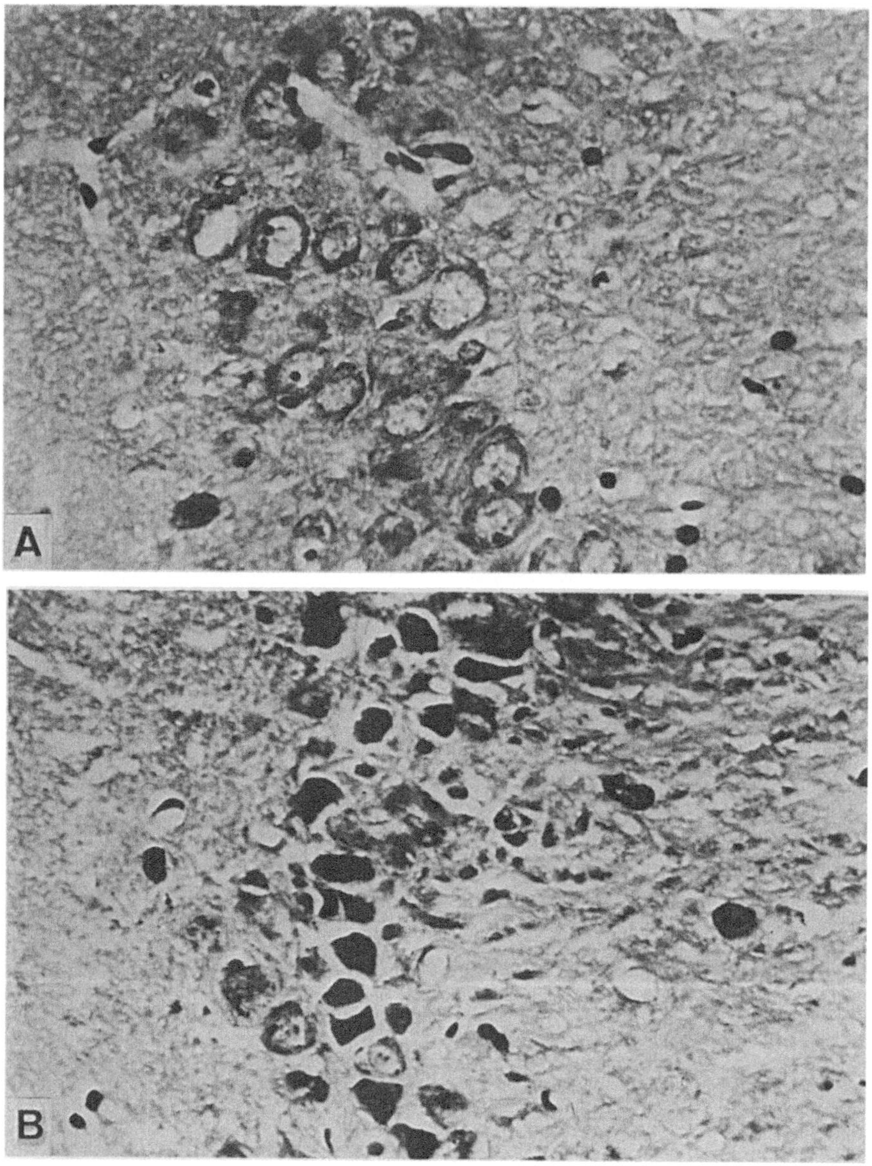

Fig. 6. Sections through CA3 hippocampal region of non-impaired (**A**) and impaired 12 months old Fischer rats. Pyknotic cells are clearly seen in **B**. H&E staining. Original magnification ×160

increase in the number of damaged cells (Fig. 7). Morpho-behavioral correlation revealed high correlation-coefficient between the number of damaged cells and escape latencies scores ($r = 0.94$ for CA3 and 0.84 for CA1).

Receptor density

Brains of adult rats showed a general decline in QNB binding, in all brain areas examined, already at the age of 12 months. No further decline was

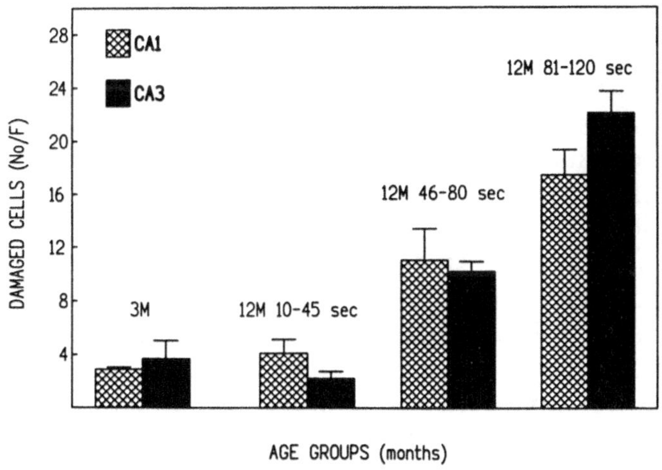

Fig. 7. Number of damaged cells within the hippocampus of Fischer rats, divided to non-impaired, intermediate and impaired subgroups according to their escape latencies

determined up to 24 months. Table 1 provides data of the decline in specifically bound 3H-QNB, for various hippocampal areas and different age groups. The most pronounced decrease in QNB binding was detected in CA1 and CA3 hippocampal regions ($P < 0.005$). In the later (CA3), 42% decrease was found at 12 months. Thus, the decline in cognitive performance, observed at the age of 12 months, was paralleled by a decrease in muscarinic receptors' density. Linear regression carried out for the quantitative autoradiographic data and the behavioral scores revealed high correlation for all regions examined (r values ranging between −0.6 and −0.7). However, data were not statistically significant.

Table 1. Age related decline in the density of muscarinic receptors labelled by 3H-QNB in the hippocampus of Wistar rats

Brain region	3 months old rats	12 months old rats
CA1 sub-field	789 ± 38	570 ± 65** (−27.8%)
CA3 sub-field	504 ± 27	291 ± 65** (−42.4%)
Dentate gyrus in.	777 ± 54	547 ± 84* (−29.7%)
Dentate gyrus ext.	727 ± 40	485 ± 58* (−33.4%)
CA4	571 ± 36	413 ± 49** (−27.7%)

The data is presented in means ± SEM of specific 3H-QNB bound to brain sections, in fmole/mg protein (precent change — 12 months vs. 3 months) (n = 6 in each age group). **$p < 0.005$ vs. 3 months old rats. *$p < 0.01$ vs. 3 months old rats

Discussion

The results of this study suggest a close relationship between age-related memory impairments and hippocampal structural degeneration in both Wistar and Fischer rats. We have demonstrated that animals having intact hippocampus perform successfully like young rats, irrespective of their age. Thus, chronological age by itself may not serve as an adequate parameter for grouping animals in aging studies. In addition, assessment of behavioral and neurobiological functions in the same individual, may provide a powerful experimental strategy for revealing brain-behavior relationships, especially regarding age-related phenomena.

The closest correlation between memory decline and structural alterations during normal aging was found for CA3 pyramidal cells, suggesting an involvement of this region in memory processes and cognitive performance. This finding was corroborated by Heimrich and his colleagues (1989), who demonstrated high correlation between synaptic efficacy in the CA3 mossy fibers projection and the learning ability of mice. The importance of CA3 cells for hippocampal circuitry and as a junction point was also demonstrated by Handelmann et al. (1981) using kainic acid lesions.

Several studies have suggested that the variation in spatial learning performance of aged rats could reflect dysfunction of the central cholinergic system, a finding which is consistent with the cholinergic hypothesis of memory, associated with neuro-degenerative disorders. Based on muscarinic receptors' autoradiography, we have shown during aging a similar pattern of decline in cognitive performance on the one hand, and a decrease in the density of muscarinic receptors on the other. However, we were not able to show individual correlation between the number of muscarinic receptors and memory scores. This finding is in agreement with other studies which were unable to detect individual correlation between cholinergic transmission and cognitive impairment. It is possible that the density of muscarinic receptors does not provide a sensitive marker for cholinergic function. Similar findings were reported for ChAT activity (Luine, 1990) and acetylcholine (ACh) release (Fischer et al., 1991). In the later microdialysis study, similar levels of extracellular ACh were found in the hippocampus of aged impaired and non-impaired Sprague-Dawley rats.

The fact that we discovered significant memory deficits already at the age of 12 months in two different strains of rats might indicate that this phenomenon is more general. This led us to the suggestion that "impaired" 12 months old (middle-aged) rats may serve as a better model for the effect of aging on cognitive deficits, since they exhibit the typical memory impairment without any physical incapacitation which is often associated with old age.

In conclusion, this study provides further support to the importance of hippocampal integrity for cognitive performance, by demonstrating that the degree of memory impairment is highly correlated with degenerative changes in this region, in particular among CA3 pyramidal cells.

References

Bartus RT, Dean RL, Beer B, Lippa AS (1982) The cholinergic hypothesis of geriatric memory dysfunction. Science 217: 408–417

Clark AS, Magnusson KR, Cotman CW (1992) In vitro autoradiography of hippocampal excitatory amino acid binding in aged Fischer 344 rats: relationship to performance on the Morris water maze. Behav Neurosci 106: 324–335

Fischer W, Nilsson OG, Bjorklund A (1991) In vivo acetylcholine release as measured by microdialysis is unrelated in the hippocampus of cognitively impaired aged rats with degenerative changes in the basal forebrain. Brain Res 556: 44–52

Gage FH, Kelley PAT, Bjorklund A (1984) Regional changes in brain glucose metabolism reflect cognitive impairments in aged rats. J Neurosci 4: 2856–2865

Gallagher M, Burwell RD, Kodsi MH, McKinney M, Southerland S, Vella-Rountree L, Lewis MH (1990) Markers for biogenic amines in the aged rat brain: relationship to decline in spatial learning ability. Neurobiol Aging 11: 507–514

Geinisman Y, Toledo-Morrell L, Morrell F (1986) Aged rats need a preserved complement of perforated axospinous synapses per hippocampal neuron to maintain good spatial memory. Brain Res 398: 266–275

Handelmann GE, Olton DS (1981) Spatial memory following damage to hippocampal CA3 pyramidal cells with kainic acid: impairment and recovery in preoperative training. Brain Res 217: 41–58

Heimrich B, Claus H, Schwegler H, Haas HL (1989) Hippocampal mossy fiber distribution and long-term potentiation in two inbred mouse strains. Brain Res 490: 404–406

Hellweg R, Fischer W, Hock C, Gage FH, Bjorklund A, Thoenen (1990) Nerve growth factor levels and choline acetyltransferase activity in the brain of aged rats with spatial memory impairments. Brain Res 537: 123–130

Henriksson BG, Soderstorm S, Gower AJ, Ebendal T, Winbald B, Mohammed AH (1992) Hippocampal nerve growth factor levels are related to spatial learning ability in aged rats. Behav Brain Res 48: 15–20

Kadar T, Silbermann M, Brandeis R, Levy A (1990) Age-related structural changes in the rat hippocampus: correlation with working memory deficiency. Brain Res 512: 113–120

Kadar T, Silbermann M, Weissman BA, Levy A (1990a) Age-related changes in the cholinergic components within the central nervous system 2. Working memory impairment and its relation to hippocampal muscarinic receptors. Mech Ageing Dev 55: 139–149

Landfield PW (1988) Review: hippocampal neurobiological mechanisms of age-related memory dysfunction. Neurobiol Aging 9: 571–579

Luine V, Heaens M (1990) Spatial memory deficits in aged rats: contributions of the cholinergic system assessed by ChAT. Brain Res 523: 321–324

Morris RGM, Garrard P, Rawlins JNP, O'Keefe J (1978) Place navigation impaired in rats with hippocampal lesions. Nature 297: 681–683

Olton DS, Walker JA, Gage FH (1978) Hippocampal connections and spatial discrimination. Brain Res 139: 295–308

Rapp PR, Amaral DG (1992) Individual differences in the cognitive and neurobiological consequences of normal aging. TINS 15: 340–345

Toledo-Morrell L, Geinisman Y, Morrell F (1988) Age-dependent alterations in hippocampal synaptic plasticity: relation to memory disorders. Neurobiol Aging 9: 581–590

Wree A, Kaever C, Birgel B, Schleicher A, Horvarth E, Zilles K (1991) Local cerebral glucose utilization in the brain of old learning impaired rats. Histochemistry 95: 591–603

Authors' address: Dr. T. Kadar, Department of Pharmacology, Israel Institute for Biological Research, Ness-Ziona 70450, Israel

J Neural Transm (1994) [Suppl] 44: 145–158

The function of the NMDA-receptor during normal brain aging

W. E. Müller, S. Stoll, K. Scheuer, and **A. Meichelböck**

Department of Psychopharmacology, Central Institute of Mental Health, Mannheim, Federal Republic of Germany

Summary. Age-related changes of N-methyl-D-aspartate (NMDA) receptors have been found in cortical areas and in the hippocampus of many species. On the basis of a variety of experimental observations it has been suggested that the decrease of NMDA-receptor density might be one of the causative factors of the cognitive decline with aging. Based on these findings several strategies have been developed to improve cognition by compensating the NMDA-receptor deficits in aging. The most promising approaches are the indirect activation of glutamatergic neurotransmission by agonists of the glycine site or the restoration of the age-related deficit of receptor density by several nootropics.

Introduction

N-Methyl-D-aspartate (NMDA) receptor-mediated glutamatergic neurotransmission appears to be of extreme importance for learning, memory and other cognitive functions in animals and man (Collingridge, 1987; Straubli and Lynch, 1991). Accordingly, the blockade of NMDA-receptor activation by competitive and non-competitive antagonists profoundly impairs learning and memory in a variety of animal models (Morris, 1989; Morris et al., 1986). Since some similarities exist between NMDA receptor antagonist induced memory impairments and age-related cognitive deficits, it has been suggested that age-related alterations of NMDA-receptor properties are involved in the cognitive decline with increasing age (Ingram et al., 1992; Pelleymounter et al., 1990). Moreover, qualitatively similar, but quantitatively more pronounced alterations of NMDA receptor properties are probably involved in the pathology of Alzheimer's disease (Deutsch and Morihisa, 1988; Palmer and Gershon, 1990; Palmer, 1991).

Previous work from this laboratory has demonstrated an age-related decline of NMDA receptor density in the forebrain of the female NMRI mouse (Cohen and Müller, 1992). The present report summarizes our previous work, adds several new findings and overviews comparable age-related alterations of NMDA receptor properties in other species. It also addresses the question if age-related deficits at the level of the NMDA receptor are relevant for the cognitive decline in aging.

Experimental section

Materials

Young (3 months), middleaged (12 months) and aged (23 months) female NMRI mice were obtained from Interfauna (Tuttlingen, FRG). ^{3}H-MK-801 (tritiated (+)-5-methyl-10,11-dihydro-5H-dibenzo(a,d)-cyclopotenten-5,10-imine hydrogen maleat) was obtained from New England Nuclear (Dreieich, FRG) (specific activity 25.7 Ci/mmol). Unlabeled (+)-MK-801 was purchased from Research Biochemicals Inc. (Natick, MA). All other chemicals were obtained from commercial suppliers.

Behavioral testing

The plexiglass passive avoidance (PA) box was divided into two compartments (dark and bright) by a sliding guilloutine door. Both compartments had a metal grid floor. The mice were habituated to the box for 2 minutes. An hour later they were placed on the bright side of the PA box and the door was opened after 30 s (habituation). The latency to enter was measured (L-naive). In a training trial on the next day the procedure was repeated, again measuring the latency (L-before). Right after entering the dark chamber, an electrical stimulus of 0.8 mA was delivered through the grid floor for 1 s. The animal remained in the dark compartment for 30 s until it returned to the illuminated compartment. After 30 s, the door was opened again and the latency to enter was recorded as a measure of short-term memory respectively acquisition (L-60 s). The retention test of the acquired memory was performed 24 hours after the training trail and step-through latency (max. 300 s) was measured (the higher the entry latency the better the retention performance).

Binding assay

The mice were sacrificed by decapitation. The brains were removed and immediately frozen at −20°C for at least 24 h. After thawing, the forebrain (including striatum, hippocampus and overlying cortex) was homogenized in 15 ml of 5 mmol/l Tris-HCl buffer, pH 7.4, at 4°C and centrifugated for 20 min at 48,000 × g. The supernatant was discarded and the pellet was washed twice under the same conditions and frozen for at least 24 h. For the final incubation, the pellet was resuspended in the same buffer to give a tissue concentration of about 4.6 mg wet weight per ml. Saturating levels of L-glutamate and glycin (100 µmol/l each) were added to all binding assays to eliminate possible binding variability attributable to changes in the concentration of these amino acids, which are known to affect the accessibility of NMDA-associated channel blockers like MK-801 to their binding site (Javitt and Zukin, 1989; Reynold et al., 1987). Triplicate aliquots of 350 µl tissue-suspension, 50 µl radioligand, 50 µl buffer or displacer and 50 µl glutamate/glycin solution (each 1 mmol/l) were incubated at 21°C for 90 min after which time binding was in equilibrium. Binding was terminated by rapid filtration through Whatman GF-C filters under slight vacuum. Non-specific binding was defined by 100 µmol/l (final concentration) of unlabelled (+)-MK-801 and accounted for 15–20% of total binding at low ligand concentrations.

The saturation experiments were always carried out using eight different ^{3}H-MK-801 concentrations ranging from 1–25 nmol/l. Data for K_D (dissociation constant) and B_{max} (maximal number of binding sites) were calculated from Scatchard plots by linear regression analysis using the EBDA-LIGAND program (Munson and Rodbard, 1980).

Protein concentration was determined according to the Lowry method (Lowry et al., 1951) using bovine serum albumin as standard.

Membrane fluidity measurements

The individual forebrain homogenates, already prepared for the ^{3}H-MK-801 binding assay, were used for determination of membrane fluidity by fluorescence spectroscopy using the hydrophobic 1,6-diphenyl-1,3,5-hexatriene (DPH) as fluorescence probe (Scheuer et al., 1993). The protein content of each sample was determined by the method of Lowry et al. (1951). The samples were diluted with 5 mmol/l Tris-HCl buffer pH 7.4 to give 30 µg protein in 100 µl.

Samples (450 µl) were incubated with 4,050 µl 5 mmol/l Tris-HCl buffer pH 7.4, 4,500 µl of a 1:150 DPH-solution and 1,000 µl buffer for 60 min at 37°C after which time anisotropy was stable. The DPH solution was prepared from a stock solution of 5 mmol/l DPH in tetrahydrofuran by 1:150 dilution with 5 mmol/l Tris-HCl buffer pH 7.4. The anisotropy was directly measured in a SLM 4,800 C Aminco spectrofluorometer using excitation and emission wavelenght of 360 nm and 450 nm respectively. The mean of 100 fluorescence intensity measurements was determined in triplicates on each sample. The steady-state fluorescence polarization (P_s) was also expressed as the anisotropy (r_s) of the probe, using the following equation (Lakowicz, 1983): $r_s = 2P_s/3 - P_s$.

Statistics

For statistical analysis (two-tailed t-test, ANOVA, correlational analysis), the SAS-package was used. Behavioral testing, binding assays and fluidity measurement were done by different experimentators, who were blinded to the other experimental data.

Age-related changes of NMDA-receptor properties

Receptor density

In accordance with previous findings (Cohen and Müller, 1992) we observed a highly significant age-related decrease of NMDA-receptor density as indicated by the specific ^{3}H-MK-801 binding to forebrain membrane of the NMRI-mouse (Table 1). Hill coefficents close to unity indicate one single population of ^{3}H-MK-801 binding sites with a dissociation constant of about 3 nmol/l in all three age groups which was only slightly higher in the aged animals. We confirmed a decline of B_{max} between young and aged animals by 22% and additionally found that the maximal number of sites of the middleaged animals is also already reduced by about 19% of maximal binding found in young animals.

Similar age-related changes of NMDA-receptor density have been reported for many species including man and are mainly found in cortical areas and the hippocampus (Table 2).

Table 1. Effect of aging on the density of NMDA receptors in the forebrain of the female NMRI mouse

Age	B_{max}	K_D	n_H
3 months (n = 12)	1.69 ± 0.08	2.89 ± 0.5	1.04 ± 0.08
12 months (n = 8)	1.38 ± 0.09***	2.82 ± 0.6	1.07 ± 0.07
23 months (n = 10)	1.31 ± 0.05***	3.73 ± 1.2*	1.03 ± 0.04

*** $p < 0.001$ (vs 3 mo)
* - < 0.05 (vs 3/12 mo)
Data for dissociation constants (K_D), maximal numbers of sites (B_{max}), and Hill coefficients (n_H) were obtained from saturation experiments using ^{3}H-MK-801 as radioligand as described in Materials and Methods. All experiments were performed in the prescence of L-glutamate and glycin (each 100 µmol/l). Data are means ± SD of (n) experiments, each representing an individual animal (from Scheuer et al., 1993)

Table 2. NMDA receptor density is reduced in the aging brain

Species	Region	References
Mouse	Whole brain	Peterson and Coleman (1989)
	Forebrain	Cohen and Müller (1992)
(SAM)	Cortex	Kitamara et al. (1992)
Rat	Hippocampus	Pelleymounter et al. (1990)
	Hippocampus	Tamara et al. (1991)
	Cortex	
	Hippocampus	Fiore and Rampello (1989)
	Hippocampus	Ingram et al. (1992)
	Hippocampus	Davis et al. (1989)
Human	Cortex	Kornhuber et al. (1989)
	Cortex	Pigott et al. (1993)

Summary or age-related reductions of NMDA-receptor density in several species. In all cases, decreases of density ranging from 10% to 30% have been observed. SAM refers to the senescent accelerated mouse model of aging

Receptor function

While the decline of NMDA-receptor density with aging seems to represent a robust finding over several species, less clear evidence is presently available on how much this decline of density is accompanied by similar changes of receptor function (Table 3). A possible explanation for the divergent findings (Table 3) might be the observation that agonist affinity of the NMDA receptor probably is up-regulated in aging as an response to reduced density (Cohen and Müller, 1992). Depending on the specific relevance of reduced density or enhanced agonist binding in each individual model,

Table 3. NMDA receptor function is reduced in the aging brain

Species	Function	References
	Decreased function	
Rat	Long-term potentation	Baskys et al. (1990)
Rat	Inhibition of carbachol stimulation IP hydrolysis	Gonzales et al. (1991)
Rat	Stimulation of norepinephrine and dopamine release	Pittaluga et al. (1993)
	No effect	
Mouse	IP hydrolysis	Hartmann and Müller (unpublished)
	Enhanced function	
Mouse	Stimulation of ^{3}H-MK-801 binding	Cohen and Müller (1992)

Summary of observations on the effects of aging on NMDA-receptor function in several animal models

Table 4. NMDA receptor deficits are involved in the age-related decline of cognitive functions

1. Cognitive deficits induced in young animals by NMDA antagonists mirror age-related cognitive deficits
   (Pelleymounter et al., 1990; Ingram et al., 1992; Venable and Kelly, 1990; this paper)

2. Sensitivity to NMDA antagonists is enhanced in aging
   (Ingram et al., 1992)

3. Positive relation between learning performance and NMDA density in rats showing an inborn high or low learning ability
   (Keller et al., 1992)

4. Reciprocal relationships between NMDA-receptor density and cognitive impairment

Rat	(Pelleymounter et al., 1990)
Rat	(Davis et al., 1993)
Mouse	(Scheuer et al., 1993)

Summary of observations supporting the assumption of an causal relation between NMDA-receptor loss and cognitive deficits

different patterns of age-related changes of NMDA-receptor responses might emerge.

Receptor density changes and cognitive deficits

Several experimental observations suggest that the decline of NMDA-receptor density is directly involved in the cognitive decline in aging (Table 4). Cognitive deficits induced in young animals by the administration of

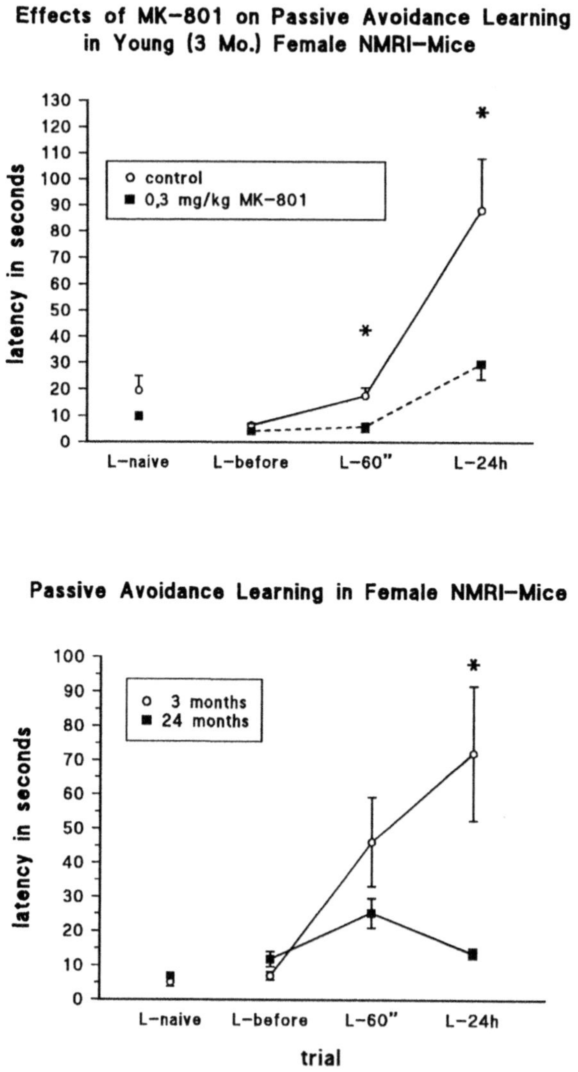

Fig. 1. L-naive is the step-through latency of experimentally naive animals without shock experience. L-before is the latency before the shock. L-60″ is the latency 60 seconds after the shock. L-24 h is the latency 24 hours after the shock. A significant effect of treatment ($p < 0.0018$, $F = 11.22$, 1 DF), repetition of measures ($p < 0.0001$, $F = 23.20$, 2 DF), and a significant interaction between treatment and repetition of measures ($p < 0.0032$, $F = 6.21$, 2 DF) was found in the ANOVA of L-before, L-60″, and L-24 h for young mice treated with 0.3 mg/kg MK-801 with $n = 20$ for treatment and control group. Post hoc Bonferroni t-tests revealed significant differences between groups for L-60″ and for L-24 h ($t > 2.02$). For young (3 mo.) and aged (22 mo.) mice there was a significant effect of age ($p < 0.0198$, $F = 6.36$, 1 DF), and a significant interaction between age and repetition of measures ($p < 0.0054$, $F = 5.92$, 2 DF) with $n = 12$ for young and $n = 11$ for aged mice. Post hoc Bonferroni t-tests revealed significant age differences for L-24 h ($t > 2.08$)

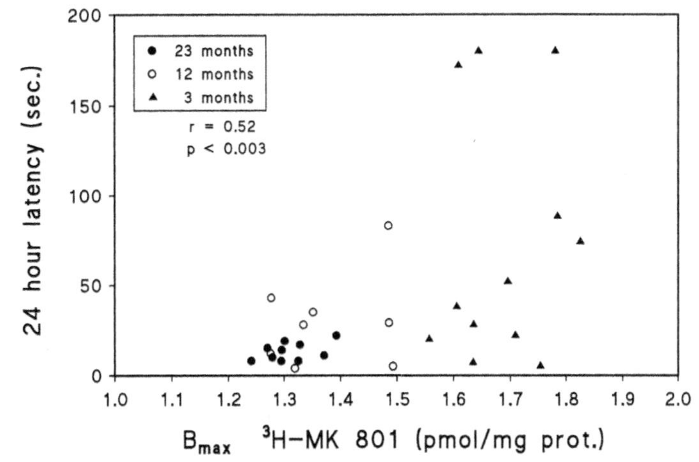

Fig. 2. Pearson correlation between 24 h latency of the passive avoidance aquisition task (Fig. 1) and B_{max} of specific ^{3}H-MK-801 binding in the forebrain for young, middleaged and aged NMRI-mice. Each point represents an individual animal. Data are taken from Scheuer et al. (1993)

NMDA-receptor antagonists mirror in many ways age-associated cognitive deficits in the same species and the same behavioral paradigm (Table 4). A specific example is the passive avoidance learning model in the mouse, were both, the antagonist MK-801 and aging affect 24 h latency (long-term memory) much more than 60 sec latency (short-term memory) (Fig. 1). The assumption of a direct role of NMDA-receptor deficits for the cognitive decline in aging is further supported by the enhanced sensitivity of aged animals to NMDA antagonists (Ingram et al., 1992), by low NMDA densities in rats with an inborn low learning ability (Keller et al., 1992), but mainly by several findings on reciprocal relationships between NMDA density and cognitive performance (Table 4). A specific example for such an relationship in the mouse is given in Fig. 2. By contrast, a negative relationship between learning performance and NMDA-receptor number has been reported by others in aged rats (Ingram et al., 1992), although there was also an age-related decrease of specific L-glutamate binding. However, in these studies NMDA receptor binding was determined using only one single concentration of ^{3}H-L-glutamate. As already mentioned, some evidences suggest an increase of the affinity of L-glutamate for the NMDA-receptor in aging, probably as an adaptive response to the reduced density. Thus, the results cited above using only one radioligand concentration for receptor determination might be explainable by the age-related decline of NMDA-receptor density superimposed by the enhanced affinity of L-glutamate during aging. Nevertheless, taking all data together, fairly good evidence suggests an inverse relationship between NMDA-receptor density and learning performance in aging. However, we can not completely refuse the alternative explanation that aging affects NMDA-receptor density and learning performance in a similar fashion, without both parameters being related.

Relation to membrane fluidity

The biochemical basis by which aging affects NMDA-receptor number and cognitive functions is not known. Studies on the biochemical mechanisms of age-related changes of density and function of central α_1-adrenergic, β-adrenergic, dopaminergic, and muscarinic cholinergic receptors have suggested a crucial role of the fluidity of the neuronal membrane (Cimino et al., 1984; Henry and Roth, 1986; Joseph et al., 1991; Muccioli et al., 1992). This links the age-related loss of neuroceptors with the "membrane hypothesis of aging" which assumes that a variety of changes of the composition of brain membranes like enhanced lipid peroxidation, enhanced cholesterol to phospholipid ratios, and enhanced cross-linking of structural proteins finally affect brain function by altering the properties of transmembrane proteins. While all the above mentioned changes of membrane composition might alter fluidity, the specific mechanisms relevant for the receptor deficits are not yet specifically identified. Observations indicating that receptor deficits can be easily reversed in vitro by membrane fluidization (Joseph et al., 1991; Muccioli et al., 1992) suggest that fluidity changes per se, rather than changes in the composition of membrane fractions of different fluidity are involved.

In agreement with the previously mentioned observations, our data (Fig. 3) also suggest an important role of membrane fluidity for the age-related decline of NMDA-receptor density. This conclusion, however, probably holds only true for the small changes of membrane fluidity between young and middleaged animals, which correlate highly significantly with receptor number. There is an additional large decrease of membrane fluidity in the

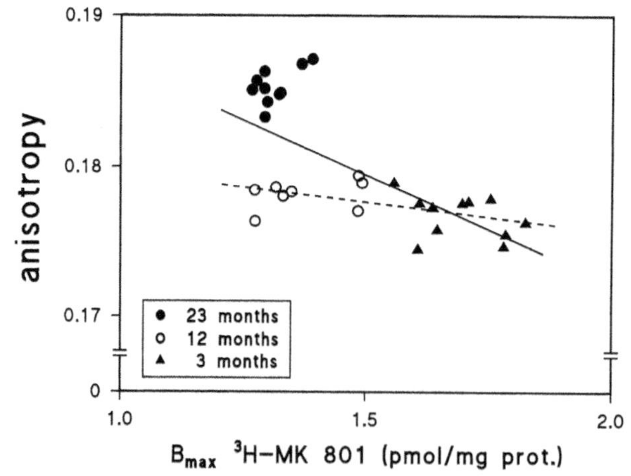

Fig. 3. Influence of age on the correlation between B_{max} of specific ^{3}H-MK-801 binding and membrane fluidity. Correlation was done over all three age groups ($r = -0.70$, $p < 0.001$; solid line) as well as for young and middleaged animals ($r = -0.49$, $p < 0.03$; dashed line). Data are taken from Scheuer et al. (1993)

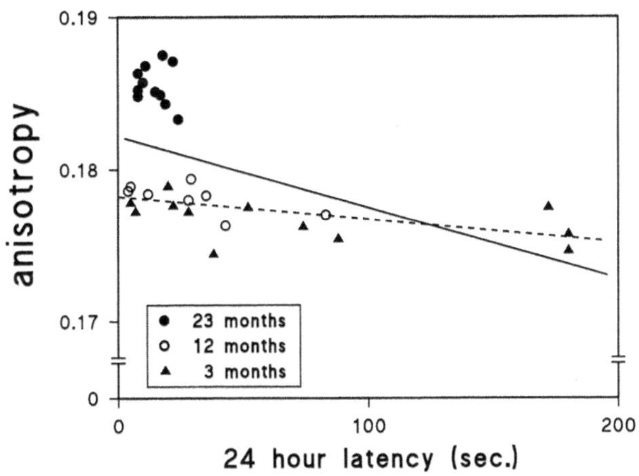

Fig. 4. Influence of age on the correlation between passive avoidance learning (24 h latency) and membrane fluidity. Correlation was done over all three age groups ($r = -0.52$, $p < 0.03$; solid line) as well as for young and middleaged animals ($r = -0.58$, $p < 0.01$; dashed line). Data are taken from Scheuer et al. (1993)

aged animals, which, however, does not seem to have further effects on receptor density. Accordingly, the small changes of the overall membrane fluidity already present in the middleaged animals probably must be attributed to different changes of membrane properties than the pronounced changes in the aged animals.

24 h latency correlated similarly with membrane fluidity for the young and middleaged, but not for the aged animals (Fig. 4). This could simply confirm our initial hypothesis that changes of membrane fluidity by decreasing the number of NMDA-receptors affect the learning performance. However, we cannot completely refuse the alternative explanation that age-related alterations of membrane fluidity affect learning and receptor number independently, without that the reduced NMDA-receptor number per se is relevant for the cognitive impairment.

In summary, deficits of learning and NMDA-receptor density are already present in middleaged mice and correlate highly significantly with small changes of membrane fluidity. An additional pronounced change of membrane fluidity in aged mice is not accompanied by additional changes of the two other parameters.

NMDA-receptor related treatment strategies for cognitive deficits in aging

Because of the possibly relevant role of NMDA-receptor deficits for the cognitive decline in aging, several strategies have been developed to improve cognition by enhancing NMDA-receptor deficits (Table 5).

Table 5. NMDA-receptor related treatment strategies for cognitive deficits in aging

1. NMDA agonist L-glutamate	Deutsch and Morihisa (1988) Vogel et al. (1966)
NMDA	Davis and Flood (1987) Jones et al. (1989) Parada-Tursha and Turski (1991) (NMDA improves learning deficits induced by NMDA antagonists and improves learning after intracerebroventricular administration but not after i.p. administration)
2. Glycine agonists	D-cycloserine Monahan et al. (1989) (improves learning in rats) Jones et al. (1991) (reverses scopolamine induced cognitive impairment in humans) Fishkin et al. (1993) (reverses scopolamine induced cognitive impairment in rats)
3. Nootropics	enhance NMDA density in aging (Table 7)

Table 6. Evidences for an enhanced sensitivity of the glycine recognition site of the NMDA receptor complex in the aged brain

Enhanced affinity of glycine binding in rat hippocampus (Jablonka and Schindler, 1990)
Enhanced stimulation of ^{3}H-MK-801 binding by glycine in the forebrain of the mouse (Cohen and Müller, 1992)
Enhanced stimulation of the effect of NMDA on norepinephrine release in rat hippocampus (Pittaluga et al., 1993)

NMDA-receptor agonists

Although recent metaanalyses of many clinical studies done with L-glutamate many years ago do suggest positive effects of L-glutamate on cognitive deficits (Deutsch and Morihisa, 1988; Vogel, 1966), this approach has not been further evaluated because of the possible risk of neurotoxic side effects. This is certainly also the reasons that only very few studies are available on the possible effects of NMDA itself on cognition (Table 5).

Agonists at the glycine site

Indirect activation of NMDA-receptor function by glycine-like agonists (e.g. D-cycloserine) probably avoids neurotoxic effects since using this strategy function NMDA-receptor is elevated only to its maximum physiological activity. The relevance of this approach is further supported by observations of an enhanced sensitivity of the glycine site for agonists in aging (Table 6).

Table 7. Nootropics enhance NMDA receptor density in the aged brain

Drug	Region	Species	References
L-Acetylcarnitine	Hippocampus	Rat	Fiore and Rampallo (1989)
L-Acetylcarnitine	Hippocampus	Rat	Davis et al. (1993)
Phosphatidylserine	Forebrain	Mouse	Cohen and Müller (1992)
Piracetam	Forebrain	Mouse	Cohen and Müller (1993)
Pyritinol	Forebrain	Mouse	Hartmann et al. (1993)
Bifemelane	Hippocampus	Rat	Ogawa et al. (1992)

Only significant effects of subchronic treatment with several nootropic drugs on NMDA receptor densities in the brain are given. In all cases, elevations of NMDA receptor densities between 10 and 25% were observed

Accordingly, some initial observations in animals and man indicate the usefulness of the administration of D-cycloserine to enhance cognition (Table 5). However, much more data are needed to decide if this strategy seems to be promising to treat age-related memory disorders including Alzheimer's disease, where the glycine site of the NMDA-receptor seems to be specifically disturbed (Palmer, 1991).

Nootropics enhance NMDA-receptor density

Subchronic treatment with several nootropics has been demonstrated to enhance NMDA-receptor density in aged animals (Table 7). This effect was not observed in young animals (Cohen and Müller, 1992). Since the decline of NMDA-receptor density correlates with the cognitive decline in aging (see 3.3) and since the elevation of NMDA-receptor density in aged animals seems to be accompanied by cognitive improvement (Davies et al., 1993), it seems possible that this effect is relevant for the cognition enhancing properties of nootropics.

Conclusions

Today, a variety of experimental evidences indicate that deficits of NMDA-receptor mediated glutamatergic neurotransmission are involved in age-related memory disorders. At present, possible treatment strategies related to these biochemical deficits are to enhance glycine modulation or to restore age-related deficits of NMDA-receptor density.

Acknowledgements

This paper was supported by a grant from the Dr. E. Neele Stiftung, Frankfurt. The excellent technical assistance of S. Koch is gratefully acknowledged.

References

Baskys A, Reynolds JM, Carlen PL (1990) NMDA depolarizations and long-term potentation are reduced in the aged rat neocortex. Brain Res 530: 142–146

Cimino M, Vantinin G, Algeri S, Curatola G, Pezzoli C, Stramentinoli G (1984) Age-related modification of dopaminergic and β-adrenergic receptor system: restoration to normal activity by modifying membrane fluidity with S-adenosylmethionine. Life Sci 34: 2029–2039

Cohen SA, Müller WE (1992) Age-related alterations of NMDA-receptor properties in the mouse forebrain: partial restoration by chronic phosphatidylserine treatment. Brain Res 584: 174–180

Cohen SA, Müller WE (1993) Effects of piracetam on N-methyl-D-asparate receptor properties in the aged mouse brain. Pharmacology 47: 217–222

Collingridge G (1987) The role of NMDA receptors in learning and memory. Nature 330: 604–608

Davis JL, Klood JS (1987) Excitatory amino acids and memory processing. Soc Neurosci Abstr 13: 658

Davis S, Markowska AL, Wenk GL, Barnes CA (1993) Acetyl-L-carnitine: behavioral, electrophysiological, and neurochemical effects. Neurobiol Aging 14: 107–115

Deutsch SI, Moruhisa JM (1988) Glutamatergic abnormalities in Alzheimer's disease and a rationale for clinical trials with L-glutamate. Clin Neuropharmacol 1: 11–35

Fiore L, Rampello L (1989) L-Acetylcarnitine attenuates the age-dependent decrease of NMDA-sensitive glutamate receptors in rat hippocampus. Acta Neurol 11: 346–351

Fishkin RJ, Ince ES, Carlezon Jr WA, Dunn RW (1993) D-cycloserine attentuates scopolamine-induced learning and memory deficits in rats. Behav Neural Biol 59: 150–157

Gonzales RA, Brown LM, Jones TW, Trent RD, Westbrook SL, Leslie SW (1991) N-methyl-D-asparate mediated respones decrease with age in Fischer 344 rat brain. Neurobiol Aging 12: 219–225

Hartmann H, Cohen SA, Müller WE (1993) Effects of subchronic administration of pyritinol on receptor deficits and phosphatidylinositol metabolism in the brain of the aged mouse. Neuropharmacology 32: 119–125

Henry JM, Roth GS (1986) Modulation of rat striatal membrane fluidity: effects on age related differences in dopamine receptor concentrations. Life Sci 39: 1223–1229

Ingram DK, Garofalo P, Spangler EL, Mantione CR, Odano I, London ED (1992) Reduced density of NMDA receptors and increased sensitivity to Dizocilpine-induced learning impairment in aged rats. Brain Res 580: 273–280

Jablonka B, Schindler U (1990) Changes of strychnine-insensitive glycine sites in the hippocampus of senescent rats. Eur J Pharmacol 183: 476–477

Javitt DC, Zukin SR (1989) Biexponential kinetics of ^{3}H-MK-801 binding: evidence for access to closed and open N-methyl-D-asparate receptor channels. Mol Pharmacol 35: 378–393

Joseph JA, Yamagami K, Roth GS (1991) Loss of muscarinic responsiveness in senscence may be the result of decreased membrane fluidity. Soc Neurosci Abstr 17: 53

Jones RW, Schaeffer CL, DeNoble VJ (1989) Systemically administrated N-methyl-D-asparate interferes with acquisition of a passive avoidance response in rats. Pharmacol Biochem Behav 34: 181–185

Jones RW, Wesnes KA, Kirby J (1991) Effects of NMDA modulation in scopolamine dementia. Ann NY Acad Sci 640: 241–244

Keller EA, Borghese CM, Carrer HF, Ramirez OA (1992) The learning capacity of high or low performance rats is related to the hippocampus NMDA receptors. Brain Res 576: 162–164

Kitamura Y, Zhao XH, Ohnuki T, Makiko T, Nomura Y (1992) Age-related changes in transmitter glutamate and NMDA receptor/channels in the brain of senescence-accelerated mouse. Neuroscience 137: 169–172

Kornhuber J, Mack-Burkhardt F, Konradi C, Fritze J, Riederer P (1989) Effect of antemortem and postmortem factors on ^{3}H-MK-801 binding in the human brain: transient elevation during early childhood. Life Sci 45: 745–749

Lakowicz JR (1983) Principles of fluorescence spectroscopy. Plenum Press, New York

Lowry O, Rosenbrough NI, Farr AL, Randall RJ (1951) Protein measurement with the Folin phenol reagent. J Biol Chem 193: 265–275

Morris RGM, Anderson E, Lynch G, Baudry M (1986) Selective impairment of learning and blockade of long-term potentation by an N-methyl-D-asparate receptor antagonist, AP5. Nature 319: 774–776

Morris RGM (1989) Synaptic plasticity and learning: selective impairment of learning in rats and blockade of long-term potentation in vivo by the N-methyl-D-asparate receptor antagonist AP5. J Neurosci 9: 3040–3057

Monahan JB, Handelmann GE, Hood WF, Cordi AA (1989) D-cycloserine, a positive modulator of the N-methyl-D-asparate receptor, enhances performance of learning tasks in rats. Pharmacol Biochem Behav 34: 649–653

Muccioli G, Scordamaglia A, Bertacco S, Di Carlo R (1992) Effect of S-adenosyl-L-methionine on brain muscarinic receptors of aged rats. Eur J Pharmacol 227: 293–299

Munson P, Rodbard D (1980) Ligand: a versatile computerized approach for characterization of ligand-binding systems. Anal Biochem 107: 220–239

Ogawa N, Mizukawa K, Haba K, Asanuma M, Mori A (1992) Effects of chronic bifemelane hydrochloride administration on receptors for N-methyl-D-asparate in the aged-rat brain. Neurochem Res 17: 687–691

Palmer AM (1991) Excitatory amino acid neurons and receptors in Alzheimer's disease. In: Kozilowski AP, Barrinuevo G (eds) Neurobiology of the NMDA receptor. VCH, New York, pp 203–237

Palmer AM, Gershon S (1990) Is the neuronal basis of Alzheimer's disease cholinergic or glutamatergic? FASEB J 4: 2745–2752

Parada-Turska J, Turski WA (1990) Excitatory amino acid antagonists and memory: effect of drugs acting at N-methyl-D-aspartate receptors in learning and memory tasks. Neuropharmacology 29: 1111–1116

Pelleymounter MA, Beatty G, Gallagher M (1990) Hippocampal ^{3}H-CPP binding and spatial learning deficits in aged rats. Psychobiology 18: 298–304

Peterson C, Cotman CW (1989) Strain-dependent decrease in glutamate binding to the N-methyl-D-aspartatic acid receptor during aging. Neurosci Lett 104: 309–313

Pigott MA, Perry EK, Perry RH, Court JA (1992) 3H-MK-801 binding to the NMDA-receptor complex, and its modulation in human frontal cortex during development and aging. Brain Res 588: 277–286

Pittaluga A, Fedele E, Risiglione C, Raiteri M (1993) Age-related decrease of the NMDA receptor-mediated noradrenaline release in rat hippocampus and partial restoration by D-cycloserine. Eur J Pharmacol 231: 129–134

Reynolds IJ, Murphy SN, Miller RJ (1987) ^{3}H-labeled MK-801 binding to the excitatory amino acid receptor complex from rat brain is enhanced by glycine. Proc Natl Acad Sci USA 84: 7744–7748

Scheuer K, Stoll S, Paschke U, Weigel R, Müller WE (1994) NMDA receptor density and membrane fluidity as major determinats of the cognitive decline in aging. Pharmacol Biochem Behav (submitted)

Spangler EL, Bresnahan EL, Garofalo P, Muth NJ, Heller B, Ingram DK (1991) NMDA receptor channel antagonism by dizocilipine (MK-801) impairs performance of rats in aversively motivated complex maze tasks. Pharmacol Biochem Behav 40: 949–958

Straubli U, Lynch G (1990) NMDA receptors and memory: evidence from pharmacological and correlational studies. In: Kozikowski AP, Barrionuevo G (eds) Neurobiology of the NMDA receptor. VCH, New York, pp 129–148

Tamaru M, Yoneda Y, Ogita K, Shimizu J, Nagata Y (1991) Age-related decreases of the N-methyl-D-asparate receptor complex in the rat cerebral cortex and hippocampus. Brain Res 542: 83–90

Venable N, Kelly PH (1980) Effect of NMDA-receptor antagonists on passive avoidance learning and retrieval in rats and mice. Neurobiol Aging 1: 59–63

Vogel W, Broverman DM, Draguns JG (1966) The role of glutamic acid in cognitive behaviors. Psychol Bull 65: 367–382

Authors' address: Prof. Dr. W.E. Müller, Department of Psychopharmacology, Central Institute of Mental Health, D-68159 Mannheim, Federal Republic of Germany

J Neural Transm (1994) [Suppl] 44: 159–171

Age-related alterations by chronic intermittent hypoxia on cerebral synaptosomal ATPase activities

G. Benzi, A. Gorini, R. Arnaboldi, B. Ghigini, and **R. F. Villa**

Institute of Pharmacology, State University of Pavia, Pavia, Italy

Summary. The age-related alterations in the plasticity of synaptic energy-requiring ATPases [Na^+,K^+-ATPase, low- and high-affinity Ca^{2+}-ATPase, Mg^{2+}-ATPase, and Ca^{2+},Mg^{2+}-ATPase] were assayed in synaptosomes and synaptosomal subfractions [namely, synaptosomal plasma membranes and synaptic vesicles] in the cerebral cortex from 3- and 24-month-old normoxic rats and rats subjected to either mild or severe chronic (four weeks) intermittent normobaric hypoxia.

With the exception of the high-affinity Ca^{2+}-ATPase, aging induced a decrease in the ATPase activities from normoxic rats. The adaptation to *mild* hypoxia was characterized by an increase in the activity of Mg^{2+}-ATPase in 3-month-old rats, concomitant with a decrease in the activities of: (i) Na^+,K^+-ATPase and high-affinity Ca^{2+}-ATPase in both 3- and 24-month-old rats, and (ii) Ca^{2+},Mg^{2+}-ATPase in 3-month-old ones.

The adaptation to chronic intermittent *severe* hypoxia was characterized by a decrease in the activities of: (i) Na^+,K^+-ATPase, Ca^{2+},Mg^{2+}-ATPase and high-affinity Ca^{2+}-ATPase in both 3- and 24-month-old rats, and (ii) low-affinity Ca^{2+}-ATPase only in 24-month-old ones. The effect on Mg^{2+}-ATPase activity was characterized by a decrease in the activity of the enzymatic form located in the synaptic plasma membranes [involved in ATP hydrolysis to adenosine production], concomitant with an increase in the activity of the form located in the synaptic vesicles [involved in the turnover of transmitters, e.g., glutamate].

Introduction

The age-related alteration in the plasticity of cerebral enzyme proteins may be carefully studied by means of the chronic sequential cycles of hypoxia/normoxia inducing alternative conditions of either absolute hypoxemia or relative hyperoxemia characterized by some adaptive modifications, namely: (a) selective alterations of enzymes having different half-lives (Dunlop, 1978); (b) decrease in the cytochrome oxidase activity in synaptosomal fractions from cerebral cortex, hippocampus, striatum, and cerebellum (Marzatico et al., 1986); (c) decrease in the malate dehydrogenase activity in the synaptosomal fraction from cerebellum (Marzatico et al., 1986); (d) change in the composition of specific proteins in synaptic plasma membranes isolated from rat cerebral cortex (Villa et al., 1991); (e) decrease in the incorporation of labelled precursor into deoxyribonucleic acid, ribonucleic acid, and proteins from various cerebral region subfractions (Serra et al., 1981); (f) decrease in the incorporation of labelled precursors into lipids in different subfractions purified from some brain regions (Alberghina and Giuffrida, 1981); (g) activation of cerebral mitochondrial phospholipase A_2 (Alberghina et al., 1982).

ATPases associated with the synaptic membranes are affected by post-hypoxic or post-ischemic recovery (Goldberg et al., 1985; Mrsulja et al., 1985; Palmer et al., 1985; Taylor et al., 1985; Schwartz et al., 1976), calcium having widespread actions on membrane phospholipase stimulation (Raichle, 1983; Chan et al., 1984; White et al., 1984), xanthine oxidase activation (Chan et al., 1984; McCord, 1985), and metabolic enzyme system alteration (Weishaar, 1984).

Synaptic non-mitochondrial ATPases [namely, Na^+,K^+-ATPase, Ca^{2+}-ATPase, Mg^{2+}-ATPase and Ca^{2+},Mg^{2+}-ATPase] play a pivotal role in the regulation of presynaptic nerve ending homeostasis. Na^+,K^+-ATPase, an enzyme located in the synaptic plasma membranes, functions as an electrogenic ion pump and maintains high intracellular K^+ concentration as this cation is moved against its concentration gradient (Albers, 1967; Schwartz et al., 1975; Skou, 1957, 1965). Thus, Na^+,K^+-ATPase is responsible for the maintenance of Na^+ and K^+ gradient and is involved in the establishment of the resting transmembrane potential required for the electrical activity of neurons in the CNS (Skou, 1986).

Na^+,K^+-ATPase is involved in transmitter presynaptic release (Garcia and Kirpekar, 1973; Paton et al., 1971; Vizi, 1972), in postsynaptic activities (Jariashvili, 1975; Logua and Kometiani, 1970), and in transmitter reuptake (Logan and O'Donovan, 1975), suggesting an important role in the turnover of the neurotransmitters in nerve endings (Kometiani et al., 1984). There is also an indirect influence of the enzyme via the ion gradients as transmitter turnover is not affected, even if the Na^+/K^+-pump is inhibited but the ion gradients remain constant.

Ca^{2+}-ATPase and Ca^{2+},Mg^{2+}-ATPase located in synaptic plasma membrane pump Ca^{2+} ions help to maintain the synaptoplasmic Ca^{2+} homeostasis. Moreover, Ca^{2+}-ATPase located in synaptic vesicles modulates the intracellular concentration of Ca^{2+} that initiates the interaction between ATP and ATPase to promote both the attachment of secretory granules to plasmalemma and the extrusion of their neurotransmitters to the nerve terminal exterior by exocytosis (Blitz et al., 1977; Douglas, 1968; Duncan, 1976; Lee and Shin, 1969; Lin, 1979, 1981; Lin and Way, 1982a,b; Michaelis et al., 1983; Schatzmann and Vincenzi, 1969). The Lineweaver-Burk plots of the velocity-activator curves for Ca^{2+}-ATPase indicates that the enzyme has two binding sites of different affinities for the Ca^{2+}: the higher affinity Ca^{2+}-ATPase binding site has an apparent K_m of 0.06 μM Ca^{2+} and the lower one an apparent K_m of 7.5 μM Ca^{2+} (Lin and Way, 1982a).

Synaptic plasma membranes from mammalian brain tissues have a Mg^{2+}-ATPase which has externally oriented active sites and, therefore, can be classified as an ecto-enzyme (Sorensen and Mahler, 1982) involved in the hydrolysis of ATP to generate, together with adenylate kinase and 5'-nucleotidase, adenosine (Sorensen and Mahler, 1982; Trams, 1980; White, 1978). In contrast, the enzymatic form located in synaptic vesicles modulates the turnover of different neurotransmitters according to the different nerve terminals involved (Mayanil et al., 1982; Wyllie and Gilbert, 1979).

The aim of the study is to evaluate in rat cerebral cortex synaptosomes and synaptosomal subfractions [namely synaptosomal plasma membranes and synaptic vesicles] the modification induced by aging in: (a) the ATP-consuming non-mitochondrial ATPases, and (b) ATPase plasticity, using the model of the adaptation to chronic [four weeks] normobaric intermittent mild or severe hypoxia.

Material and methods

Animals and experimental plan

The experiments were carried out on albino rats aged either from 2 to 3 months or from 23 to 24 months. The animals were kept under constant environmental conditions (temperature: 22 ± 1°C; relative humidity: 60 ± 5%; circadian rhythm: 12-hr light and 12-hr dark) and fed normal laboratory diet as pellets.

The rats were divided into six groups: (a) *groups 1 and 2* (normoxic animals): 3- or 24-month-old rats breathing room air; (b) *groups 3 and 4* (mildly-hypoxic rats): 2- or 23-month-old rats kept under mild normobaric hypoxia in a chamber flushed with $N_2:O_2$ mixture (90:10) for 12 hr/day (from 8:30 a.m. to 8:30 p.m.) for four weeks (seven days/week); (c) *groups 5 and 6* (severely-hypoxic rats): 2- or 23-month-old rats kept under severe normobaric hypoxia in a chamber flushed with $N_2:O_2$ mixture (91.5:8.5) for 12 hr/day (from 8:30 a.m. to 8:30 p.m.) for four weeks (seven days/week). Excess CO_2 was removed from the chamber with soda.

Cerebral tissue drawing and biochemical evaluation

At the scheduled time, the rats were killed by decapitation between 9:00 and 10:30 [24 h after the last hypoxia/normoxia cycle]. The brain of each animal was quickly removed (15 s) and immersed in a cold isolation medium [IM = 0.32 M sucrose; 1.0 mM EDTA-K^+; 10 mM Tris-HCl (pH 7.4)]. The dissection of the cerebral cortex was performed in a precooled box (0°–4°C) and the tissue was quickly immersed in IM.

The subcellular fractions tested were: (i) synaptosomes; (ii) synaptosomal plasma membranes; (iii) synaptic vesicles. The activity of the following enzymes was tested: (a) Na^+,K^+-ATPase; (b) low and high affinity Ca^{2+}-ATPase; (c) Mg^{2+}-ATPase; and (d) Ca^{2+},Mg^{2+}-ATPase.

Preparation of synaptosomes and synaptosomal subfractions

Cerebral cortex homogenate was prepared in IM by a Teflon-glass Potter-Elvehjem homogenizer (electronic control of the revolution speed) with a loosely fitted Teflon pestle (total clearance = 0.1 mm) rotating at 800 rpm for 1 min (five up-and-down passes of the pestle). The homogenate (7–10%) was centrifuged at 1,000 × g for 5 min to remove the nuclear fraction and the resulting supernatant was isolated and preserved. The pellet was then resuspended in IM and again centrifuged as above. This step was repeated once again. The three supernatants were then centrifuged at 15,000 × g for 20 min to obtain the crude mitochondrial fraction (P_2). The subfraction was resuspended in IM and layered over a density gradient of 7.5% and 12% Ficoll (wt/vol) in a suspension medium [SM = 0.32 M sucrose; 50 µM EDTA-K^+; 10 mM Tris-HCl (pH 7.4)].

The suspension was centrifuged at 70,000 × g for 20 min in a ultracentrifuge. Myelin fraction was suckled off and synaptosomes were then collected from the interface of the Ficoll layers, diluted with IM, and spun down by centrifugation at 15,000 × g for 25 min. The pellet obtained was then suspended in 0.32 M sucrose solution (pH 7.4) and an aliquot was used within 2 h for the assay of enzyme activities and protein related to the synaptosomal subfraction.

An aliquot of the synaptosomal subfraction was lysed in 6 mM Tris-HCl (pH 8.1) for 1 h at 0–4°C and layered over a discontinuous sucrose-HEPES gradient (SH = 0.32; 0.64; 0.80; 1.0 M). The suspension was centrifuged at 60,000 × g for 2 h and the separated bands were collected with a glass capillary pipette. The subfraction of synaptic vesicles was at the interface between 0.32 M and 0.64 M SH. The subfraction was centrifuged at 110.000 × g for 30 min. The pellet was resuspended in 0.32 M sucrose solution (pH 7.4) and it was used for the evaluation of enzyme activities and protein related to the synaptic vesicles.

The synaptosomal plasma membrane subfractions — SPM_1 and SPM_2 — were collected between 0.64 M and 0.80 M and between 0.80 M and 1.0 M SH, respectively. The subfractions were pooled and centrifuged at 110.000 × g for 30 min. The pellet was resuspended in 0.32 M sucrose solution (pH 7.4) and was used for enzyme and protein assays related to synaptosomal plasma membranes.

Biochemical assay

ATPase activities were determined by measuring the P_i released from ATP hydrolysis in the presence of different ions and enzymes. ATPase activities were expressed as µmoles of P_i released/hour/mg protein of the subfraction tested (Sawas and Gilbert, 1981; Lin and Way, 1982a; Sorensen and Mahler, 1981, 1982; Shallom and Katyare, 1985).

Na^+,K^+-ATPase activity was calculated from the difference between the Na^+,K^+, Mg^{2+}-ATPase activity and Mg^{2+}-ATPase activity. Na^+,K^+,Mg^{2+}-ATPase activity was assayed in 1.0 ml of a medium containing: 50 mM imidazole-HCl buffer pH 7.4; 120 mM NaCl; 10 mM KCl; 5 mM $MgCl_2$; 4 mM ATP. Vanadium free Na_2-ATP solutions were made fresh prior to use by neutralizing ATP to pH 7.4 with Tris. At the end of the incubation period [10–15 min; 37°C] with the sample of the subfraction tested [24–48 µg of protein], the reaction was stopped by cooling it to 4°C and adding 0.5 ml of 10% (wt/vol) of TCA solution.

Mg^{2+}-ATPase activity was evaluated in 1.0 ml of a medium containing: 50 mM imidazole-HCl buffer pH 7.4; 5 mM $MgCl_2$; 4 mM ATP. Vanadium free Na_2-ATP solutions were made fresh prior to use by neutralizing ATP to pH 7.4 with Tris. At the end of the incubation period [10–15 min; 37°C] with the sample tested [24–48 µg of protein], the reaction was stopped by cooling it to 4°C and adding 0.5 ml of 10% (wt/vol) of TCA solution.

Ca^{2+},Mg^{2+}-ATPase activity was evaluated in 1.0 ml of a medium containing: 50 mM imidazole-HCl buffer pH 7.5; 100 mM KCl; 2 mM $MgCl_2$; 120 µM $CaCl_2$; 100 µM EGTA; 100 µM freshly prepared ouabain; 10 µg of freshly prepared oligomycin in ethanol solution; 1.23 mM ATP. Vanadium free Tris-ATP solutions were made fresh prior to use by neutralizing ATP to pH 7.4 with Tris. At the end of the incubation period [15 min; 37°C] with the sample of the subfraction tested [32–48 μg of protein], the reaction was stopped by cooling it to 4°C and adding 0.5 ml of 10% (wt/vol) of TCA solution.

Low and high affinity Ca^{2+}-ATPase activities were evaluated in 1.0 ml of a medium containing: 50 mM imidazole-HCl buffer pH 7.5; 200 μM sucrose; 200 μM $CaCl_2$ for the high affinity form or 2 mM for the low affinity form; 100 μM EGTA; 1.84 mM ATP. Vanadium free Tris-ATP solutions were made fresh prior to use by neutralizing ATP to pH 7.4 with Tris. At the end of the incubation period [15 min; 37°C] with the sample of the subfraction tested [32–48 μg of protein], the reaction was stopped by cooling it to 4°C and adding 0.5 ml of 10% (wt/vol) of TCA solution. The Ca^{2+}-ATPase activity was calculated by the difference between the activity in the presence and absence of Ca^{2+}.

For the enzyme activities tested, after centrifugation of the TCA-sample, 1.0 ml of the clear supernatant was assayed for P_i by adding and mixing [5 min]: (a) 3.0 ml of a solution containing both 4.6% of sodium acetate trihydrate [$CH_3COONa.3H_2O$] and 0.25% of copper sulphate pentahydrate [$CuSO_4.5H_2O$] in 2N acetic acid [CH_3COOH]; and (b) 0.5 ml of 5% ammonium molybdate [$(NH_4)_6Mo_7O_{24}.4H_2O$]. Then, 0.5 ml of 2% MAPS [$(HOC_6H_4NHCH_3)_2.H_2SO_4$] in 5% sodium sulphite [Na_2SO_3] was added and mixed. After 10 min the absorbance was read at 870 nm versus a blank sample containing the subfraction tested and all the reagents without cations. A titration curve for P_i was made with anhydrous KH_2PO_4.

The protein content of the subfractions tested was determined using crystalline bovine serum albumin as a standard (Lowry et al., 1951).

Statistical analysis

Two statistical tests (ANOVA and Dunnett's tests) were applied to the results ($p < 0.01$), after checking the homogeneity of variance by the Bartlett's test.

Results

In synaptosomes from the cerebral cortex of rats in normoxic condition, the values of the enzyme activities markedly differ according to the individual ATPases tested. With the exception of the high-affinity Ca^{2+}-ATPase, aging induced a marked decrease in the ATPase values (Fig. 1).

In synaptosomes from cerebral cortex, the adaptation to chronic (four weeks) intermittent *mild* hypoxia was characterized (Fig. 2) by an increase in the activity of Mg^{2+}-ATPase in 3-month-old rats, concomitant with a decrease in the activities of: (i) Na^+,K^+-ATPase and high-affinity Ca^{2+}-ATPase in both 3- and 24-month-old rats; (ii) low-affinity Ca^{2+}-ATPase in 24-month-old; (iii) Ca^{2+},Mg^{2+}-ATPase in 3-month-old ones. The age-related Mg^{2+}-ATPase modifications were different according to the different subcellular fractions tested (Fig. 4).

In synaptosomes from cerebral cortex, the adaptation to chronic intermittent *severe* hypoxia was characterized (Fig. 3) by a decrease in the activities of: (i) Na^+,K^+-ATPase, Ca^{2+},Mg^{2+}-ATPase and high-affinity Ca^{2+}-ATPase in both 3- and 24-month-old rats, and (ii) low-affinity Ca^{2+}-ATPase only in 24-month-old ones. The effect on the Mg^{2+}-ATPase activity

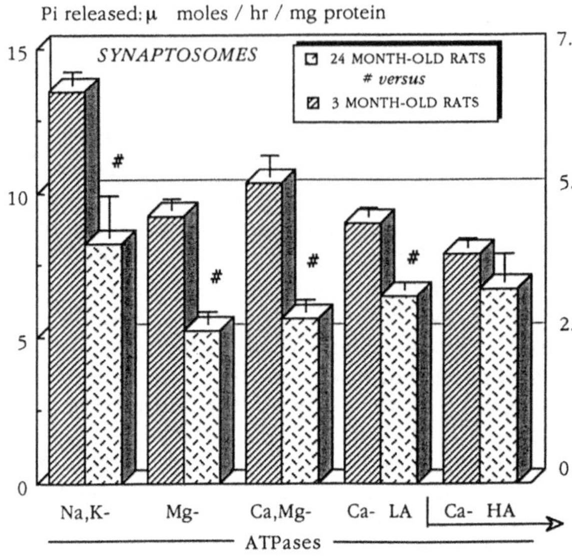

Fig. 1. Na^+,K^+-ATPase, Mg^{2+}-ATPase, Ca^{2+},Mg^{2+}-ATPase, low- [LA] and high-affinity [HA] Ca^{2+}-ATPase activities evaluated in synaptosomes from the cerebral cortex of 3- and 24-month-old normoxic rats. Each column represents the mean ± SEM; n = 5–6. # differs from 3-month-old rats: $p < 0.01$

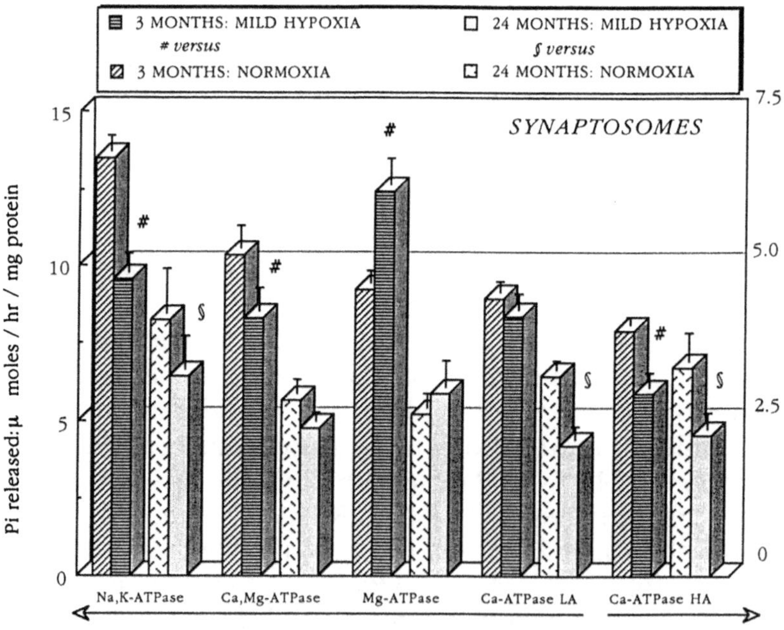

Fig. 2. Na^+,K^+-ATPase, Ca^{2+},Mg^{2+}-ATPase, Mg^{2+}-ATPase, low-affinity [LA] and high-affinity [HA] Ca^{2+}-ATPase activities evaluated in synaptosomes from the cerebral cortex of 3- and 24-month-old rats subjected to intermittent normobaric mild hypoxia = 12 hr daily exposure to $N_2:O_2$ (90:10) for four weeks. Each column represents the mean ± SEM; n = 5. # differs from 3-month-old normoxic rats: $p < 0.01$, § differs from 24-month-old normoxic rats: $p < 0.01$

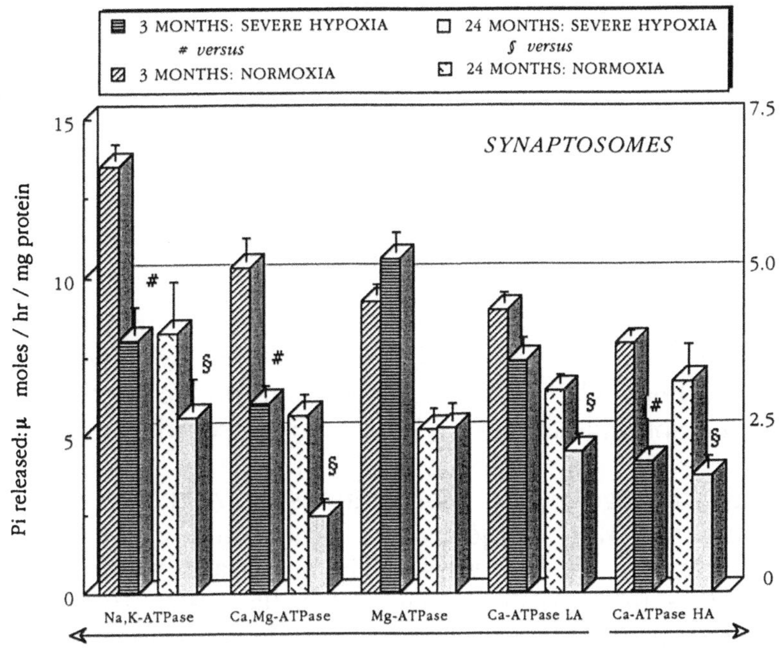

Fig. 3. Na^+,K^+-ATPase, Ca^{2+},Mg^{2+}-ATPase, Mg^{2+}-ATPase, low-affinity [LA] and high-affinity [HA] Ca^{2+}-ATPase activities evaluated in synaptosomes from the cerebral cortex of 3- and 24-month-old rats subjected to intermittent normobaric severe hypoxia = 12 hr daily exposure to $N_2:O_2$ (91.5:8.5) for four weeks. Each column represents the mean ± SEM; n = 5. # differs from 3-month-old normoxic rats: $p < 0.01$, § differs from 24-month-old normoxic rats: $p < 0.01$

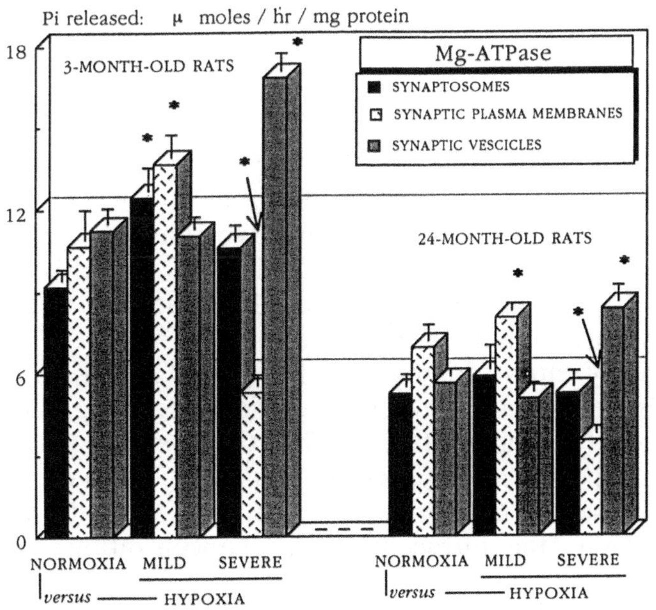

Fig. 4. Mg^{2+}-ATPase activity evaluated in synaptosomes, synaptosomal plasma membranes and synaptic vesicles from the cerebral cortex of 3- and 24-month-old rats in normoxia or subjected to either mild or severe intermittent normobaric hypoxia = 12 hr daily exposure to $N_2:O_2$ (90:10 or 91.5:8.5) for four weeks. Each column represents the mean ± SEM; n = 5. * differs from normoxic rats [first column in each group of three columns]: $p < 0.01$

was characterized by the decrease in the enzymatic form located in the synaptic plasma membranes, concomitant with an increase in the form located in the synaptic vesicles (Fig. 4).

Discussion

In the cerebral synaptic compartment the role of the individual ATPases is conditioned by their subcellular localizations. Moreover, with the exception of the high-affinity Ca^{2+}-ATPase, in rat cerebral cortex aging markedly decreases the specific activities of the energy-requiring non-mitochondrial ATPases.

To study the age-related modifications of ATPase plasticity, 2- and 23-month-old rats were subjected to either mild or severe intermittent normobaric hypoxia for four weeks. The cerebral synaptic compartment is highly sensitive to hypoxia (Benzi and Villa, 1976; Benzi et al., 1981; Folbergrovà et al., 1981; Gardiner et al., 1981; Gibson et al., 1975; Norberg and Siesjö, 1974a,b; Rafalowska et al., 1980) and the biochemical alterations are related to both severity and duration of the hypoxic insults (Smialek and Hamberger, 1970; Wood et al., 1968), to the age of the individuals (Benzi et al., 1984), and to the mitochondrial compartment involved (Marzatico et al., 1986).

As a daily normoxic period (12 hr) was scheduled after the hypoxic assault (12 hr), in this research the cerebral enzymatic adaptation is also related to repeated full reoxygenation rather than only to oxygen deprivation.

Age influences the adaptation to the chronic intermittent mild hypoxia which was characterized by an increase in the activity of Mg^{2+}-ATPase in 3-month-old rats, concomitant with a decrease in the activities of: (i) Na^+,K^+-ATPase and high-affinity Ca^{2+}-ATPase in both 3- and 24-month-old rats, and (ii) Ca^{2+},Mg^{2+}-ATPase in 3-month-old ones. The adaptation to chronic intermittent severe hypoxia was characterized by a decrease in the activities of: (i) Na^+,K^+-ATPase, Ca^{2+},Mg^{2+}-ATPase and high-affinity Ca^{2+}-ATPase in both 3- and 24-month-old rats, and (ii) low-affinity Ca^{2+}-ATPase only in 24-month-old ones.

The decrease by chronic intermittent hypoxia in Na^+,K^+-ATPase activity is also possibly related to the reoxygenation occurring during the post-hypoxic recovery (Schwartz et al., 1976; Goldberg et al., 1985; Palmer et al., 1985; Taylor et al., 1985; Mrsulja et al., 1985) and may be of primary importance because the membrane enzyme interferes with the Na^+ transmembrane gradient, the intracellular K^+ concentration and the related electrical events.

The decrease by chronic intermittent hypoxia in the activities of both Ca^{2+},Mg^{2+}-ATPase and high-affinity Ca^{2+}-ATPase interfaces with the regulation of the intracellular Ca^{2+} concentration that modulates the polarization/depolarization state of the presynaptic nerve ending membrane, together with many other mechanisms (Lin and Way, 1982b). The decrease in the high affinity Ca^{2+}-ATPase activity suggests an interference already at Ca^{2+} concentration comparable to the normal intracellular level of the ion (10^{-7} M).

The effect of severe hypoxia/normoxia cycles on the Mg^{2+}-ATPase activity is characterized by opposite changes in the enzymatic forms located either in synaptosomal plasma membranes [decrease in activity] or in synaptic vesicles [increase in activity]. Many cerebral alterations (including hypoxia) involve the release of excitatory amino acids acting at level of kainate-, quisqualate-, and NMDA-receptors, and leading to both Ca^{2+}-influx and cell bursting (Dragunow and Faull, 1988; Phillis and Wu, 1981; Rothman and Olney, 1986; Snyder, 1985). These events might be related to glutamate mobilization by the electrochemical proton gradient generated by the vesicular Mg^{2+}-ATPase (Christensen et al., 1990; Fykse and Fonnum, 1988, 1991; Maycox et al., 1988; Naito and Ueda, 1985).

In contrast, Mg^{2+}-ATPase located in the synaptic plasma membranes has externally oriented active sites and it is classified as an ecto-enzyme that carries out the ATP hydrolysis to generate, together with adenylate kinase and 5′-nucleotidase, adenosine (Nagy et al., 1983, 1986; Sorensen and Mahler, 1982; Trams, 1980; White, 1978) which induces neuronal hyperpolarization, blocks NMDA-receptors, inhibits both Ca^{2+} and Na^{+} uptake, and reduces the excitatory amino acid release. Both adenosine and adenosine-receptors are present in the cortical deeper layers, especially in the pyramidal cells in layer V (Braas et al., 1986; Goodman and Snyder, 1982; Lewis et al., 1981). There is a similar distribution of adenosine A_1-receptors and NMDA binding sites (Cotman et al., 1987), while the distribution of 5′-nucleotidase is rather different from that of adenosine A_1-receptors. Thus, it seems to be rather tenuous to relate the synaptic alteration by severe hypoxia/normoxia cycles basically to the increase in the vesicular Mg^{2+}-ATPase activity, concomitant with the decrease in the activity of the ecto-enzymatic form located in the synaptic plasma membranes.

In conclusion, the activities of the components in the synaptosomal non-mitochondrial ATPase system on their plasticity may be dependent on the age of the individuals, although the enzyme subcellular localization can play a fundamental role. Of course, it is difficult to unequivocally relate the moderate modifications in ATPase activities to the only promotion of the important age-related changes on the neural levels of ions, transmembrane gradients, neurotransmitter release, etc. There are, however, logical reasons to suggest that the changed ATPase activities and their modifications by chronic hypoxia/normoxia cycles support some age-related alterations in neural transmission.

Acknowledgements

The technical assistance of L. Maggi, G. Coscia and D. Spizzi is gratefully acknowledged. We thank G. Garlaschi, G. Corbellini and D. Minisci for their secretarial work.

References

Alberghina M, Giuffrida AM (1981) Effect of hypoxia on the incorporation of [1-^{14}C] palmitate on lipids of various brain regions. J Neurosci Res 6: 403–419

Alberghina M, Viola M, Giuffrida AM (1982) Changes in enzyme activities of glycerolipid metabolism of guinea pig cerebral hemispheres during experimental hypoxia. J Neurosci Res 7: 147–154

Albers RW (1967) Biochemical aspects of active transport. Ann Rev Biochem 36: 727–756

Benzi G, Villa RF (1976) Adenyl cyclase system and cerebral energy state. J Neurol Neurosurg Psychiatry 39: 77–83

Benzi G, Arrigoni E, Pastoris O, Raimondo S, Fulle D, Curti D, Villa RF (1981) Metabolic changes induced by acute hypoxia on the synaptosomes from dog brain. Eur Neurol 20: 235–244

Benzi G, Agnoli A, Giuffrida AM (1984) Influence of aging on cerebral derangement by acute severe hypoxia during hypovolemic hypotension. Neurobiol Aging 5: 513–519

Blitz AL, Fine RE, Toselli PA (1977) Evidence that coated vesicles isolated from brain are calcium-sequestering organelles resembling sarcoplasmic reticulum. J Cell Biol 75: 135–147

Braas KM, Newby AC, Wilson VS, Snyder SH (1986) Adenosine-containing neurons in the brain localized by immunocytochemistry. J Neurosci 6: 1952–1961

Chan PH, Schmidley JW, Fishman RA, Longar SM (1984) Brain injury, edema, and vascular permeability changes induced by oxygen-derived free radicals. Neurology 34: 315–320

Christensen H, Fykse EM, Fonnum F (1990) Uptake of glycine into synaptic vesicles isolated from rat spinal cord. J Neurochem 54: 1142–1147

Cotman CW, Monaghan DT, Ottersen OP, Storm-Mathisen J (1987) Anatomical organization of excitatory amino acid receptors and their pathways. Trends Neurosci 10: 273–280

Douglas WW (1968) Stimulus-secretion coupling: the concept and clues from chromaffin and other cells. Br J Pharmacol 34: 451–474

Dragunow M, Faull RLM (1988) Neuroprotective effects of adenosine. TIPS 9: 193–194

Duncan CJ (1976) Properties of Ca^{2+}-ATPase activity of mammalian synaptic membrane preparations. J Neurochem 27: 1277–1279

Dunlop DS (1978) Measuring protein synthesis and degradation rates in CNS tissue. In: Marks K, Rodnight R (eds) Research methods in neurochemistry, vol 4. Plenum Press, New York, pp 91–138

Folbergrová J, Nilsson B, Sakabe T, Siesjö BK (1981) The influence of hypoxia on the concentration of cyclic nucleotides in the rat brain. J Neurochem 36: 1670–1674

Fykse EM, Fonnum F (1988) Uptake of γ-aminobutyric acid by a synaptic vesicle fraction isolated from rat brain. J Neurochem 50: 1237–1242

Fykse EM, Fonnum F (1991) Transport of γ-aminobutyrate and L-glutamate into synaptic vesicles. Effect of different inhibitors on the vesicular uptake of neurotransmitters and on the Mg^{2+}-ATPase. Biochem J 276: 363–367

Garcia AG, Kirpekar SM (1973) Release of noradrenaline from cat spleen by sodium deprivation. Br J Pharmacol 47: 729–747

Gardiner M, Nilsson B, Rehncrona S, Siesjö BK (1981) The influence of hypoxia on the concentration of cyclic nucleotides in the rat brain. J Neurochem 36: 1500–1505

Gibson GE, Jope R, Blass JP (1975) Decreased synthesis of acetylcholine accompanying impaired oxidation of pyruvic acid in rat brain slices. Biochem J 14: 17–23

Goldberg WJ, Dorman RV, Dabrowiecki Z, Horrocks LA (1985) The effects of ischemia and CDP amines on Na^{+},K^{+}-ATPase and acetylcholinesterase activities in rat brain. Neurochem Pathol 3: 237–248

Goodman RR, Snyder SH (1982) Autoradiographic localization of adenosine receptors in rat brain using [^{3}H]cyclohexyladenosine. J Neurosci 2: 2130–2141

Jariashvili TYa (1975) Materials elucidating the specificity of the action of biogenic amines on the activity of transport ATPase. Bull Acad Sci Georgian SSR 79: 701–704

Kometiani ZP, Tsakadze LG, Jariashvili TYa (1984) Functional significance of the effects of neurotransmitters on the Na^+,K^+-ATPase system. J Neurochem 42: 1246–1250

Lee KS, Shin BC (1969) Studies on the active transport of calcium in human red cells. J Gen Physiol 54: 713–729

Lewis ME, Patel J, Moon Edley S, Marangos PJ (1981) Autoradiographic visualization of rat brain adenosine receptors using N_6-cyclohexyl [^{3}H]adenosine. Eur J Pharmacol 73: 109–110

Lin SC (1979) Ca^{2+}-ATPase of nerve ending particles. Possible role in regulating Ca^{2+} level in cytosol. Fed Proc 38: 376

Lin SC (1981) Ca^{2+}-activated or Mg^{2+}-activated ATP hydrolysis in synaptosomes from mouse brain. Fed Proc 40: 301

Lin SC, Way EL (1982a) A high affinity Ca^{2+}-ATPase in enriched nerve-ending plasma membranes. Brain Res 235: 387–392

Lin SC, Way EL (1982b) Calcium-activated ATPases in presynaptic nerve endings. J Neurochem 39: 1641–1651

Logan JG, O'Donovan DJ (1975) The effects of noradrenaline, 5-hydroxytryptamine and dopamine on synaptic membrane ATPases. J Physiol (Lond) 250: 47–49

Logua GSh, Kometiani ZP (1970) On the effect of acetylcholine on the transportable ATPase of rat brain synaptic fractions. Bull Acad Sci Georgian SSR 60: 709–712

Lowry OH, Rosebrough NJ, Farr AL, Randall RJ (1951) Protein measurement with the Folin phenol reagent. J Biol Chem 193: 265–275

Marzatico F, Curti D, Dagani F, Taglietti M, Benzi G (1986) Brain enzyme adaptation to mild normobaric intermittent hypoxia. J Neurosci Res 16: 419–428

Mayanil CSK, Kazmai SMI, Baquer NZ (1982) Na^+,K^+-ATPase and Mg^{2+}-ATPase activities in different regions of rat brain during alloxan diabetes. J Neurochem 39: 903–908

Maycox PR, Deckwerth T, Hell JW, Jahn R (1988) Glutamate uptake by brain synaptic vesicles. J Biol Chem 263: 15423–15428

McCord JM (1985) Oxygen-derived free radicals in postischemic tissue injury. N Engl J Med 312: 159–163

Michaelis EK, Michaelis ML, Chang HH, Kitos TE (1983) High affinity Ca^{2+}-stimulated Mg^{2+}-dependent ATPase in rat brain synaptosomes, synaptic membranes, and microsomes. J Biol Chem 258: 6101–6108

Mrsulja BB, Micic DV, Stefanovich V (1985) Propentofylline and postischemic brain edema: relation to Na^+-K^+-ATPase activity. Drug Dev Res 6: 339–344

Nagy À, Shuster TA, Rosenberg MD (1983) Adenosine triphosphatase activity at the external surface of chicken brain synaptosomes. J Neurochem 40: 226–234

Nagy AK, Shuster TA, Delgado-Escueta AV (1986) Ecto-ATPase of mammalian synaptosomes: identification and enzymatic characterization. J Neurochem 47: 976–986

Naito S, Ueda T (1985) Characterization of glutamate uptake into synaptic vesicles. J Neurochem 44: 99–109

Norberg K, Siesjö BK (1974a) Cerebral metabolism in hypoxic hypoxia. I. Pattern of activations of glycolysis: a re-evaluation. Brain Res 86: 31–44

Norberg K, Siesjö BK (1974b) Cerebral metabolism in hypoxic hypoxia. II. Citric acid cycle intermediates and associated amino acids. Brain Res 86: 45–54

Palmer GC, Palmer SJ, Christie-Pope BC, Callahan III AS, Taylor MD, Eddy LJ (1985) Classification of ischemic-induced damage to Na^+-K^+-ATPase in gerbil forebrain. Neuropharmacology 24: 509–516

Paton WDM, Vizi ES, Zar Aboo M (1971) The mechanism of acetylcholine release from parasympathetic nerves. J Physiol (Lond) 215: 819–848
Phillis JW, Wu PH (1981) The role of adenosine and its nucleotides in central synaptic transmission. Prog Neurobiol 16: 187–239
Rafalowska U, Erecinska M, Wilson D (1980) The effect of acute hypoxia on synaptosomes from rat brain. J Neurochem 34: 1160–1165
Raichle ME (1983) The pathophysiology of brain ischemia. Ann Neurol 13: 2–10
Rothman SM, Olney JW (1986) Glutamate and the pathophysiology of hypoxic ischemic brain damage. Ann Neurol 19: 105–111
Sawas AH, Gilbert JC (1981) Effects of adrenergic agonists and antagonists and of the catechol nucleus on the Na^+-K^+-ATPase and Mg^{2+}-ATPase activities of synaptosomes. Biochem Pharmacol 30: 1799–1803
Schatzmann JJ, Vincenzi FF (1969) Calcium movements across the membrane of human red cells. J Physiol (Lond) 201: 369–395
Schwartz A, Lindenmayer GE, Allen JC (1975) The sodium-potassium adenosine triphosphatase: pharmacological physiological and biochemical aspects. Pharm Rev 27: 3–134
Schwartz JP, Mrsulja BB, Mrsulja BJ, Passonneau JV, Klatzo I (1976) Alterations of cyclic nucleotide-related enzymes and ATPase during unilateral ischemia and recirculation in gerbil cerebral cortex. J Neurochem 27: 101–107
Serra I, Alberghina M, Viola M, Giuffrida AM (1981) Effect of hypoxia on nucleic acids and protein synthesis in different brain regions. Neurochem Res 6: 595–605
Shallom JM, Katyare SS (1985) Altered synaptosomal ATPase activity in rat brain following prolonged in vivo treatment with nicotine. Biochem Pharmacol 34: 3445–3449
Skou JC (1957) The influence of some cations on an adenosine triphosphatase from peripheral nerves. Biochim Biophys Acta 23: 394–401
Skou JC (1965) Enzymatic basis for active transport of Na^+ and K^+ across cell membrane. Physiol Rev 45: 596–617
Skou JC (1986) The sodium potassium pump. Scand J Clin Lab Invest 46 [Suppl] 180: 11–23
Smialek M, Hamberger A (1970) The effect of hypoxia and ischemia on cytochrome oxidase activity and protein synthesis in brain mitochondria. Brain Res 17: 369–371
Snyder SH (1985) Adenosine as a neuromodulator. Ann Rev Neurosci 8: 103–124
Sorensen RG, Mahler HR (1981) Calcium-stimulated adenosine triphosphatases in synaptic membranes. J Neurochem 37: 1407–1418
Sorensen RG, Mahler HR (1982) Localization of endogenous ATPases at the nerve terminal. J Bioenerg Biomembr 14: 527–547
Taylor MD, Mellert TK, Parmentier JL, Eddy LJ (1985) Pharmacological protection of reoxygenation damage to in vitro brain slice tissue. Brain Res 347: 268–273
Trams EG (1980) A proposal for the role of ecto-enzymes and adenylates in traumatic shock. J Theor Biol 87: 609–621
Villa RF, Turpeenoja L, Magrì G, Gorini A, Ragusa N, Giuffrida-Stella AM (1991) Effect of hypoxia on protein composition of synaptic membranes from cerebral cortex during aging. Neurochem Res 16: 827–832
Vizi ES (1972) Stimulation by inhibition of (Na^+,K^+,Mg^{2+})-activated ATPase of acetylcholine release in cortical slices from rat brain. J Physiol (Lond) 226: 95–117
Weishaar RE (1984) "Calcium antagonists and calcium agonists": is there a place in pharmacology for these two misnomers? Life Sci 35: 455–462
White BC, Wiegenstein JG, Winegar CD (1984) Brain ischemic anoxia. Mechanisms of injury. JAMA 25: 1586–1590
White TD (1978) Release of ATP from a synaptosomal preparation by elevated extracellular potassium and by veratridine. J Neurochem 30: 329–336
Wood JD, Watson WJ, Ducker AJ (1968) The effect of growth of hypoxia on brain gamma-aminobutyric acid levels. J Neurochem 15: 603–608

Wyllie MG, Gilbert JL (1979) An investigation into the role of synaptic vesicular Mg^{2+}-ATPase in neurotransmitter release, using benzhydryl piperazines. Arch Int Pharmacodyn Ther 241: 4–15

Authors' address: Dr. G. Benzi, Institute of Pharmacology, State University of Pavia, 11 piazza Botta, I-27100 Pavia, Italy.

J Neural Transm (1994) [Suppl] 44: 173–187

Impairment in memory function and neurodegenerative changes in the cholinergic basal forebrain system induced by chronic intake of ethanol

T. Arendt

Department of Neurochemistry, Paul Flechsig Institute of Brain Research, University of Leipzig, Leipzig, Federal Republic of Germany

Summary. Chronic intake of ethanol both in human and rat results in a substantial impairment in memory function associated with a reduction in the number of cholinergic neurons in the basal forebrain which give rise to the cholinergic afferentation of the cortical mantle. Degenerative changes in the basal forebrain are paralleled by the concomitant reduction of presynaptic cholinergic markers (synthesis, content and release of acetylcholine) in the neocortex and hippocampus. Cognitive dysfunction in rat induced by ethanol treatment can be ameliorated by the pharmacological manipulation of central cholinergic neurotransmission by physostigmine, arecoline or nicotine. Furthermore, fetal brain transplants rich in cholinergic neurons are able to restore cognitive function which argues in favour of a cholinergic aspect of alcohol-induced behavioural dysfunction. The observation, that implantation of purified astrocytes results in a similar restoration of learning and memory abilities associated with a recovery of cholinergic function, further indicates that behavioural sequelae of alcohol intake can largely be ameliorated by a trophic stimulation directed towards the cholinergic basal forebrain system.

Regarding the cholinergic basal forebrain system as a component of the ascending reticular activation system, the involvement of the cholinergic afferentation of the cortical mantle in the mediation of memory processes and their dysfunction under neurodegenerative conditions can be explained on the basis of the "Hippocampal Memory Indexing Theory" of Teyler and DiScenna. The hypothesis is formulated in the present paper that mental dysfunction observed after chronic ethanol consumption can largely be attributed to a degeneration of the cholinergic pathway of the ascending activation system resulting in an impairment of cortical activation, clinically appearing as the "syndrome of partial cholinergic deafferentation of the cortical mantle".

Introduction

The development of successful strategies for the clinical management of dementing disorders requires a detailed knowledge on the pathogenetic events under these conditions. The direct link between molecular pathology and neuropsychology, however, remains difficult to study. The identification of both common and specific steps in the pathogenetic chains of different dementing disorders, and the investigation of the relationship between these events and cognitive disturbances might, therefore, be one approach to define the neurobiological substrate of the major symptomatology of these disorders. Major progress in developing a better insight into the mechanisms which are primarily responsible for the symptoms of Alzheimer's disease and related dementing disorders has been considerably restricted by the lack of a valid animal model.

A potentially unifying framework for modelling the critical step in the pathomechanism of Alzheimer's disease might be provided by the evidence of a degeneration in the cholinergic basal forebrain system as a perhaps universal mechanism underlaying the development of impaired cognitive function in a certain subgroup of dementia (Bigl et al., 1990). Experiments on rodent and non-human primate brain using electrolytic or excitotoxic lesion techniques have provided much insight into the significance of this cholinergic basal forebrain projection system in learning and memory (Wenk, 1989). Brain damage induced by placing acute lesions to certain brain structures, however, is different with respect to its temporal-spatial spreading of degeneration, compensation and reparation as compared to the process of chronic neuronal degeneration such as observed in Alzheimer's disease and related conditions. The availability of an animal model which more closely mimics the process of chronic degeneration would, therefore, be of some value for the investigation of the functional significance of cell loss in the cholinergic basal forebrain as well as for the development of strategies for the prevention and amelioration of symptoms associated with this degenerative process. The hypothesis presented in this paper suggests that mental dysfunction observed after both acute and chronic ethanol consumption can largely be attributed to an impairment of the cholinergic pathway of the ascending activation system, clinically appearing either acutely as an "anticholinergic syndrome" or chronically as a "syndrome of partial cholinergic deafferentation of the cortical mantle". Longterm intake of ethanol in rodent might, therefore, be a useful model to mimic this process of degeneration of the cortical cholinergic afferentation, a constant feature of Alzheimer's disease and a number of related neuropsychiatric disorders.

Table 1. Pattern of neuropsychological dysfunction in mental disorders and after "blockade" of cholinergic synapses in young healthy volunteers by scopolamine

	Scopolamine	Korsakoff's syndrome	Alzheimer's disease	Huntington's chorea
Memory				
Primary memory	—	—	↓	—
Secondary anterograde memory				
Explicit component				
Semantic memory	↓	↓	↓	—
Spatial memory	↓	↓	↓	—
Implicit component				
Procedural and skill learning	—	—	—	↓
Priming	—	—	—	↓
Secondary retrograde memory	—	—	—/↓	↓
Conceptual ability and behavioural regulation	↓	↓	↓	—
Attention	↓	—	↓	↓

↓ impaired; — no impairment. Data adapted from Brown and Marsden (1988) and Kopelman and Corn (1988)

The pharmacological blockade of cholinergic neurotransmission by ethanol: "The central anticholinergic syndrome"

Neuropsychology

Comparative neuropsychological studies indicate that early Alzheimer's disease and postalcoholic Korsakoff syndrome show a similar pattern of memory impairment which, in turn, could be differentiated from the memory impairment in Huntington's disease (Butters, 1985; Butters et al., 1983; Moss et al., 1986) (Table 1). These similarities between Alzheimer's disease and Korsakoff syndrome might be restricted to the initial memory dysfunction of both disorders. Beyond the very early stages, patients with Alzheimer's disease develop an additional disruption of language and constructional abilities which contrast with the amnesia in Korsakoff syndrome and might be related to a more widespread cerebral damage in Alzheimer's disease. However, after a long drinking history, a proportion of severe alcoholics proceed to intellectual impairment, exceeding those with Korsakoff syndrome, developing the clinical picture of presenile dementia (Lishman, 1986; King, 1986). Moreover, there are indications that even many patients with Korsakoff syndrome may have cognitive dysfunctions which extent beyond those of memory deficits, and should be regarded as demented (Lishman, 1986). An unknown number of alcohol abusers suffering from alcoholic dementia might be misdiagnosed as having Alzheimer's disease (King, 1986; Dufour, 1993). The difficulty in establishing alcoholic dementia as a distinct disorder and in distinguishing it from Alzheimer's disease might be explained by a crucial neurobiological substrate underlaying both alcohol-induced brain damage and Alzheimer's disease.

The anterograde amnesic deficit demonstrated for Alzheimer's disease and alcoholic Korsakoff syndrome most likely reflects primarily an acquisition or learning deficit, and in this respect resembles the amnesia induced by the application of anticholinergic drugs to healthy subjects (Kopelman, 1985). In particular, on tests of secondary memory, "cholinergic blockade" by scopolamine produces a pattern similar to that seen in both disorders, namely a pronounced impairment in learning verbal and visuo-spatial material, a "normal" forgetting rate once learning had been accomplished, and relative preservation of the response to priming and of skill learning (procedural memory) (Table 1).

Clinically, this central pharmacological blockade of cholinergic neurotransmission appears as a syndrome, known as the "central anticholinergic syndrome" (Schneck and Rupreht, 1989). Behaviourally, the syndrome is characterized by agitation, including seizures, restlessness, hallucinations, disorientation or signs of depression such as stupor, coma and respiratory depression. From anaesthesia and intensive care where observations on the amnesic effects of drugs with anticholinergic action, frequently used as a premedication for surgical operation, date back to the beginning of the century (Gauss, 1906; Thomson and Cotterill, 1906; Pandit and Dundee, 1970), it is known that this central anticholinergic syndrome can be prevented by administration of physostigmine (Schneck and Rupreht, 1989). As, similarly, physostigmine had been proven to be effective in the clinical management of several alcohol withdrawal states, especially alcohol delirium (Rupreht et al., 1989), one might assume an anticholinergic syndrome as being part of the clinical picture of acute alcohol intoxication.

Neurochemistry

Consumption of ethanol affects a variety of neurotransmitter functions in the brain (Hunt and Majchrowicz, 1979; Shanley and Wilce, 1993). The most consistent data, however, have been obtained on the action of ethanol on cholinergic neurotransmission (Massarelli, 1979) (Table 2). The release of acetylcholine had been reported to be most sensitive to the inhibitory action of ethanol as compared to the release of serotonin, dopamine, noradrenaline, glutamate, and GABA, which might probably suggest a direct effect of ethanol on cholinergic neurons (Carmichael and Israel, 1975). The effects of ethanol on acetylcholine release are paralleled by its behavioural effects as well as by an ethanol-induced EEG synchrony being regarded as an indicator of the CNS depression (Sauerland and Harper, 1970; Erickson and Graham, 1973). Furthermore, the observation that the mesencephalic reticular formation appears to be even more sensitive with respect to the depressant actions of ethanol on acetylcholine release than

Table 2. Effects of chronic ethanol administration on central cholinergic systems

		Species	Brain region
Acetylcholine concentration	↓	Rat/Mice	Neocortex, hippocampus, basal forebrain, caudate nucleus, brain stem
Acetylcholine release (K^+-evoked)	↓	Rat (in vitro)	Neocortex, hippocampus, basal forebrain
Acetyl-CoA availability (Acetyl-CoA content/ pyruvate dehydrogenase activity)	↓	Rat/Mice	Neocortex, hippocampus, basal forebrain
Choline uptake	↓	Rat	Neocortex, hippocampus
Acetylcholine synthesis/ choline acetyltransferase activity	↓	Rat/Mice	Neocortex, hippocampus, basal forebrain, caudate nucleus, hypothalamus
Acetylcholinesterase activity	↓	Rat/Mice	Neocortex, hippocampus, basal forebrain, caudate nucleus, brain stem
Muscarinic receptors	↑	Rat/Mice	Neocortex, hippocampus, forebrain, caudate nucleus

↓ decreased; ↑ increased. For references see Arendt (1993)

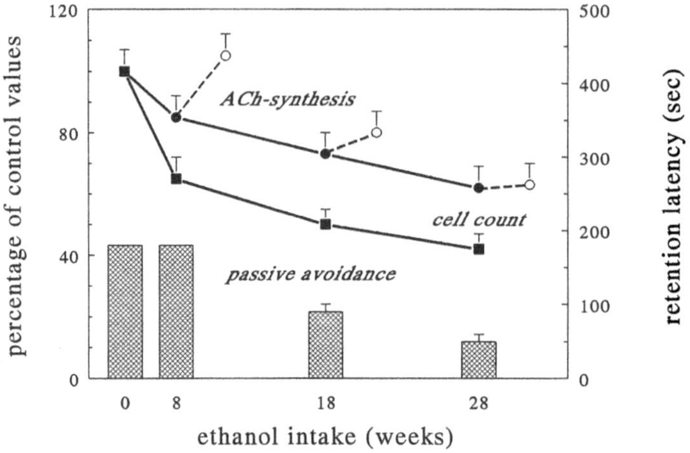

Fig. 1. Degeneration of the cholinergic afferentation of the cerebral cortex and memory impairment as a function of the duration of ethanol intake in rat. Animals treated for 18 or 28 weeks show an impaired performance in the retention of a passive avoidance task (columns, right scale). The synthesis of acetylcholine in the frontal cortex (●) and the number of cholinergic neurons in the basal forebrain (■) determined after completion of the task progressively declines over the duration of ethanol intake (left scale). Following an ethanol-free period of 4 weeks, changes in ACh-synthesis are fully, partially, or not reversible, depending on the duration of previous ethanol intake (○). Data are mean (+SD) of n = 20 animals, expressed as a percentage of control values of non-treated animals

the cerebral cortex (Erickson and Graham, 1973) might suggest that the central depressant action of ethanol is largely mediated by a depression of the cholinergic pathway of the ascending reticular activating system.

A progressive decrease of the synthesis, content and release of ACh as a function of the duration of ethanol intake has been demonstrated in a long-term follow up study in rat (Arendt et al., 1989a) (Fig. 1). Ethanol-induced changes appear to be fully reversible when the ethanol is eliminated from the blood early during the course of treatment. After continuation of the ethanol application for a longer period of time, however, these changes are only partially reversible or even become irreversible, independently of the presence of ethanol in the blood. Thus, there appears to be a critical "point of no return" where the "pharmacological blockade" by ethanol progresses into a "neurotoxic lesion".

The chronic neurotoxic lesion induced by chronic consumption of ethanol: "The syndrome of partial cholinergic deafferentation of the cortical mantle"

Postmortem human studies

The chronic intake of ethanol in human (Arendt et al., 1983, 1989b) results in a degeneration of the cholinergic basal forebrain projection system as

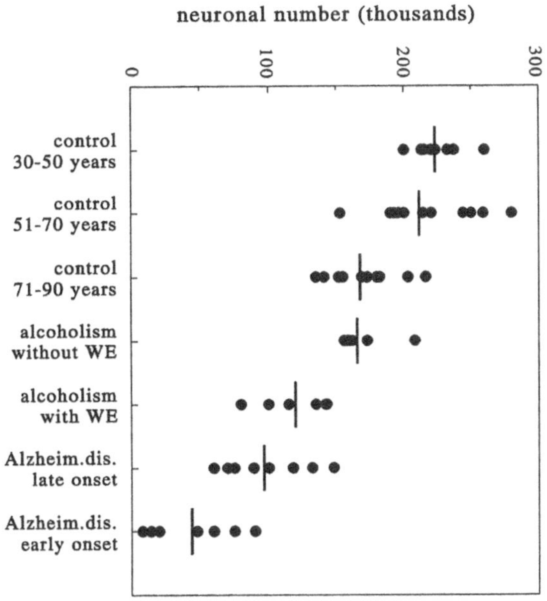

Fig. 2. Changes in the total number of neurons in the cholinergic basal forebrain nuclei in normal aging, chronic alcoholism and Alzheimer's disease. Data indicate the neuron number in one hemisphere of individual cases and the mean of each group. *WE* Wernicke's encephalopathy

revealed by a decrease in neuronal number in the cholinergic basal forebrain nuclei (Fig. 2) and a depletion of cortical ChAT activity accompanied by an impairment in memory function.

Loss of cholinergic basal forebrain neurons in chronic alcoholism is most pronounced in cases with a Wernicke encephalopathy, where it reaches an extent of about 40% to 50%, whereas cases without Wernicke encephalopathy are less severely affected being only marginally different from normal elderly (Fig. 2). In comparison, normal aging is related to a loss of about 20% of neurons without being accompanied by mental impairment, while dementia of the Alzheimer type is characterized by a loss of more than 50% of cells in cases with late onset which even reaches about 80% in cases with early onset. Thus, a continuum of degenerative changes is formed induced by aging, chronic alcohol and Alzheimer's disease.

Animal studies

The degeneration of the cholinergic afferentation of the cerebral cortex induced by chronic ethanol consumption in human can be mimicked in rodent (Arendt et al., 1988a,b, 1989a; Beracochea et al., 1987). Treatment of rat with a 20% ethanol solution applied in the drinking water results in a time-dependent degeneration of the basal forebrain cholinergic projection system as revealed by a progressive cell loss in the cholinergic basal forebrain paralleled by deficiencies in synthesis, content and release of ACh in target structures such as the cerebral cortex (Arendt et al., 1987, 1989a) (see Fig. 1 and Table 2). The degeneration is accompanied by a progressive impairment in memory function when the cell loss in the basal forebrain exceeds a certain threshold; i.e. when irreversible changes come into play which in the present paradigm occurs after about 18 weeks of intake (Fig. 1).

The impairment of cortical activation after ethanol-induced degeneration of the cholinergic basal forebrain

Damage inflicted to the cholinergic neurons of the basal forebrain nuclei is paralleled by deficits in a variety of functional aspects, including cortical arousal, selective attention, sleep-wake cycle and learning and memory (Bartus et al., 1985; Collerton, 1986). The behavioural deficits described after such lesions appear to affect virtually all aspects of cognitive function that the cerebral cortex is assumed to subserve. Thus, it can be hypothesized that the physiological mechanisms of the behavioural impairment is due not so much to the death of a circumscribed set of basal forebrain neurons per se, but must be sought in the malfunction of their cortical target areas. In other terms, the behavioural impairment might be related to a functional

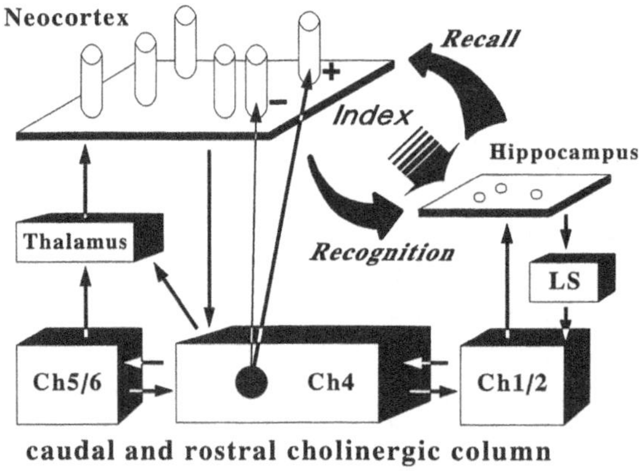

Fig. 3. Hypothetical involvement of the cholinergic afferentation of the cortical mantle in the processing and storage of information, based on the Hippocampal Memory Indexing Theory of Teyler and DiScenna (1986). Groups of cholinergic projection neurons of the basal forebrain and upper brain stem are reciprocally connected with the thalamus, neocortex, and hippocampus as well as with one another and can be viewed as components of the ascending reticular activation system of Moruzzi and Magoun (1949). Abbreviations Ch refer to the nomenclature according to Mesulam (1983a,b): *Ch1* medial septal nucleus; *Ch2* vertical limb of the diagonal band nucleus; *Ch4* nucleus basalis of Meynert; *Ch5* nucleus pedunculopontinus tegmenti; *Ch6* nucleus laterodorsalis tegmenti; *LS* lateral septum

impairment of the neocortex and hippocampus due to an erroneous gaiting or arousing mechanism normally provided by the cholinergic basal forebrain which can be viewed as the anatomical substrate of the ascending activation system (Moruzzi and Magoun, 1949; Steriade and Buzsaki, 1990).

Sequelae of a partial cholinergic deafferentation of the cortical mantle with respect to the impairment in learning and memory function can be explained on the basis of the "Hippocampal Memory Indexing Theory" of Teyler and DiScenna (1986) (Fig. 3). According to this theory, incoming environmental information is relayed by thalamocortical pathways and engages the activities of numerous cortical modules distributed in space and time over the neocortical surface. This spatio-temporal pattern of active cortical modules is subsequently stored as an index in the hippocampus by means of long-term potentiation through neocortico-limbic pathways. According to this theory, only the location and temporal sequencing of activated cortical modules are encoded and not the neuronal transformation of the experiential event itself. Reactivation of neocortical modules in the appropriate spatio-temporal sequence will stimulate the original experience (Recognition Memory). If the hippocampal activation exceeds a certain threshold level, the hippocampal index will be activated and will in turn, reactivate the pattern of cortical modules which match the original event (Recall Memory). As a prerequisite for this processing and encoding of information, a subtle mechanism of tuning the thalamic, neocortical and hippocampal activity corresponding to the demands of sensory inputs is needed which is met by the anatomical and electrophysiological properties of the cholinergic projection system of the basal forebrain and upper brainstem.

The complex of symptoms which after cortical cholinergic deafferentation results from the dysfunction of cortical information processing might be modified by the intrinsic properties and connections of those particular cortical areas which are affected by the deafferentation. Thus, the partial cholinergic deafferentation of the cortical mantle leads to a dysfunction of cognitive abilities which is not only quantitatively but also qualitatively different depending on the particular site of involvement during the course of the disease. As a complex of symptoms with an underlaying common neurobiological substrate, this neuropsychological impairment which forms a constant component of the clinical picture of a multitude of dementing disorders, such as postalcoholic Korsakoff syndrome, Alzheimer's disease, Parkinsons's disease, Down's syndrome, progressive supranuclear palsy and others, can be described as the "syndrome of partial cholinergic deafferentation of the cortical mantle".

Defining the "syndrome of partial cholinergic deafferentation of the cortical mantle" as an essential part of mental dysfunction occurring subsequently to chronic alcoholism may help to understand the neurobiological substrate of the major symptomatology and might, therefore, be of heuristic value for both the development of strategies for the management of this particular disorder and the conceptualization of the neuronal organisation of memory processes under normal and pathological conditions.

Strategies for the management of the behavioural impairment induced by chronic ethanol consumption

Pharmacological intervention in cholinergic neurotransmission after chronic ethanol

The ethanol-induced impairment in memory function can be ameliorated by pharmacological manipulation of cholinergic neurotransmission (Arendt et al., 1990; Beracochea et al., 1986; Hodges et al., 1991; Martin and Nimmerrichter, 1993). The inhibition of acetylcholine degradation by physostigmine and the application of arecoline or nicotine as direct muscarinic and nicotinic receptor agonists are equally effective in partially restoring ethanol-induced memory deficiency while application of neostigmine appears to be behaviourally ineffective (Fig. 4). These results are in agreement with other studies reporting effects of physostigmine on behavioural impairment in rodent and human induced by ethanol consumption (Beracochea et al., 1986; Erickson and Chai, 1976; Erickson and Burnam, 1971; Daunderer, 1978; Laurent et al., 1981).

Transplantation of neurons and glial cells

Intracortical grafting of fetal brain tissue derived from the primordial basal forebrain has been shown to ameliorate cognitive deficits caused by aging (Gage et al., 1984), excitotoxic lesions of the basal forebrain (Dunnett et al., 1985) or chronic ethanol treatment (Arendt et al., 1988a). As such grafts necessarily contain a multiplicity of cellular and non-cellular con-

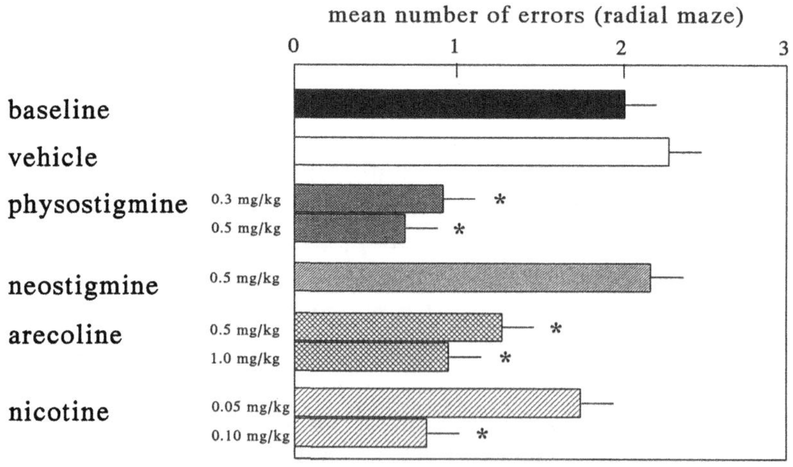

Fig. 4. Effects of pharmacological manipulation of cholinergic neurotransmission on the asymptotic performance in the radial arm maze of rats treated with ethanol (20% v/v) for 28 weeks. Drugs were applied (i.p.) 15 minutes before testing. Data are mean (+SD) of 12 animals. * Values are significantly different from baseline, $p < 0.01$ (Student's t-test)

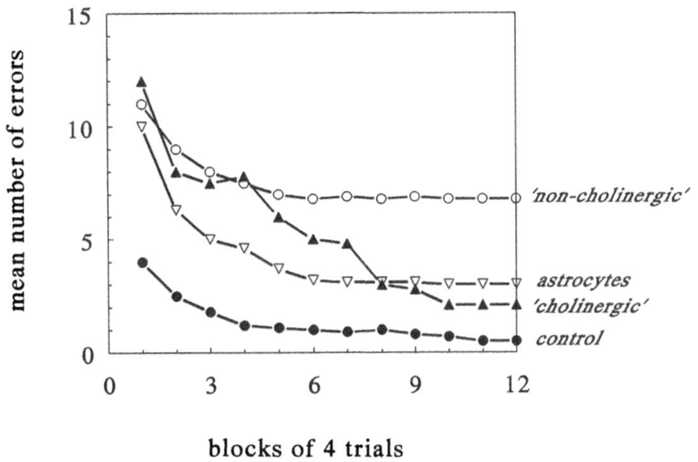

Fig. 5. Effects of grafts of different cellular origin on the recovery of ethanol-induced memory-impairment. Mean number of errors over four blocks of trials in the 8-arm radial maze. Animals received grafts of fetal brain tissue derived from the basal forebrain ("cholinergic" graft), from the hippocampus ("non-cholinergic" graft) or purified astrocytes at 4 sites in the neocortex and hippocampus (group size n = 8 animals)

stituents, it might be difficult to determine the critical factors underlying restoration of function. One approach to solve this problem is to compare grafts of different origin and/or type, of which some contain but others lack potentially relevant factors. In a number of experiments (Arendt et al., 1988a, 1989a; Brückner and Arendt, 1991, 1992) we have applied this approach to investigate the relationship between cholinergic deafferentation of the cortical mantle after chronic ethanol intake and cognitive dysfunction.

Tissue taken at embryonic day 16 either from the primordial basal forebrain or from the hippocampus was placed to the cerebral cortex, to the hippocampus or to both locations of rats previously been treated with ethanol. As tissue taken from the basal forebrain is rich in cholinergic neurons while those from the hippocampus is largely devoid of cholinergic neurons, grafting of the former is designated in the following as "cholinergic" transplants and grafting of the latter as "non-cholinergic" transplant. Animals which had received "cholinergic" transplants improved their performance in the radial arm maze while those receiving "non-cholinergic" transplants remained impaired (Fig. 5). Improvement was significant for both reference and working memory.

Improvement in memory function induced by "cholinergic" transplants derived from the basal forebrain could probably be attributed to the supply of acetylcholine by the graft or to some other factor associated with cholinergic neurons. Behavioural improvement after transplantation was quantitatively related to the increase in the acetylcholine content, while no transplant related effects were observed for the diminished content of noradrenaline, dopamine or serotonin (Table 3).

After grafting "cholinergic" transplants to the cerebral cortex, a recovery in synthesis and release of ACh, previously reduced by chronic ethanol

Table 3. Correlations between the content of various neurotransmitters in the neocortex and the pre-/posttreatment differences in maze performance

	Acetylcholine	Noradrenaline	Serotonin	Dopamine
"cholinergic" transplants	0.96*	0.09	0.11	0.21
"non-cholinergic" transplants	0.13	0.08	0.21	0.16

The rank correlation coefficient was calculated from the difference between asymptotic performance of 5 animals, 13 weeks after having received either cholinergic or non-cholinergic transplants in the neocortex and their asymptotic maze performance prior to transplantation. * $p < 0.01$ (Mann-Whitney u test)

treatment, has also been detected in the basal forebrain (Arendt et al., 1989a) (Fig. 6). This observation probably indicates that graft-induced behavioural recovery might at least partially be due to trophic mechanisms rather than to the establishment of synapses between graft and host.

Trophic factors which might be able to induce a behavioural recovery appear to be largely derived from reactive astrocytes. Therefore, we have investigated the effects of intracortical and intrahippocampal grafting of purified astrocytes on the impaired memory function induced by chronic intake of ethanol (Brückner and Arendt, 1992). The results demonstrate that grafting of purified astrocytes is equally effective in improving the performance in the eight-arm radial maze as is the transplantation of embryonic tissue from the primordial basal forebrain (Fig. 5). In comparison, the rate of behavioural recovery is even more accelerated after grafting of astrocytes. The time course of the improvement in memory function which can be detected as early as two weeks after operation (see

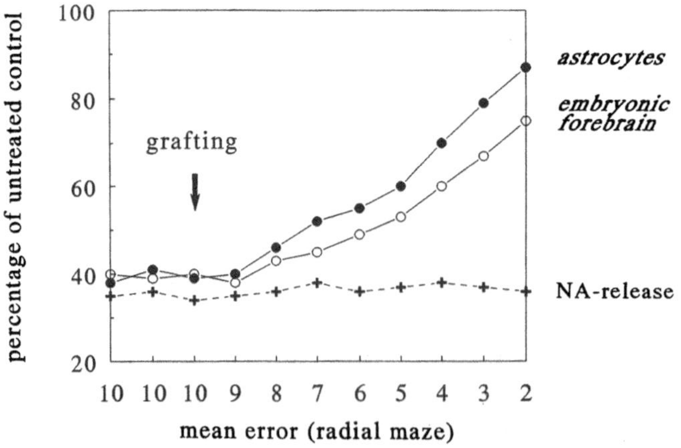

Fig. 6. Effects of grafts of embryonic basal forebrain or of purified astrocytes placed to the neocortex and hippocampus on the K^+-evoked release of acetylcholine. The release of noradrenaline (*NA*; K^+-evoked) is indicated for comparison. Each point represents the mean value of a group of 5 animals matched for performance in the radial maze at different intervals after operation

Fig. 5) as well as the increase of synthesis and release of acetylcholine in the basal forebrain (see Fig. 6), make it most likely that graft-induced behavioural recovery in the present paradigm is due to a neurotrophic influence of astrocytes mediated via the cholinergic basal forebrain system.

On the basis of the present data, it is suggested that brain damage in rodents induced by ethanol may serve as a model to investigate the brain behaviour relationship in a group of dementing disorders which are all associated with a partial cholinergic deafferentation of the cortical mantle supposed to be crucial for the development of cognitive dysfunction.

References

Arendt Th (1993) The cholinergic deafferentation of the cerebral cortex induced by chronic consumption of alcohol: reversal by cholinergic drugs and transplantation. In: Hunt WA, Nixon SJ (eds) Alcohol induced brain damage. NIH Publication No 93-3549. US Department of Health and Human Services, Rockville, pp 431–460

Arendt Th, Allen Y, Marchbanks RM, Schugens MM, Sinden J, Lantos PL, Gray JA (1989a) Cholinergic system and memory in the rat: effects of chronic ethanol, embryonic basal forebrain brain transplants and excitotoxic lesions of the cholinergic basal forebrain projection system. Neuroscience 33: 435–462

Arendt Th, Allen Y, Sinden J, Schugens MM, Marchbanks RM, Lantos PL, Gray JA (1988a) Cholinergic-rich brain transplants reverse alcohol-induced memory deficits. Nature 332: 448–450

Arendt Th, Bigl V, Arendt A, Tennstedt A (1983) Loss of neurons in the nucleus basalis of Meynert in Alzheimer's disease, paralysis agitans and Korsakoff's disease. Acta Neuropathol 61: 101–108

Arendt Th, Bigl V, Schugens MM (1989b) Ethanol as a neurotoxin — A model of the syndrome of cholinergic deafferentation of the cortical mantle. Brain Dysfunct 2: 169–180

Arendt Th, Hennig D, Gray JA, Marchbanks RM (1988b) Loss of neurons in the rat basal forebrain cholinergic projection system after prolonged intake of ethanol. Brain Res Bull 21: 563–570

Arendt Th, Marchbanks RM, Gray JA (1987) Effects of prolonged ethanol consumption on cholinergic function in the basal forebrain and memory. Biochem Soc Trans 15: 499–500

Arendt Th, Schugens MM, Bigl V (1990) The cholinergic system and memory: amelioration of ethanol induced deficiency by physostigmine in rat. Acta Neurobiol Exp 50: 251–262

Bartus RT, Dean RL, Pontecorvo MJ, Flicker CH (1985) The cholinergic hypothesis: a historical overview, current position and future directions. Ann NY Acad Sci 444: 332–358

Beracochea D, Durkin TP, Jaffard R (1986) On the involvement of the central cholinergic system in memory deficits induced by long term ethanol consumption in mice. Pharmacol Biochem Behav 24: 519–524

Beracochea D, Lescaudron L, Verna A, Jaffard R (1987) Neuroanatomical effects of chronic ethanol consumption on dorsomedial and anterior thalamic nuclei and on substantia innominata in mice. Neurosci Lett 73: 81–84

Bigl V, Arendt Th, Biesold D (1990) The nucleus basalis of Meynert during ageing and in dementing neuropsychiatric disorders. In: Steriade M, Biesold D (eds) Brain cholinergic systems. Oxford University Press, Oxford, pp 364–386

Brown RG, Marsden CD (1988) "Subcortical dementia": the neuropsychological evidence. Neuroscience 25: 363–387

Brückner MK, Arendt Th (1991) Astroglia: expression of amyloid β-protein and source of trophic activity. In: Elsner N, Penzlin H (eds) Synapse-transmission, modulation. Thieme, Stuttgart, p 536

Brückner MK, Arendt Th (1992) Intracortical grafts of purified astrocytes ameliorate memory deficits in rat induced by chronic treatment with ethanol. Neurosci Lett 14: 251–254

Butters N (1985) Alcoholic Korsakoff's syndrome: some unresolved issues concerning etiology and cognitive deficits. J Clin Exp Neuropsychol 7: 181–210

Butters N, Albert M, Sax DS, Miliotis P, Nagode J, Sterste A (1983) The effect of verbal mediators on the pictorial memory of brain-damaged patients. Neuropsychologia 21: 307–323

Carmichael FJ, Israel Y (1975) Effects of ethanol on neurotransmitter release by rat brain cortical slices. J Pharmacol Exp Ther 193: 824–834

Collerton D (1986) Cholinergic function and intellectual decline in Alzheimer's disease. Neuroscience 19: 1–28

Daunderer M (1978) Akute Alkohol-Intoxikation: Physostigmin als Antidot gegen Äthanol. Fortschr Med 96: 1311–1312

Dufour MC (1993) The epidemiology of alcohol-induced brain damage. In: Hunt WA, Nixon SJ (eds) Alcohol induced brain damage. NIH Publication No 93-3549. US Department of Health and Human Services, Rockville, pp 3–14

Dunnett SB, Toniolo G, Fine A, Ryan CN, Björklund A, Iversen SD (1985) Transplantation of embryonic ventral forebrain neurons to the neocortex of rats with lesions of nucleus basalis magnocellularis. II. Sensorimotor and learning impairments. Neuroscience 16: 787–797

Erickson CK, Burnam WL (1971) Cholinergic alteration of ethanol-induced sleep and death in mice. Agents Act 2: 8–13

Erickson CK, Chai KJ (1976) Cholinergic modification of ethanol-induced electroencephalographic synchrony in the rat. Neuropharmacology 15: 39–43

Erickson CK, Graham DT (1973) Alterations of cortical and reticular acetylcholine release by ethanol in vivo. J Pharmacol Exp Ther 185: 583–593

Gage FH, Björklund A, Stenevi U, Dunnett SB, Kelly PAT (1984) Intrahippocampal septal grafts ameliorate learning impairments in aged rats. Science 222: 533–536

Gauss CJ (1906) Geburten in künstlichem Dämmerschlaf. Arch Gynäkol 78: 579–631

Hodges H, Allen Y, Sinden J, Mitchell SN, Arendt Th, Lantos PL, Gray JA (1991) The effects of cholinergic drugs and cholinergic-rich foetal neuronal transplants on alcohol-induced deficits in radial maze performance in rats. Behav Brain Res 43: 7–28

Hunt WA, Majchrowicz E (1979) Alterations in neurotransmitter function after acute and chronic treatment with ethanol. In: Majchrowicz E, Noble EP (eds) Biochemistry and pharmacology of ethanol, vol 2. Plenum Press, New York, pp 167–185

King MB (1986) Alcohol abuse and dementia. Int J Geriatr Psychiatry 1: 31–36

Kopelman MD (1985) Rates of forgetting in Alzheimer-type dementia and Korsakoff's syndrome. Neuropsychologia 23: 623–638

Kopelman MD, Corn TH (1988) Cholinergic "blockade" as a model for cholinergic depletion. A comparison of the memory deficits with those of Alzheimer-type dementia and the alcoholic Korsakoff-syndrome. Brain 111: 1079–1110

Laurent B, Hibert-Kuntzler O, Chazot G, Michel D, Schott B (1981) Effets de la physostigmine sur les syndromes amnesiques. Rev Neurol 137: 649–660

Lishman WA (1986) Alcoholic dementia: a hypothesis. Lancet i: 1184–1185

Martin PR, Nimmerrichter AA (1993) Pharmacological treatment of alcohol-induced brain damage. In: Hunt WA, Nixon SJ (eds) Alcohol induced brain damage. NIH Publication No 93-3549. US Department of Health and Human Services, Rockville, pp 461–477

Massarelli R (1979) Effect of ethanol on the cholinergic system. In: Majchrowicz E, Noble EP (eds) Biochemistry and pharmacology of ethanol, vol 2. Plenum Press, New York, pp 223–240

Mesulam M-M, Mufson EJ, Levey AI, Wainer BH (1983a) Cholinergic innervation of the cortex by the basal forebrain: cytochemistry and cortical connections of the septal area, diagonal band nuclei, nucleus basalis (substantia innominata) and hypothalamus in the rhesus monkey. J Comp Neurol 214: 170–197

Mesulam M-M, Mufson EJ, Wainer BH, Levey AI (1983b) Central cholinergic pathways in the rat: an overview based on an alternative nomenclature (Ch1–Ch6). Neuroscience 10: 1185–1201

Moruzzi G, Magoun HW (1949) Brain stem reticular formation and activation of the EEG. Electroencephalogr Clin Neurophysiol 1: 455–473

Moss MB, Albert MS, Butters N, Payne M (1986) Differential patterns of memory loss among patients with Alzheimer's disease, Huntington's disease and alcoholic Korsakoff's syndrome. Arch Neurol 43: 239–246

Pandit SK, Dundee JW (1970) Pre-operative amnesia. The incidence following the intramuscular injection of commonly used premedicants. Anaesthesia 25: 493–499

Rupreht J, Schneck HJ, Dworacek B (1989) Physostigmine — recent pharmacologic data and their significance for practical use. Anaesthesiol Reanim 14: 235–241

Sauerland EK, Harper RM (1970) Effects of ethanol on EEG spectra of the intact brain and isolated forebrain. Exp Neurol 27: 490–496

Schneck HJ, Rupreht J (1989) Central anticholinergic syndrome (CAS) in anesthesia and intensive care. Acta Anaesthesiol Belg 40: 219–228

Shanley BC, Wilce PA (1993) Receptor changes associated with long-term ethanol consumption. In: Hunt WA, Nixon SJ (eds) Alcohol induced brain damage. NIH Publication No 93-3549. US Department of Health and Human Services, Rockville, pp 299–324

Steriade M, Buzsaki G (1990) Parallel activation of thalamic and cortical neurons by brainstem and basal forebrain cholinergic systems. In: Steriade M, Biesold D (eds) Brain cholinergic systems. Oxford University Press, Oxford, pp 3–62

Teyler TJ, DiScenna P (1986) The hippocampal memory indexing theory. Behav Neurosci 100: 147–154

Thompson HT, Cotterill D (1906) Note on the use of scopolamine-morphine combination as an anaesthetic adjunct. Edinburgh Med J (New Series) 3: 548–554

Wenk H (1989) The nucleus basalis magnocellularis Meynert (NbmM) complex — a central integrator of coded "limbic signals" linked to neocortical modular operation? A proposed (heuristic) model of function. J Hirnforsch 30: 127–151

Author's address: Dr. T. Arendt, Department of Neurochemistry, Paul Flechsig Institute of Brain Research, University of Leipzig, Jahnallee 59, D-04109 Leipzig, Federal Republic of Germany.

J Neural Transm (1994) [Suppl] 44: 189–194

Dysfunction of the brain cholinergic system during aging and after lesions of the nucleus basalis of Meynert

G. Pepeu and **I. Marconcini Pepeu**

Department of Preclinical and Clinical Pharmacology, University of Florence, Florence, Italy

Summary. In order to establish how closely the models mimic aging and Alzheimer's disease, a comparison was made, using the extensive literature available, between brain cholinergic dysfunction in aging animals and man, and between that in animals with lesions of the nucleus basalis of Meynert and in subjects affected by Alzheimer's disease. It is concluded that cholinergic dysfunction in the aging rat closely resembles that in aging man. A similarity can also be found between the cortical dysfunction induced by lesions of the nucleus basalis and that occurring in patients with Alzheimer's disease. However, cholinergic dysfunction only represents a limited aspect of the neurotransmitter deficits and neuropathological alterations of the disease.

Since the observation that senile dementias of Alzheimer and Alzheimer's type (AD) are associated with a large decrease in cortical choline acetyltransferase activity (ChAT) (Davies and Maloney, 1976), resulting from the degeneration of the cholinergic neurons of the nucleus basalis of Meynert (NBM) (Whitehouse et al., 1982), the cholinergic hypofunction of the aging brain has been the target of many preclinical and clinical investigations. Perhaps too many if one considers that several other neuronal systems are affected by aging and AD (Hardy et al., 1985). Nevertheless, it must be remembered that the cholinergic system is the main target of tacrine, the only treatment which has thus far been considered effective in AD patients (Farlow et al., 1992).

Investigations on brain cholinergic dysfunction have several aims: to define its nature, and its pathogenetic role in the cognitive impairment which characterizes, with different severity AD and aging, and the etiology of the loss of cholinergic neurons. Their final goal remains the identification of drugs for correcting the hypofunction and the evaluation of their neurochemical and clinical efficacy. Many of the investigations have been carried out on patients, as well as on biopsies and postmortem brain samples. However, much information has also been obtained from aging animals and efforts have been devoted to developing animal models which mimic senile dementias.

The usefulness and limitations of animal models in the study of the aging brain and AD have been repeatedly and extensively discussed (Olton, 1985; Dean and Bartus, 1985; Pepeu et al., 1986, 1990; Smith, 1988), leaving little to be added. This presentation will therefore reassess the resemblance between the cholinergic deficits of the "models" and the conditions which they are supposed to mimic, in the light of information derived from recent work from our laboratory and the literature.

Comparison between brain cholinergic dysfunctions in aging animals and aging men

The comparison is made difficult by the disproportion between the amount of information available in animals, largely rodents, and in man, and the inconsistencies due to differences in species and strains, and methodological approaches. These inconsistencies clearly appear in all recent reviews describing the effects of aging on brain cholinergic mechanisms (Decker, 1987; Sherman and Friedman, 1990; Pepeu et al., 1992). As a further example of variability even within the same strain, it can be mentioned that, while investigating cortical ACh release by the microdialysis technique in our laboratory, a 43–48% decrease was found in 20 to 24 month-old male Wistar rats from two sources and a 10% decrease only in Wistar rats from a third breeder (unpublished results). Nevertheless, Table 1 shows that, by critically selecting the data, usually from rats 24 to 28 months of age, several analogies can be demonstrated between the age-associated changes in the central cholinergic system of the rat and man. Table 1 only reveals two differences, namely that in the rat no consistent decrease in ChAT activity, and in man no decrease in the number of nicotinic receptors in normal aging have been found. However, it cannot be excluded that the decrease in ChAT activity reported in normal aged humans might be attributable to undetected cases of AD. Conversely, it is possible that the determination of ChAT activity under optimal substrate and co-factor conditions

Table 1. Comparison between brain cholinergic dysfunction in the aging rat and man

Changes	Rat	Ref.	Man	Ref.
Loss of cholinergic forebrain neurons	+	[1]	+	[2]
Decrease in cortical ChAT activity	−	[3]	+	[4]
Decrease in cortical ACh release	+	[5]	+	[6]
Decrease in cortical AChE activity	+	[7]	+	[4]
Decrease in muscarinic receptor density	+	[8]	+	[9]
Decrease in nicotinic receptor density	+	[8]	−	[9]

+ = yes; − = no
1 Smith et al. (1993); 2 deLacalle et al. (1991); 3 Decker (1987); 4 McGeer et al. (1971); 5 Pedata et al. (1983); 6 Feuerstein et al. (1992); 7 Meneguz et al. (1992); 8 Araujo et al. (1990); 9 Nordberg et al. (1990)

may mask the age-associated decrease, particularly since a reduction in ACh synthesis has been shown in aging rats (Vannucchi et al., 1987).

It must be added that cholinergic dysfunction in the rat is not confined to the forebrain since a significant impairment in ACh release from the hippocampus and striatum of aging rats has been reported (Wu et al., 1988; Pepeu et al., 1993). No information on ACh release in these regions is available in man but a decrease in ChAT and acetylcholinesterase activity in the hippocampus of normally aging men has been demonstrated (Perry et al., 1992).

Comparison between the cholinergic dysfunction induced by lesions of the NBM in animals and by AD in man

The degeneration of NBM cholinergic neurons and their projections to the cerebral cortex observed in AD (Whitehouse et al., 1982), and in other neurological diseases associated with dementia (Bigl et al., 1990), is considered an important pathogenetic element. For this reason, degeneration of NBM neurons has been induced in experimental animals, mostly rats, and in a few cases monkeys, in order to obtain a model of the cholinergic dysfunction occurring in AD as a tool for both understanding the role of the forebrain cholinergic system in cognitive processes and behavior, and investigating drugs potentially active in AD. Information on the procedures used for placing the lesions and the extent of the cholinergic dysfunction which can be achieved are found in Pepeu and Casamenti (1991). Both unilateral and bilateral lesions have been used, but a more severe, longer-lasting dysfunction can be obtained with bilateral than unilateral lesions (Casamenti et al., 1988). The morphological, neurochemical and behavioral changes following NBM lesions have been reviewed by Pepeu et al. (1990),

Table 2. Comparison between brain cholinergic dysfunction in animals with lesion of the nucleus basalis of Meynert and subjects affected by Alzheimer's disease

Changes	Lesions	AD
Loss of cholinergic forebrain neurons	Confined to NBM	Involving other forebrain nuclei [1]
Decrease in cortical ChAT activity	+ [2]	+ [3]
Decrease in hippocampal ChAT activity	− [4]	+ [5]
Decrease in cortical ACh synthesis	+ [2]	+ [6]
Decrease in cortical AChE activity	+ [2]	+ [7]
Decrease in muscarinic receptor density	+ [2]	+ [8]
Decrease in nicotinic receptor density	+ [9]	+ [8]

+ = yes; − = no
References: 1 Perry et al. (1992); 2 Pepeu and Casamenti (1990); 3 Davies and Maloney (1976); 4 Hepler et al. (1985); 5 Bartus et al. (1982); 6 Sims et al. (1980); 7 Geula and Mesulam (1989); 8 Nordberg et al. (1990); 9 Tilson et al. (1989)

and for a critical, and somewhat skeptical, evaluation of the possibility to correlate the cognitive impairment with cholinergic lesion see Fibiger (1991). In Table 2 a comparison is made between the cholinergic dysfunction in AD and in rats with NBM lesions. In AD patients cholinergic dysfunction involves the cortex and hipocampus while in lesioned animals only the cortical cholinergic system is affected, particularly in the frontal areas. An attempt to place a double lesion in NBM and septum has been made (Hepler et al., 1985) in order to reduce ChAT activity in both the cortex and hippocampus, but the low survival rate of the rats limits the use of this model. If we confine ourselves to cortical cholinergic dysfunction, a good similarity between the model and spontaneous pathology can be demonstrated. However, AD is characterized by the presence of amyloid plaques and neurofibrillary tangles and the dysfunction of several other neurotransmitter systems (Hardy et al., 1985). None of these modifications occur in animals with NBM lesions.

Conclusions

The comparison between the cholinergic dysfunction in the aging rat and the aging man reveals that the animal closely mimics man. The mechanisms involved in the aging process are presumably the same and result in comparable consequences for the cholinergic neurons. Therefore rats can be profitably used for studying the aging of the cholinergic system and attempting to either delay or correct the dysfunction by pharmacological treatments. In this regard, a remarkable recovery of cholinergic dysfunction has been obtained in the rat with the administration of NGF (Fischer et al., 1987), and phosphatidylserine (Casamenti et al., 1991). The recovery is associated with an improvement of the cognitive impairment, a finding demonstrating the relationship between cholinergic system and memory. To which extent similar results can be obtained in man still needs appropriate investigations.

Even if the cortical dysfunction induced by NBM lesions in the rats resembles that observed in AD patients, it only represents a small piece of the intricate puzzle formed by AD. Rats and even monkeys with NBM lesions cannot be considered a model of AD, but may be used for investigating the spontaneous and drug-stimulated recovery of the cholinergic system, its role in cognitive processes, and for detecting drugs which may correct memory impairment. For the latter purpose see Sarter et al. (1992).

Acknowledgements

Supported by a CNR grant (Target Project on Aging).

References

Araujo DM, Lapchack PA, Meaney MJ, Collier B, Quirion R (1990) Effects of aging on nicotinic and muscarinic autoreceptor function in the rat brain: relationship to presynaptic cholinergic markers and binding sites. J Neurosci 10: 3069–3078

Bartus RT, Dean RL, Beer B, Lippa AS (1982) The cholinergic hypothesis of geriatric memory dysfunction. Science 217: 408–417

Bigl V, Arendt T, Biesold D (1990) The nucleus basalis of Maynert during ageing and in dementing neuropsychiatric disorders. In: Steriade M, Biesold D (eds) Brain cholinergic system. Oxford University Press, Oxford, pp 364–386

Casamenti F, Di Patre PL, Bartolini L, Pepeu G (1988) Unilateral and bilateral nucleus basalis lesions: differences in neurochemical and behavioral recovery. Neuroscience 24: 209–215

Casamenti F, Scali C, Pepeu G (1991) Phosphatidylserine reverses the age-dependent decrease in cortical acetylcholine release: a microdialysis study. Eur J Pharmacol 194: 11–16

Davies P, Maloney AJR (1976) Selective loss of cholinergic neurons in Alzheimer's disease. Lancet ii: 1403

Dean RL, Bartus RT (1985) Animal models of geriatric cognitive dysfunction: evidence for an important cholinergic involvement. In: Traber J, Gispen WH (eds) Senile dementia of Alzheimer's type. Springer, Berlin Heidelberg New York Tokyo, pp 269–287

de Lacalle, Iraizos I, Gonzalos LM (1991) Differential changes in cell size and number in topographic subdivisions of human basal nucleus in normal aging. Neuroscience 43: 445–456

Decker MW (1987) The effect of aging on hippocampal and cortical projections of the forebrain cholinergic system. Brain Res Rev 12: 423–438

Farlow M, Gracon SI, Hershey LA, Lewis KW, Sadowski CH, Dolan-Ureno J (1992) A controlled trial of tacrine in Alzheimer's disease. JAMA 268: 2523–2529

Feuerstein J, Lehman J, Sauermann W, van Velthoven V, Jackisch R (1992) The autoinhibitory feedback control of acetylcholine release in human neocortex tissue. Brain Res 572: 64–71

Fibiger HC (1991) Cholinergic mechanisms in learning, memory and dementia: a review of recent evidence. TINS 14: 220–223

Fischer W, Wictorin K, Bjorklund A, William LR, Varon S, Gage FH (1987) Amelioration of cholinergic neuron atrophy and spatial memory impairment in aged rats by nerve growth factor. Nature 329: 65–68

Geula C, Mesulam MM (1989) Cortical cholinergic fibers in ageing and Alzheimer's disease. A morphological study. Neuroscience 33: 469–476

Hardy J, Adolfsson R, Alafuzoff I, Bucht G, Marcusson J, Nyberg P, Perdhal E, Wester P, Winblad B (1985) Transmitter deficits in Alzheimer's disease. Neurochem Int 7: 345–363

Hepler DJ, Olton DA, Wenk GL, Coyle JT (1985) Lesions of the nucleus basalis magnocellularis and medial septal area of rats produce qualitatively similar memory impairments. J Neurosci 5: 866–873

McGeer E, Fibiger HC, McGeer PL, Wickinson (1971) Aging and the brain enzymes. Exp Gerontol 6: 391–396

Meneguz A, Bisso GM, Michalek H (1992) Age-related changes in acetylcholinesterase and its molecular forms in various brain areas of rats. Neurochem Res 8: 785–790

Nordberg A, Alafuzoff I, Winblad B (1992) Nicotinic and muscarinic subtypes in the human brain: changes with aging and dementia. J Neurosci Res 31: 103–111

Olton DS (1985) Strategies for the development of animal models of human memory impairment. In: Olton DS, Gamzu E, Corkin S (eds) Memory dysfunctions: an integration of animal and human research from preclinical and clinical perspectives. Ann NY Acad Sci 444: 113–121

Pedata F, Slavikova J, Kotas A, Pepeu G (1983) Acetylcholine release from cortical slices during postnatal development and aging. Neurobiol Aging 6: 31–34
Pepeu G, Casamenti F (1991) Lesioning the nucleus basalis. In: Conn PM (ed) Methods in neuroscience, vol 7. Lesions and transplantation. Academic Press, San Diego, pp 139–150
Pepeu G, Casamenti F, Pedata F, Bartolini L (1986) Usefulness, constraint and limits of animal investigations in the study of aging brain and senile dementia. In: Courtois Y, Facheaux B, Forette B, Knook DL, Tréton JA (eds) Modern trends in aging research. John Libbey, London, pp 487–495
Pepeu G, Di Patre PL, Casamenti F (1991) Spontaneous and drug-stimulated recovery of cortical cholinergic function after lesion of the nucleus basalis. In: Steriade M, Biesold D (eds) Brain cholinergic system. Oxford University Press, Oxford, pp 357–363
Pepeu G, Marconcini Pepeu I, Casamenti F (1990) The validity of animal models in the search for drugs for the aging brain. Drug Res Del 7: 1–10
Pepeu G, Casamenti F, Marconcini Pepeu I, Scali C (1993) The brain cholinergic system in ageing mammals. J Reprod Fert [Suppl] 46: 155–162
Perry EK, Johnson M, Kerwin JM, Piggott MA, Court JA, Shaw PJ, Ince PG, Brown A, Perry RH (1992) Convergent cholinergic activities in aging and Alzheimer's disease. Neurobiol Aging 13: 393–400
Sarter M, Hagan J, Dudchenko P (1992) Behavioral screening for cognition enhancers: from indiscriminate to valid testing, part I. Psychopharmacology 107: 144–159
Sherman KA, Friedman E (1990) Pre- and postsynaptic cholinergic dysfunction in aged rodent brain regions: new findings and interpretative review. Int J Dev Neurosci 8: 689–708
Sims NR, Bowen DM, Smith CCT, Flack RHA, Davison AN, Snowden JS, Neary D (1980) Glucose metabolism and acetylcholine synthesis in relation to neuronal activity in Alzheimer's disease. Lancet i: 333–335
Smith G (1988) Animal models of Alzheimer's disease: experimental cholinergic denervation. Brain Res Rev 13: 103–118
Smith ML, Deadwyler SA, Booze RM (1993) 3-D Reconstruction of the cholinergic basal forebrain system in young and aged rats. Neurobiol Aging 14: 389–392
Tilson HA, Schwartz RD, Ali SF, McLamb RL (1989) Colchicine administered into the area of the nucleus basalis decreases cortical nicotinic cholinergic receptors labelled by [3H]-acetylcholine. Neuropharmacology 8: 855–861
Vannucchi G, Pepeu G (1987) Effect of phosphatidylserine on acetylcholine release and content in cortical slices from aging rats. Neurobiol Aging 8: 403–407
Whitehouse PJ, Price DJ, Struble RG, Clark AW, Coyle JT, DeLong ML (1982) Alzheimer's disease and senile dementia: loss of neurons in the basal forebrain. Science 215: 1237–1239
Wu CF, Bertorelli R, Sacconi M, Pepeu G, Consolo S (1988) Decrease of brain acetylcholine release in aging freely moving rats detected by microdialysis. Neurobiol Aging 9: 357–361

Authors' address: Prof. Dr. G. Pepeu, Department of Preclinical and Clinical Pharmacology, University of Florence, Viale Morgagni 65, I-50134 Florence, Italy.

J Neural Transm (1994) [Suppl] 44: 195–208

Receptor function in cortical rat brain regions after lesion of nucleus basalis

R. Schliebs, T. Feist, S. Roßner, and **V. Bigl**

Paul Flechsig Institute for Brain Research, Department of Neurochemistry, University of Leipzig, Leipzig, Federal Republic of Germany

Summary. The present study was undertaken to study the interaction of cholinergic and glutamatergic mechanisms in cholinoceptive cortical target regions which is assumed to play an important role for realizing cognitive functions.

The densities of cortical muscarinic cholinergic receptor subtypes and corresponding receptor genes m_1 through m_4, as well as NMDA, AMPA and kainate glutamate receptor subtypes were measured in rats one week after unilateral mechanical lesion of the anterior part of the nucleus basalis magnocellularis (NbM) applying quantitative receptor autoradiography and in situ hybridization.

The studies revealed that in cortical regions displaying a low amount of acetylcholinesterase activity due to forebrain cholinergic lesion, NMDA receptor binding was markedly reduced in comparison to the unlesioned side, whereas AMPA and kainate binding has been significantly increased in these regions. M_1-muscarinic cholinergic receptor binding was not changed in any of the cortical regions studied, whereas M_2-receptor densities are slightly reduced in frontal and parietal cortices following lesion. These alterations in cortical M_2-muscarinic receptor binding are complemented by corresponding changes in the m_2- and m_4-mRNA transcripts.

The comparison of binding profiles through selected cortical regions of both lesioned and normal brain side revealed that lesion of the NbM affects NMDA receptors in all cortical layers of the lesioned side, whereas AMPA receptors are affected preferentially in the upper and kainate receptors preferentially in the middle and deeper cortical layers.

The differential changes in glutamate receptor subtypes following lesion might be regarded as the consequence of a cortical reorganization compensating for the reduced cholinergic presynaptic input. The data further suggest that presynaptic cortical cholinergic deficits might affect glutamatergic functions with different intensity and different directions.

Introduction

The basal forebrain cholinergic system is known to play an important role in cortical arousal and normal cognitive function. Glutamate, a major excitatory neurotransmitter in the adult central nervous system, seems to additionally be involved in a variety of physiological and pathophysiological processes including changes in synaptic plasticity and learning and memory processes (see e.g. Collingridge and Singer, 1990; Francis et al., 1993).

Cortical cholinergic dysfunction has been implicated in cognitive deficits that occur in Alzheimer's disease. The changes in markers of neocortical cholinergic system are complemented by alterations in glutamate receptors in the cerebral cortex (for review, see Nordberg, 1992) suggesting a functional link between both transmitter systems in the cerebral cortex. This is emphasized by rat brain studies demonstrating that glutamate can regulate the release of cortical acetylcholine (Lodge and Johnston, 1985). To study the impact of cholinergic hypoactivity on glutamatergic mechanisms in cholinoceptive cortical target regions, the effect of cholinergic degeneration in rat basal forebrain on cortical glutamate receptor subtypes was investigated in a complex manner by screening the whole cerebral cortex with the aim to quantify deficits in both cholinergic (M_1, M_2-muscarinic receptors and muscarinic receptor genes m_1 through m_4) and glutamatergic parameters (glutamate receptor subtypes: NMDA, AMPA, and kainate). To mimic the dysfunction of the forebrain cholinergic system, adult rats were subjected to unilateral electrolytic lesions of the nucleus basalis magnocellularis (NbM), a region which is analogous to the nucleus basalis of Meynert in human brain.

The densities of cortical muscarinic cholinergic and glutamate receptor subtypes were studied in rats one week after surgery applying quantitative receptor autoradiography and in situ hybridization.

Material and methods

Treatment of animals and tissue preparation

Male Wistar rats (275–300 g) were anesthetized with 100 mg ketamine and 15 mg xylidinothiazine/kg body weight. The rat's head was placed in a stereotaxic frame and a steel electrode (diameter 0.4 mm, which was isolated except the last 0.5 mm) was placed into the nucleus basalis magnocellularis (NbM) using the coordinates: 2.2 mm posterior to the bregma, 2.9 mm lateral to the bregma, and 7.0 mm vertically from the skull surface. Electrolytic lesions were made by allowing to flow a direct current of 60 µA for 60 sec through the electrode. 7 days following lesion brains were rapidly removed and frozen at −70°C. Twelve µm thick serial coronal sections were cut, thaw-mounted on gelatin coated (for binding studies) or poly-L-lysine coated slides (in situ hybridization), and stored at −20°C (binding studies) or at 4°C in 95% ethanol after fixation in 4% paraformaldehyde (in situ hybridization) pending biochemical analysis.

To check for the efficiency of the cholinergic lesion, a histochemical staining for acetylcholinesterase (AChE) was performed in adjacent sections in each cortical level measured. Only those animals with a properly placed lesion were considered for biochemical analysis. Sham-operated animals were considered as controls.

Receptor autoradiography

Muscarinic cholinergic M_1-subtype receptor binding was carried out as described previously (Kumar and Schliebs, 1992) using [^{3}H]pirenzepine (NEN, specific activity 3.03 TBq/mmol) as a specific radioligand. [^{3}H]AF-DX384 (NEN, spec. act. 4.44 TBq/mmol) binding to M_2-muscarinic cholinergic subtype was performed as follows: after preincubation for 30 min, slides were incubated for one hour at room temperature with Krebs buffer (pH 7.4) containing 30 nM [^{3}H]AF-DX384. After incubation, slides were given four rinses of 3 min each in ice-cold 50 mM Tris-HCl (pH 7.4) followed by a short dipping in ice-cold distilled water before drying in a light air stream.

Radioligands used to define NMDA, AMPA, and kainate receptors were [^{3}H]MK-801 (NEN, spec. act. 951 GBq/mmol), [^{3}H]CNQX (NEN, 577 GBq/mmol), and [^{3}H]kainate (NEN, 2.15 TBq/mmol), respectively; receptor binding in tissue slices was performed as described earlier (Kumar et al., 1994).

The labeled and dried tissue sections were apposed to tritium-sensitive film (^{3}H-Hyperfilm, Amersham) at 4°C, and exposed for two to five weeks (depending on the radioligand used).

In situ hybridization

In situ hybridization to detect m_1 to m_4 mAChR mRNA was carried out as recently described (Roßner et al., 1993, 1994) using ^{35}S-labeled oligonucleotides as specific probes.

Evaluation of autoradiograms

Quantitative analysis of the autoradiograms was done on video camera-based, computer assisted imaging device (SIGNUM) using the autoradiographic software package Image Pro Plus, version 1.1. For calibration of grey values (optical density) ^{3}H-microscale standards (Amersham) co-exposed with labeled sections were used.

For evaluation of cortical regions all over the brain coronal sections cryocutted at selected distances from the bregma: +2.7, +1.2, −0.8, −1.3, 3.3, −5.8 mm (according to the atlas of Zilles, 1985) were used as indicated in Fig. 1.

Optical density readings were performed in the frontal cortices (Fr1, Fr2, Fr3), parietal cortices (Par1, Par2), temporal cortex, occipital cortex (V1, V2), forelimb and hindlimb area, piriform, cingulate, entorhinal and retrosplenial cortex according to the atlas of Zilles (1985). Measurements were made on 4 consecutive sections from each animal. The corresponding data obtained from 7 animals were averaged. For some selected regions the optical density through the depth of the cortical area along a 1 mm thick band (thick profile) was determined and corrected for background. Thick profiles obtained from 7 animals were averaged (for details of quantification, see also Kumar and Schliebs, 1992).

To test for statistically significant differences between binding levels of various brain regions from control and experimental animals, the two-tailed Student's t-test was applied.

Results

To check for the efficiency of the electrolytic lesion in the NbM in destroying cholinergic projections to cortical target regions, brain sections were stained for acetylcholinesterase and semiquantitatively evaluated by image analysis. The data revealed that in experimental rats the enzyme staining was considerably lower in cortical regions (frontal, parietal cortices and hindlimb areas) ipsilateral to the lesion as compared to the corresponding areas of the opposite hemisphere. The reduction was more prominent in the more rostral cortical regions (reduction by about 60% as compared to the unlesioned side), whereas the more caudal cortical areas were less affected (by about 20%) following unilateral lesion of NbM (Fig. 1).

7 days following lesion of the rat NbM, the NMDA receptor binding is significantly reduced in the frontal and parietal cortices as well as hindlimb areas of the lesioned brain side (Fig. 2). In contrast, after 7 days of electrolytic lesion of the NbM, kainat and AMPA receptor binding is increased in the frontal and parietal cortices as well as in forelimb and hindlimb areas of the lesioned brain side (Fig. 2).

In order to elucidate whether the effects of NbM lesion on cortical glutamate receptors are due to changes in a particular cortical layer, binding profiles through the depth of the cortical area along a 1 mm thick band were determined. The comparison of binding profiles through selected cortical regions of both lesioned and normal brain side reveals that lesion of the NbM affects NMDA receptors in all cortical layers of the lesioned side, whereas AMPA receptors are affected preferentially in the upper and kainate receptors preferentially in the middle and deeper cortical layers (Fig. 3).

Following 7 days of lesion of the NbM M_1-muscarinic acetylcholine receptor (mAChR) binding was not changed in any of the cortical regions studied, whereas M_2-mAChR densities are only slightly reduced in frontal and parietal cortices (Fig. 4). In order to see whether particular mRNA mAChR subtypes are involved in situ hybridization experiments were performed. The data revealed that the changes in M_2-mAChR binding following unilateral lesion of NbM are complemented by changes in the m_2- and m_4-gene transcripts in frontal and parietal cortices (Fig. 4).

---→

Fig. 1. Effect of unilateral electrolytic lesion of NbM on acetylcholinesterase-staining. Data represent the mean values obtained from 7 animals and are given as relative changes expressed as percentage over unlesioned brain side. Brain coronal sections cryocutted at selected distances from the bregma (from top: +2.7, +1.2, −1.3, −3.3, −5.8 mm) were stained for AChE and used for evaluation of brain regions by image analysis. *Cing* cingulate cortex; *Entorhin* entorhinal cortex; *FL* forelimb area; *Fr* frontal cortex; *Pir* piriform cortex; *Par* parietal cortex; *HL* hindlimb area; *Retro* retrosplenial cortex; *Temp* temporal cortex; *V* visual cortex. $^{*}P < 0.05$, $^{**}P < 0.01$ vs. corresponding region on the unlesioned brain side

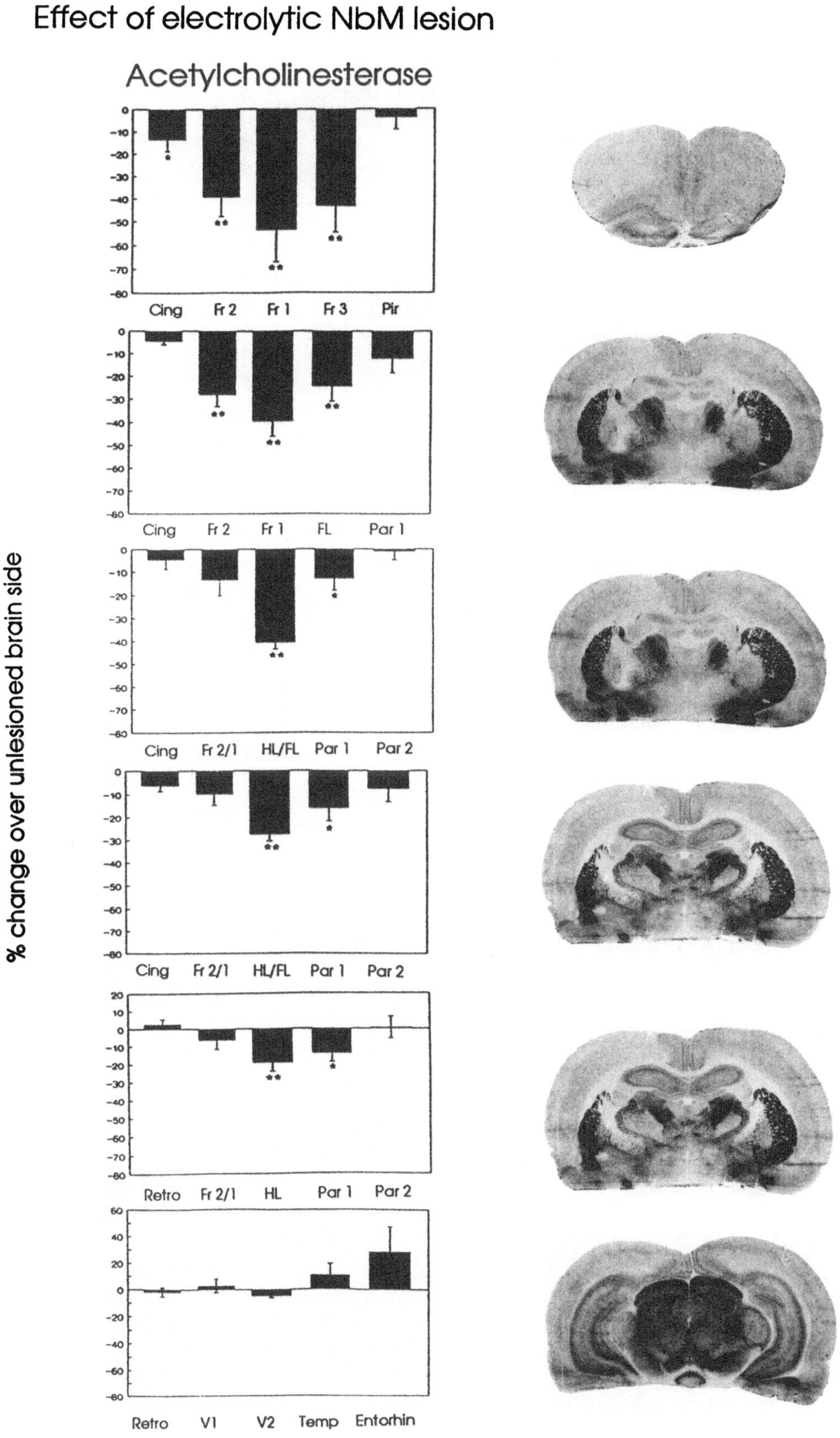
Effect of electrolytic NbM lesion
Acetylcholinesterase
% change over unlesioned brain side
Cing Fr 2 Fr 1 Fr 3 Pir
Cing Fr 2 Fr 1 FL Par 1
Cing Fr 2/1 HL/FL Par 1 Par 2
Cing Fr 2/1 HL/FL Par 1 Par 2
Retro Fr 2/1 HL Par 1 Par 2
Retro V1 V2 Temp Entorhin

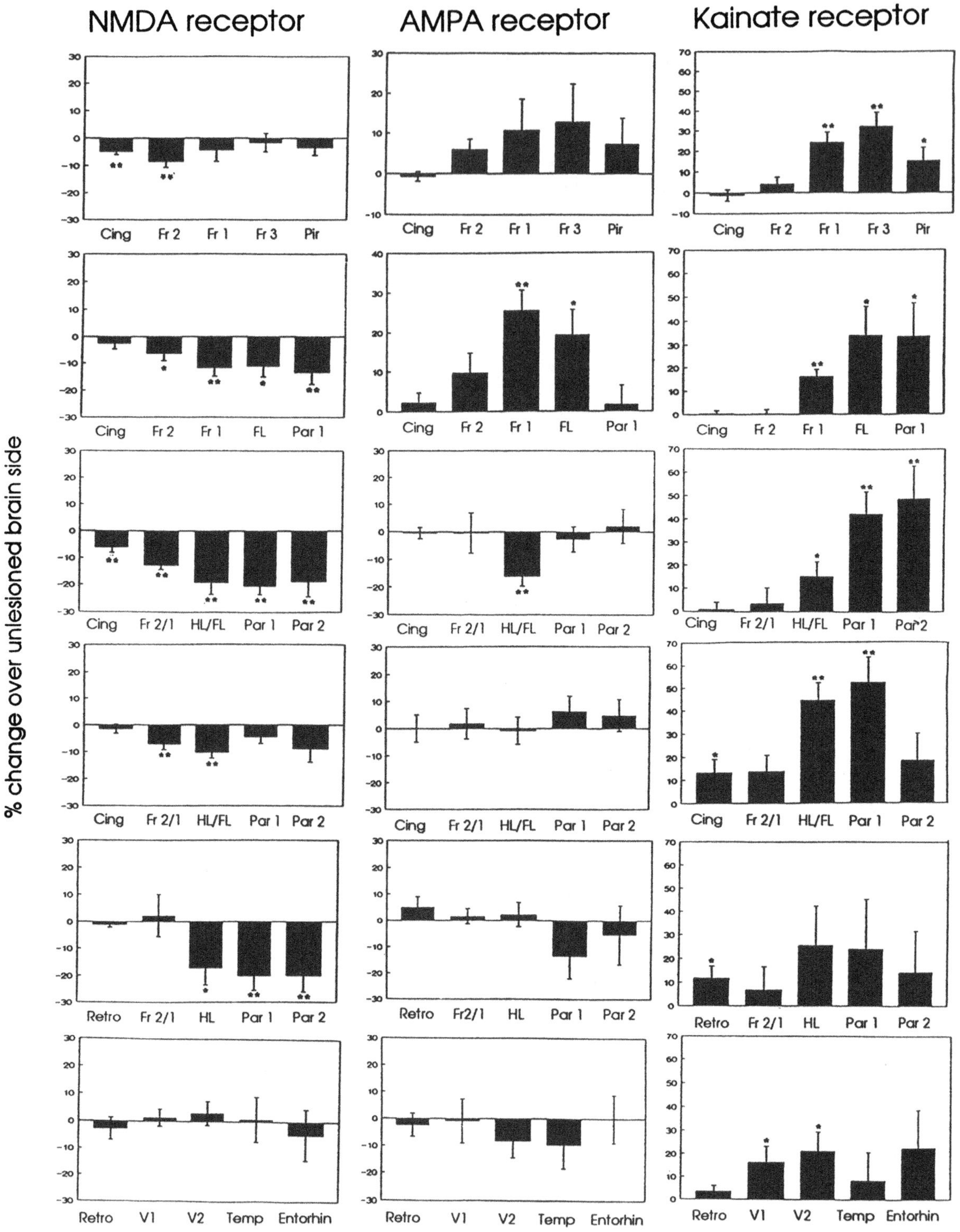

Fig. 2. Effect of unilateral electrolytic lesion of NbM on NMDA, AMPA, and kainate receptor binding. Brain coronal sections cryocutted at selected distances from the bregma (from top: +2.7, +1.2, −1.3, −3.3, −5.8 mm, see also Fig. 1) were used for receptor autoradiography, and the autoradiograms evaluated by image analysis. Data represent the mean values obtained from 7 animals and are given as relative changes expressed as percentage over unlesioned brain side. For abbreviations, see legend to Fig. 1. $^*P < 0.05$, $^{**}P < 0.01$ vs. corresponding region on the unlesioned brain side

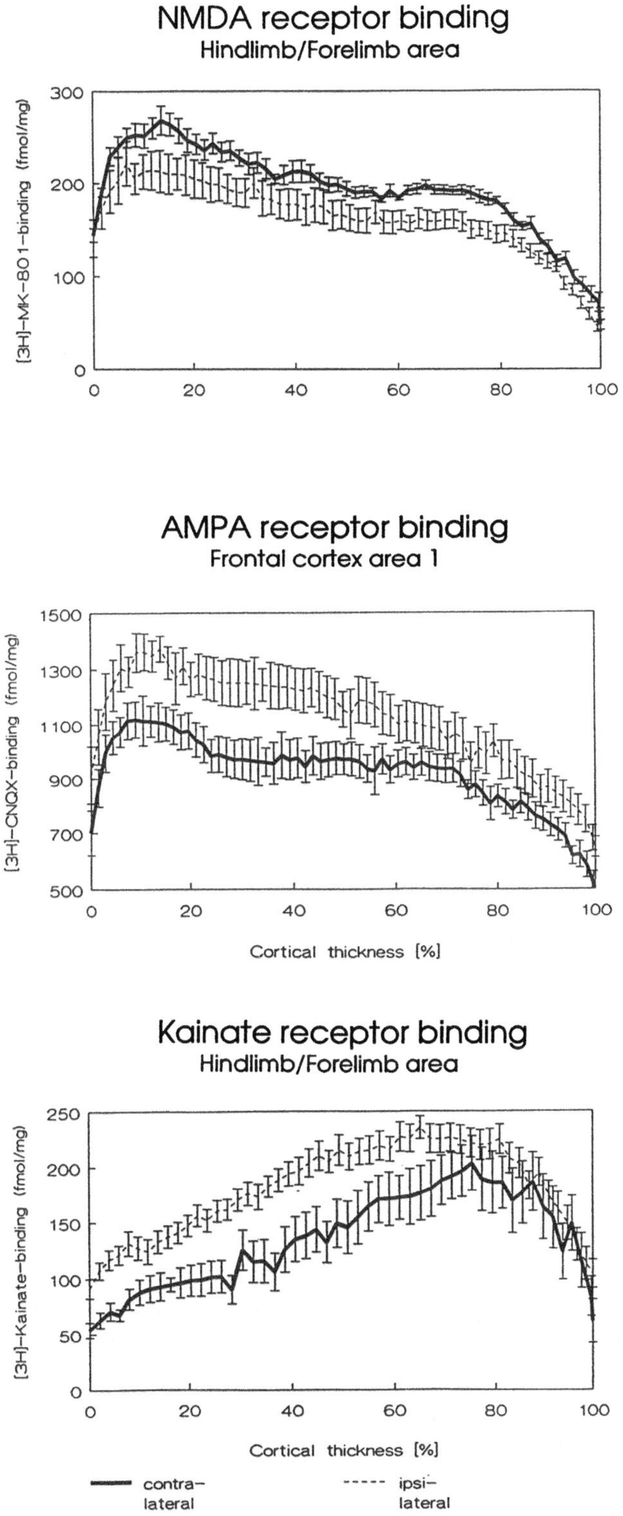

Fig. 3. Comparison of the laminar distributions of NMDA, AMPA, and kainate receptor binding sites through the whole cortical depth of selected cortical brain regions between both hemispheres from animals subjected to unilateral NbM lesion. The values are means ± SD from 7 animals

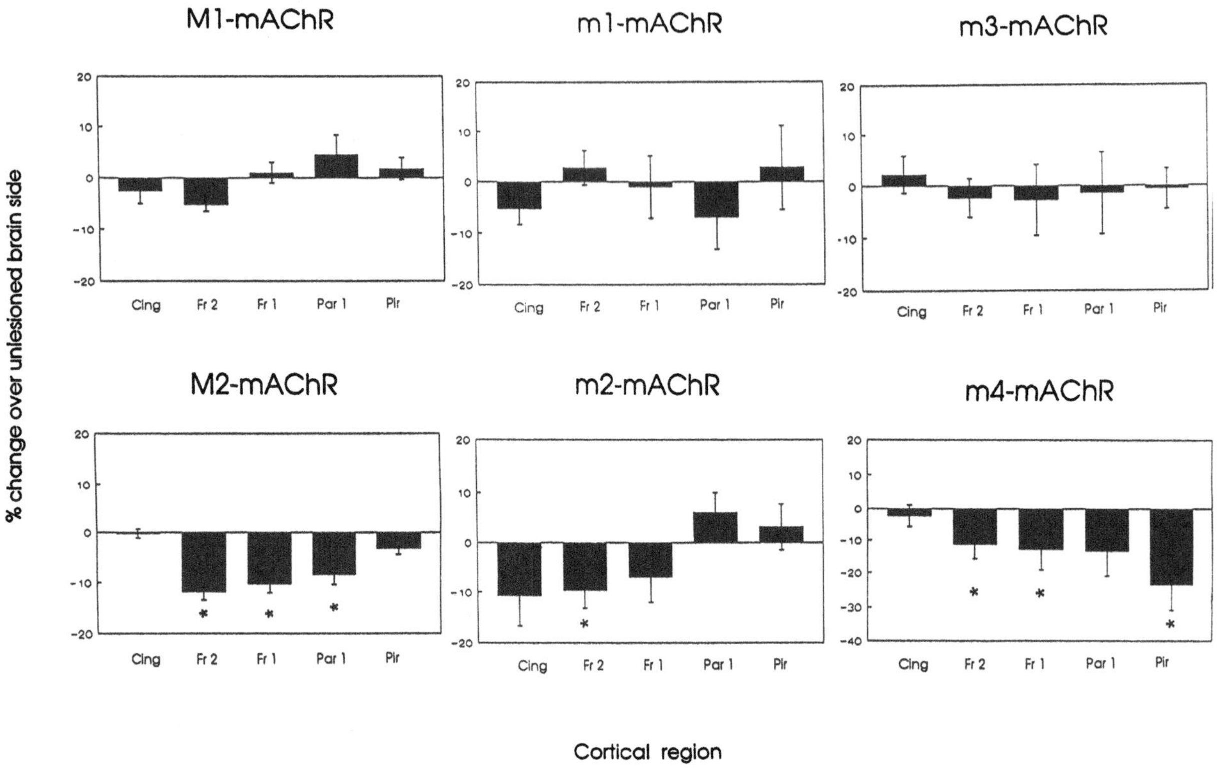

Fig. 4. Effect of unilateral electrolytic lesion of NbM on M_1- and M_2-muscarinic acetylcholine receptors and m_1 through m_4 muscarinic receptor genes. Brain coronal sections cryocutted at selected distances from the bregma were used for receptor autoradiography and in situ hybridization, and the autoradiograms evaluated by image analysis. Data represent the mean values obtained from 7 animals and are given as relative changes expressed as percentage over unlesioned brain side. For abbreviations, see legend to Fig. 1. $^*P < 0.05$ vs. corresponding region on the unlesioned brain side

Discussion

The NbM has been identified as the main origin of cholinergic innervation to the cerebral cortex (Wenk et al., 1980). The projections to the cortex are topographically organized, with frontal, temporal and parietal cortices receiving the majority of these efferents. To mimic cholinergic dysfunction an unilateral lesion was placed into the NbM resulting in a significant unilateral reduction of cortical AChE activity which is more pronounced in rostral compared to caudal cortical areas. This compares well to other studies demonstrating that differential placement of lesions within the NbM produce varying depletions of ChAT in the neocortex: the more posteriorly the lesion is placed, the greater the depletion of choline acetyltransferase (ChAT; Wenk et al., 1984). It has been shown that the distribution of AChE-positive fibres parallels that of ChAT-immunoreactive fibres in most cortical areas, except retrosplenial and cingulate cortices (Lysakowski et al., 1989) thus allowing to use AChE-staining as a marker to detect degeneration of cortical cholinergic terminals.

The lesion procedure used in this study should affect mainly cholinergic neurons because in the rat approximately 80–90% of the NbM efferents are cholinergic (see Smith, 1988). However, it cannot be excluded that to some extent other neurons (probably mostly GABA neurons as demonstrated by Lindefors et al., 1992) or fibres passing the site of lesion are also affected. The use of various excitotoxins to produce NbM lesion seems to spare crossing fibres. However, also these toxins are far from being selective for cholinergic neurons which might explain why the behavioural effects depend on the toxin used to produce lesion (see e.g. Dunnett et al., 1991; Wenk et al., 1992). The size of the lesioned area as well as the reduction of AChE activity in various cortical areas in the present study are similar to those observed when utilizing excitotoxins to destroy cells in the NbM (for review and references, see Smith, 1988).

NbM lesions produce persistent depletion in cortical high-affinity choline uptake which does not recover over time (Bartus et al., 1985; El-Defrawy et al., 1986; Nakamura and Ishihara, 1990), and which result in significant reductions of spontaneous and potassium-stimulated acetylcholine release (El-Defrawy et al., 1986). In contrast, the changes in cerebral glucose utilization after NbM ablation in rats have been found to be only of transient nature (DeMicheli and Soncrant, 1992). NbM lesions decreased significantly the density of nicotine binding sites (Zawia et al., 1992). Hemicholinium-3 binding to choline uptake sites was found to be decreased in frontolateral cortex following lesion (Wenk et al., 1992) which was not confirmed by Steckler et al. (1993). The findings on muscarinic receptors following NbM lesion are controversial: Watson et al. (1985) demonstrated a decreased number in pirenzepine and QNB binding whereas other authors (Altman et al., 1985; Bartus et al., 1985; McKinney and Coyle, 1982) did not observe any change in the total number of muscarinic binding sites as a consequence of NbM lesion. This compares with findings from Scarth et al. (1989) who were unable to find any difference in cortical phosphatidylinositol metabolism linked to M_1-mAChRs, between lesioned and control rats. In contrast, increased values in the maximum receptor number and binding constant of the total number of muscarinic receptor sites were found 28 days after basal forebrain lesion induced by ibotenic acid (Fuji et al., 1993). After unilateral NbM lesion, the density of cortical M_2-mAChRs was found to be reduced three days after lesion (Mash et al., 1985) but increased after one week postlesion (Atack et al., 1989).

7 days following unilateral NbM lesion in this study we found a small reduction in M_2-mAChR binding restricted to frontal cortices, but no change in M_1-receptor binding sites as compared to the unlesioned side. M_2-radioligands seem to recognize with high affinity both m_2 and m_4-receptors (Bonner, 1989; Buckley et al., 1989). Thus the changes in cortical M_2-mAChRs following lesion might be due to alterations in m_2- and/or m_4-receptors. In fact, the alterations in M_2-mAChRs are found to be

complemented by changes in the amount of both m_2 and m_4 mAChR mRNA.

The decrease in M_2-mAChRs following lesion can be explained when assuming that these sites are located on cholinergic nerve terminals, whereas the decrease in the level of the corresponding receptor mRNA then might be induced in the postsynaptic neuron through an intracellular signal transduction cascade as consequence of cortical cholinergic hypofunction. However, there is some evidence that also a postsynaptic population of cortical M_2-mAChRs exists (Dawson et al., 1991).

There are several studies describing the effect of NbM lesion on non-cholinergic markers like glutamic acid decarboxylase (Nakamura et al., 1992), biogenic amines (noradrenaline, dopamine, serotonin) and metabolites (Connor et al., 1992; Langlais et al., 1993) as well as various neuropeptides (Gaykema et al., 1989; Rattan and Tejwani, 1992; Moyse et al., 1993), but there is still a lack of data on the effect of NbM lesion on glutamate receptors. A lack of impairment of glutamate receptor number and affinity as well as NMDA receptors was observed following NbM lesion (Maragos et al., 1991; Smith, 1988).

However, in this study unilateral electrolytic lesion of the NbM resulted in reduced numbers of NMDA receptors in frontal cortical areas suggesting that at least some of the cortical NMDA receptors might be located on cholinergic terminals originating in the NbM. Thus, the reduced number of NMDA receptors might be due to the degeneration of cholinergic terminals. However, cortical lesion studies could not find any evidence that cortical NMDA receptors may exist on presynaptic terminals of cholinergic neurons originating in the NbM (Maragos et al., 1991). The basal forebrain magnocellular complex receives among others also a strong glutamatergic innervation from the cortex (see e.g. Martin et al., 1993) suggesting that the nucleus basalis lesion should have functional consequences also on cortical glutamatergic neurons. Therefore, the changes in the number of cortical NMDA receptors following lesion might be considered as a loss and/or down-regulation of NMDA receptor sites. In contrast, the increased kainate and AMPA binding following lesion should be considered as up-regulation of receptor sites. The functional role of up-regulation of postsynaptic glutamate receptors is assumed to compensate for reduced presynaptic input. This would suggest that cholinergic terminals directly affect glutamate transmission on presynaptic glutamatergic elements. However, the differential changes in glutamate receptor might also represent the biochemical consequences of a functional cortical reorganization as suggested for other deafferentation paradigms (Dykes and Metherate, 1989).

In a recent study it was demonstrated that electrolytic but not ibotenic acid lesion of NbM resulted in increased numbers of astrocytes in the cortex (Lescaudron et al., 1992). Thus the enhanced number of AMPA and kainate receptor sites could, at least partly, be the consequence of cortical gliosis in response to the electrolytic NbM lesion, when assuming that at least some of cortical AMPA and kainate receptors are located on astrocytes. However, there is no strong evidence that glutamate receptor subtypes are located on glia cells (Greenamyre and Young, 1989).

Glutamate is used as transmitter in corticofugal as well as cortico-cortical systems. Immunohistochemical studies revealed that glutamate-containing neurons are highly concentrated in layer V and the deep part of layer VI, whereas glutamate-containing axon terminals show the highest density in layers I through IV with peak values in layer IV of primary sensory areas (for review, see Zilles et al., 1990). Cholinergic fibres are mainly concentrated in layer I and layer V in motor and most sensory cortical areas, except in the visual cortex which additionally showed a dense band in layer IV (Eckenstein et al., 1988). This correspondence between cholinergic terminals and glutamatergic neurons supports a direct link between both transmitter systems.

The differential laminar distribution of glutamate subtypes in the cortex suggests that glutamate should exert a different influence in each particular cortical layer by inducing different cellular responses through the distinct receptor subtypes. This is emphasized by our findings that NbM lesion affects glutamate receptor subtypes mainly in those cortical layers which are highly enriched with the corresponding receptor subtype.

It is interesting to note that the same sort of alterations in glutamate receptor subtypes observed in rat cortex following NbM lesion have been detected in cortical brain areas from patients with Alzheimer's disease (Nordberg, 1992). This supports the suggestion that the receptor changes observed might indicate compensatory mechanisms due to presumably cholinergic degenerative events. But regardless of possible explanations, the data further support a glutamatergic strategy which may prove to be therapeutically useful in Alzheimer's disease (Nordberg, 1992).

Acknowledgements

The authors gratefully acknowledge the expert technical assistence of Mrs. R. Jendrek. This work was supported by a grant from the Bundesministerium für Forschung und Technik, no. FKZ 01 ZZ 9130/2.8 to R.S.

References

Altman H, Crosland RD, Jenden DJ, Berman RF (1985) Further characterizations of the nature of the behavioral and neurochemical effects of lesions to the nucleus basalis of Meynert in the rat. Neurobiol Aging 6: 125–130

Atack JR, Wenk GL, Wagster MV, Kellar KJ, Whitehouse PJ, Rapoport SI (1989) Bilateral changes in neocortical [^{3}H]pirenzepine and [^{3}H]oxotremorine-M binding following unilateral lesion of the rat nucleus basalis magnocellularis: an autoradiographic study. Brain Res 483: 367–372

Bartus RT, Flicker C, Dean RD, Pontecorvo M, Figueiredo JC, Fisher SK (1985) Selective memory loss following nucleus basalis lesions: long term behavioral recovery despite persistent cholinergic deficiencies. Pharmacol Biochem Behav 23: 125–135

Bonner TI (1989) The molecular basis of receptor diversity. Trends Neurosci 2: 148–151

Buckley NJ, Bonner TI, Buckley CM, Brann MR (1989) Antagonist binding properties of five cloned muscarinic receptors expressed in CHO-Kl cells. Mol Pharmacol 35: 469–476

Collingridge GL, Singer W (1990) Excitatory amino acid receptors and synaptic plasticity. In: Collingridge G, Lodge D (eds) TIPS Special report: pharmacology of excitatory amino acids. Elsevier, Cambridge, pp 42–53

Connor DJ, Thal LJ, Mandel RJ, Langlais, Masliah E (1992) Independent effects of age and nucleus basalis magnocellularis lesion: maze learning, cortical neurochemistry, and morphometry. Behav Neurosci 106: 776–788

Dawson VL, Hunt ME, Wamsley JK (1991) Alterations in cortical muscarinic receptors following cholinotoxin (AF64A) lesion of the rat nucleus basalis magnocellularis. Neurobiol Aging 13: 25–32

Dykes RW, Metherate R (1989) Sensory cortical reorganization following peripheral nerve injury. In: Finger S, Levere TE, Almli CR, Stein DG (eds) Brain injury and recovery: theoretical and controversial issues. Plenum, New York, pp 215–234

DeMicheli E, Soncrant TT (1992) Age-dependent cerebral metabolic effects of unilateral nucleus basalis magnocellularis ablation in rats. Neurobiol Aging 13: 687–695

Dunnett SB, Everitt BJ, Robbins TW (1991) The basal forebrain-cortical cholinergic system: interpreting the functional consequences of excitotoxic lesions. Trends Neurosci 14: 494–501

Eckenstein FP, Baughman RW, Quinn J (1988) An anatomical study of cholinergic innervation in rat cerebral cortex. Neuroscience 25: 457–474

El-Defrawy SR, Boegman RJ, Jhamandas K, Beninger R, Shipton L (1986) Lack of recovery of cortical cholinergic function following quinolinic or ibotenic acid injections into the nucleus basalis magnocellularis in rats. Exp Neurol 91: 628–633

Francis PT, Sims NR, Procter AW, Bowen DM (1993) Cortical pyramidal neurone loss may cause glutamatergic hypoactivity and cognitive impairment in Alzheimer's disease: investigative and therapeutic perspectives. J Neurochem 60: 1589–1604

Fuji K, Hiramatsu M, Hayashi S, Kameyama T, Nabeshima T (1993) Effects of propentofylline, a NGF synthesis stimulator, on alterations in muscarinic cholinergic receptors induced by basal forebrain lesion in rats. Neurosci Lett 150: 99–102

Gaykema RPA, Compaan JC, Nyakas C, Horvath E, Luiten PGM (1989) Long-term effects of cholinergic basal forebrain lesions on neuropeptide Y and somatostatin immunoreactivity in rat neocortex. Brain Res 489: 392–396

Greenamyre JT, Young AB (1989) Synaptic localization of striatal NMDA, quisqualate and kainate receptors. Neurosci Lett 101: 133–137

Kumar A, Schliebs R (1992) Postnatal development of cholinergic receptors, protein kinase C and dihydropyridine-sensitive calcium antagonist binding in rat visual cortex. Effect of visual deprivation. Int J Dev Neurosci 10: 491–504

Kumar A, Schliebs R, Bigl V (1994) Development of NMDA, AMPA, and kainate receptors in individual layers of rat visual cortex and the effect of monocular deprivation. Int J Dev Neurosci 12:31–41

Langlais PJ, Connor DJ, Thal L (1993) Comparison of the effects of single and combined neurotoxic lesions of the nucleus basalis magnocellularis and dorsal noradrenergic bundle on learning and memory in the rat. Behav Brain Res 54: 81–90

Lescaudron L, Bitran BS, Stein DG (1992) GM1 ganglioside effects on astroglial response in the rat nucleus basalis magnocellularis and its cortical projection areas after electrolytic or ibotenic lesions. Exp Neurol 116: 85–95

Lindefors N, Boatell ML, Mahy N, Persson H (1992) Widespread neuronal degeneration after ibotenic acid lesioning of cholinergic neurons in the nucleus basalis revealed by in situ hybridization. Neurosci Lett 135: 262–264

Lodge D, Johnston GAR (1985) Effect of ketamine on amino acid-evoked release of acetylcholine from rat cerebral cortex in vitro. Neurosci Lett 56: 371–375

Lysakowski A, Wainer BH, Bruce G, Hersh LB (1989) An atlas of the regional and laminar distribution of choline acetyltransferase immunoreactivity in rat cerebral cortex. Neuroscience 28: 291–336

Maragos WF, Greenamyre JT, Chu DCM, Penney JB, Young AB (1991) A study of cortical and hippocampal NMDA and PCP receptors following selective cortical and subcortical lesions. Brain Res 538: 36–45

Martin LJ, Blackstone CD, Levey AI, Huganir RL, Price DL (1993) Cellular localization of AMPA glutamate receptors within the forebrain magnocellular complex of rat and monkey. J Neurosci 13: 2249–2263

Mash D, Flynn D, Potter L (1985) Loss of M2 muscarinic receptors in the cerebral cortex in Alzheimer's disease and experimental cholinergic denervation. Science 228: 1115–1117

McKinney M, Coyle J (1982) Regulation of neocortical muscarinic receptors: effects of drug treatments and lesions. J Neurosci 2: 97–105

Moyse E, Szigethy E, Danger JM, Vaudry H, Wenk GL, Beaudet A, Epelbaum J (1993) Short- and long-term effects of nucleus basalis magnocellularis lesions on cortical levels of somatostatin and its receptors in the rat. Brain Res 607: 154–160

Nakamura S, Ishihara T (1990) Task-dependent memory loss and recovery following unilateral nucleus basalis lesion: behavioral and neurochemical correlation. Behav Brain Res 39: 113–122

Nakamura S, Tani Y, Maezono Y, Ishihara T, Ohno T (1992) Learning deficits after unilateral AF64A lesions in the rat basal forebrain: role of cholinergic and noncholinergic systems. Pharmacol Biochem Behav 42: 119–130

Nordberg A (1992) Neuroreceptor changes in Alzheimer disease. Cerebrovasc Brain Met Rev 4: 303–328

Rattan AK, Tejwani GA (1992) The neurotoxic actions of ibotenic acid on cholinergic and opioid peptidergic systems in the central nervous system of the rat. Brain Res 571: 298–305

Roßner S, Kues W, Witzemann V, Schliebs R (1993) Laminar expression of m_1-, m_3-, and m_4-muscarinic cholinergic receptor genes in the developing rat visual cortex using in situ hybridization histochemistry. Effect of monocular visual deprivation. Int J Dev Neurosci 11: 369–378

Roßner S, Kumar A, Witzemann V, Schliebs R (1994) Development of laminar expression of the m_2 muscarinic cholinergic receptor gene in rat visual cortex and the effect of monocular visual deprivation. Dev Brain Res 77: 55–61

Scarch BJ, Jhamandas K, Boegman RJ, Beninger RJ, Reynolds JN (1989) Cortical muscarinic receptor function following quinolinic acid-induced lesion of the nucleus basalis magnocellularis. Exp Neurol 103: 158–164

Smith G (1988) Animal models of Alzheimer's disease: experimental cholinergic denervation. Brain Res Rev 13: 103–118

Steckler T, Andrews JS, Marten P, Turner JD (1993) Effects of NBM lesions with two neurotoxins on spatial memory and autoshaping. Pharmacol Biochem Behav 44: 877–889

Watson M, Vickeroy T, Fibiger H, Roeske W, Yamamura H (1985) Effects of bilateral ibotenate-induced lesions of the nucleus basalis magnocellularis upon selective cholinergic biochemical markers in the rat anterior cerebral cortex. Brain Res 177: 337–345

Wenk H, Bigl V, Meyer U (1980) Cholinergic projections from magnocellular nuclei of the basal forebrain to cortical areas in rats. Brain Res Rev 2: 295–316

Wenk GL, Cribbs B, McCall L (1984) Nucleus basalis magnocellularis: optimal coordinates for selective reduction of choline acetyltransferase in frontal neocortex by ibotenic acid injection. Exp Brain Res 56: 335–340

Wenk GL, Harrington CA, Tucker DA, Rance NE, Walker LC (1992) Basal forebrain neurons and memory: a biochemical, histological, and behavioral study of differential vulnerability to ibotenate and quisqualate. Behav Neurosci 106: 909–923

Zawia N, Arendash GW, Wecker L (1992) Basal forebrain cholinergic neurons in aged rat brain are more susceptible to ibotenate-induced degeneration than neurons in young adult brain. Brain Res 589: 333–337

Zilles K (1985) The cortex of the rat. A stereotaxic atlas. Springer, Berlin Heidelberg New York Tokyo

Zilles K, Wree A, Dausch ND (1990) Anatomy of the neocortex: neurochemical organization. In: Kolb B, Tees RC (eds) The cerebral cortex of the rat. MIT Press, Cambridge, pp 113–150

Authors' address: Dr. R. Schliebs, Department of Neurochemistry, Paul Flechsig Institute for Brain Research, University of Leipzig, Jahnallee 59, D-04109 Leipzig, Federal Republic of Germany.

J Neural Transm (1994) [Suppl] 44: 209–217

Trophic factors during normal brain aging and after functional damage

R. Hellweg

Department of Psychiatry, Free University of Berlin, Berlin, Federal Republic of Germany

Summary. There is an increasing body of information concerning the physiological role of several target-derived neurotrophic proteins that are structurally and functionally related to the classical neurotrophic molecule NGF and which resemble a genetic family called neurotrophins. However apart from NGF, there is little knowledge about the pathophysiological role of these neurotrophins concerning aging or dementia. To our present knowledge, decreased NGF production does not seem to play a causal role in age-related cognitive impairment which is usually associated with neurodegenerative processes in the cholinergic basal forebrain system. However, there are several experimentally found indications that NGF might be of importance in the stimulation of compensatory changes and repair mechanisms.

Moreover, recent findings suggest that disturbances in cerebral glucose metabolism may play an important role in cognitive disabilities during normal aging and also in dementia disorders such as Alzheimer's disease. Intracerebroventricular (ICV) injection of streptozotocin (STZ) has been reported to decrease cerebral glucose utilization and energy metabolism and to impair passive avoidance learning in adult rats. One week after ICV STZ treatment, NGF content was significantly decreased in the septal region, where NGF-responsive cell bodies are known to be located and where NGF exerts its neurotrophic action after retrograde transport from NGF-producing targets. In contrast, NGF levels were increased within 3 weeks after ICV STZ treatment by about the same magnitude as has been observed for aged learning-impaired rats in the target regions for the basal forebrain cholinergic neurons. Similar to these aged rats, ChAT levels were barely altered in adult rats up to 3 weeks after ICV injection of STZ. This might be due to masking effects exerted by compensatory NGF-mediated stimulation of ChAT activity in the remaining functional neurons. Thus disturbance of the cerebral glucose and energy metabolism may produce patterns of behavioral and biochemical alterations similar to those observed in aged rats with memory impairments. This suggests that the central disturbance of glucose metabolism may be an important event for memory impairment, which ultimately affects NGF-responsive central cholinergic neurons, this effect probably taking the form of a diminished capacity to bind and/or transport NGF. Damage of the cerebral glucose and energy metabolism by ICV treatment with STZ may therefore provide an appropriate animal model for the study of the early metabolic events in cerebral aging and degenerative disorders associated with cognitive impairments.

Introduction

There is an increasing body of information concerning the physiological role of target-derived neurotrophic factors. These proteins are structurally and functionally related to the classical neurotrophic molecule called nerve growth factor (NGF). They resemble a genetic family called neurotrophins, which with target cells influence their innervating neurons (for review see Thoenen, 1991). NGF has been originally shown to be synthesized and released in limiting amounts by target tissues of the peripheral sympathetic and neural-crest derived sensory nervous system (for review see Thoenen et al., 1987). Following release NGF is taken up by the innervating nerve terminals via specific NGF receptors and transported retrogradely in an NGF-NGF receptor complex to the corresponding perikarya where it exerts its neurotrophic action (for review see Raivich et al., 1990). Interruption of the retrograde axonal transport in adult animals by surgical or pharmacological procedures has the same effect as the neutralization of endogenous NGF by specific antibodies, namely impaired neuronal function even in fully differentiated neurons (for review see Thoenen et al., 1987). Thus all pathological processes accompanied by a trophic deficiency and/or directly or indirectly affecting retrograde axonal transport of trophic factor(s) may produce alterations in these neurons. This may well represent an important pathogenetic mechanism in several neurological disorders (for review see Vantini, 1992).

There is compelling evidence that learning and memory functions are closely related to the integrity of the cholinergic neurons of the basal forebrain and their projections to the cortex and hippocampus (for review see Kesner, 1988). With age all major cholinergic cell groups in the mammalian forebrain undergo degenerative changes and several publications demonstrate a close relationship between severe cholinergic degeneration and profound spatial learning impairment in rats (Fischer et al., 1989; Koh et al., 1989).

The state of preservation of cholinergic neurons is conventionally analyzed by determining the levels of the acetylcholine synthesizing enzyme choline acetyltransferase (ChAT). In the basal forebrain system, this enzyme is known to be regulated by NGF which is synthesized in the cholinergically innervated target areas (i.e., cortex, hippocampus, olfactory bulb) and transported retrogradely to the cholinergic forebrain nuclei where it exerts neurotrophic actions (for reviews see Thoenen et al., 1987; Ebendal, 1989). Moreover, a closely coordinated increase in NGF and ChAT activity levels has been observed in the hippocampus during specific periods of early postnatal development, further supporting the physiological role of NGF for the cholinergic neurons of the basal forebrain system (Auburger et al., 1987).

More recently, it has been shown that an additional member of the genetic family of neutrotrophins, namely brain-derived neutrotrophic factor (BDNF), like NGF is capable of enhancing survival and differentiated functions of rat cholinergic neurons in culture (Alderson et al., 1990; Knusel et al., 1991). However, little is known about the role(s) of NGF or other trophic factors in normal brain aging or age-related neurodegenerative processes (cf. Ebendal, 1989). Since the available data concerning the remaining neurotrophins (i.e., neurotrophin-3 to -5) are sparse, it is not reasonable to assign any potential clinical relevance to these molecules at present (Vantini, 1992).

Materials and methods

Rats were treated as described in detail previously (Hellweg et al., 1990, 1992b; Mayer et al., 1990; Nitsch and Hoyer, 1991). Dissected Brain tissues were processed for determination of NGF and ChAT activity as detailed elsewhere (Hellweg et al., 1989 and Auburger et al., 1987, respectively).

Results and discussion

Levels of NGF and ChAT activity in the aging rat brain

In collaboration with Prof. A. Björklund's group from the University of Lund (Sweden) we have determined levels of NGF and ChAT activity in young (3-month-old) nonimpaired and aged (2-year-old) subpopulations of Sprague-Dawley rats with either slight or severe spatial learning impairments which were assessed by the Morris water maze test (see Fischer et al., 1989). Even in rats with severe spatial memory impairments, central NGF levels remained normal with advancing age or were even increased by up to 30%, notably in the cortex, olfactory bulb, septal region, and cerebellum (Hellweg et al., 1990). Thus, sustained NGF production appears to be unable to prevent atrophy and cell loss (Hellweg et al., 1990) as known to occur in the forebrain cholinergic system in behaviourally impaired aged rats (Fischer et al., 1989). As discussed elsewhere (Hellweg et al., 1990, 1992a, b), age-dependent cholinergic atrophy may be due to reduced responsiveness to NGF rather than to lack of the trophic factor. The increased NGF levels that have been observed in the aging brain of different rodent strains (Otten et al., 1989) may be a result of reactive gliosis and/or NGF accumulation due to lacking removal by functionally impaired cholinergic neurons of the basal forebrain. They may also play a role in the compensatory changes associated with the process of degeneration (see Hellweg et al., 1990, 1992a, b for further discussion; cf. Crutcher et al., 1993). In fact, ChAT activity levels in the target regions were only marginally affected even in severely impaired aged rats, despite their cell body regions having undergone substantial atrophy and cell loss (Fischer et al., 1989; Hellweg et al., 1990). This suggests a compensatory increase in ChAT activity in the remaining functional neurons. Thus the sustained NGF levels in the aged brain could be important for the maintenance of cholinergic neurotransmission in a situation where the system is undergoing progressive degenerative changes (Hellweg et al., 1990).

We are currently investigating whether this keeps true even in older rats, namely 30-month-old Wistars. As compared to 1-year-old control rats, preliminary data show a significant increase in ChAT activity of about 20% in the aged septal region and cerebellum, while ChAT activity is only slightly decreased by 10% in the aged hippocampus; ChAT activity does not differ significantly with age in the remaining brain regions investigated, namely olfactory bulb, cortex, striatum, and brainstem. It seems therefore reasonable to assume the rat brain disposes of some compensating capacities even in very advanced age. We are currently investigating whether this assumption can be corroborated by corresponding changes in endogenous NGF concentrations.

Expression of neurotrophins and their endogenous levels in Alzheimer's disease patients

In the case of pathological aging like Alzheimer's disease (AD), no changes in postmortem brain levels of NGF-mRNA have been observed (Goedert et al., 1986; Philips et al., 1991). In contrast, BDNF-mRNA is reportedly decreased in the hippocampus of patients with AD (Philips et al., 1991). Due to the lack of specific antibodies against BDNF, it is still impossible to quantify endogenous levels of BDNF protein reliably. Somewhat conflicting results have been published with respect to postmortem levels of NGF protein in AD brain. Using our improved ELISA for NGF determination (Hellweg et al., 1989) Allen et al. (1991) reported normal NGF content in AD cerebral cortex and hippocampus; NGF concentrations in some cortical regions showed a trend towards slight increases by up to 50% in AD, the difference from corresponding brain tissues of age-matched normal subjects failed to reach statistical significance though. In fact, highly significantly increased levels of NGF-like activity in postmortem cortical tissues of AD patients have been published very recently, using both ELISA and bioassay techniques (Crutcher et al., 1993). Based on our previously published experimental data (Hellweg et al., 1990, 1992b), we have suggested (Hellweg et al., 1992a) that AD is characterized by increased, rather than decreased levels of cortical NGF which may be a consequence of degenerative changes in the basal forebrain cholinergic system (cf. Crutcher et al., 1993). However, comparison of NGF levels among sera from a very limited number of patients suffering from different neurological diseases and age-matched healthy controls showed significantly reduced NGF levels both in Huntington and Parkinson sera. In AD sera only a trend towards NGF decrease was reported that failed to be significantly different from control sera (Lorigados et al., 1992). Whether these findings are of any pathophysiological relevance is not yet known (cf. Murase et al., 1993). With respect to the postmortem data so far published, one has to consider the possible dependence of neurotrophin concentrations on the stage of disease (cf. Hellweg et al., 1992b,c).

Levels of NGF and ChAT activity after experimental impairment of cerebral glucose and energy metabolism

Little is known about the *early* metabolism events in cerebral aging and age-related cerebral disorders such as AD (see Hellweg et al., 1992a,b for further discussion). Recent findings suggest disturbances in cerebral glucose metabolism playing an important role in cognitive disabilities during normal brain aging and in AD (see Wenk, 1989; Hoyer et al., 1991; Smith et al., 1992). Moreover learning impairments in aged rats are reported to be related to the extent of decreased glucose utilization in the septohippocampal system and the prefrontal cortex (Gage et al., 1984).

Prof. S. Hoyer's group from the University of Heidelberg (Germany) has developed an animal model in normal adult Wistar rats by injecting streptozotocin (STZ) in a subdiabetogenic dose by the intracerebroventricular (ICV) route without changing the blood glucose level (Mayer et al., 1990; Nitsch and Hoyer, 1991). In this model recent experimental data have shown both a disturbance of cerebral glucose utilization (Nitsch et al., 1989) and impaired brain energy metabolism within 3 weeks after a single ICV injection of STZ (Nitsch and Hoyer, 1991). Moreover retention in the passive avoidance test was significantly impaired 1 to 6 days after the learning trial performed 13 days after ICV treatment with STZ (Mayer et al., 1990). The observed deficits in learning and memory may be related to a distinct susceptibility of some specific neurotransmitters or -modulators in certain brain areas after a more general impairment of brain glucose and energy metabolism (Ding et al., 1992; Hellweg et al., 1992b). There is indication that cerebral glucose utilization may influence cholinergic biochemistry (for review see Wenk, 1989).

In fact, NGF levels were raised after ICV STZ treatment by about the same magnitude in target regions of cholinergic basal forebrain neurons of adult (1-year-old) Wistar rats (Hellweg et al., 1992b) as previously demonstrated in aged learning-impaired Sprague-Dawley rats (Hellweg et al., 1990; cf. Otten et al., 1989). In contrast, the NGF content was decreased by 30% within 1 week after ICV STZ treatment in the septal region where NGF-responsive cell bodies are located (Hellweg et al., 1992b). Our results are in agreement with the idea of an impaired retrograde NGF removal after ICV treatment with STZ and probably reflect an ongoing cholinergic deafferentation (Hellweg et al., 1992b). Interestingly there is increasing evidence that the forebrain cholinergic neurons of aged rats have a reduced capacity to bind and transport NGF from their targets (Angelucci et al., 1988; Koh et al., 1989). Thus deficiency of NGF in perikarya of the basal forebrain (after ICV STZ treatment and during aging) may ultimately lead to impaired cholinergic transmission and thereby impair learning and memory functions (Hellweg et al., 1992b). So far, our results would be consistent with a variety of reports on severe reduction in number of NGF receptor-containing neurons in AD (see Strada et al., 1992). If NGF receptors are located on cholinergic neurons, receptor loss and consequent decrease in trophic support may precede cell degeneration in AD (Strada et al., 1992). The relationship between the loss of NGF receptors and the pathogenesis of the disease remains to be investigated (Strada et al., 1992) as well as whether the increase in some trophic factors observed by several groups in AD brain might reflect compensatory mechanism(s) (see Arendt and Brückner, 1992; cf. Crutcher et al., 1993) as is suggested by our animal data.

On day 21 after ICV STZ treatment, we found a significant decrease by 15% in ChAT activity in the left hippocampus, whereas ChAT activity levels in the remaining brain regions did not change significantly up to 3 weeks after ICV STZ treatment (Hellweg et al., 1992b). Likewise, in the brain of aged and cognitively impaired rodents, ChAT activity levels are reported to be only marginally affected (Fischer et al., 1989; Hellweg et al., 1990). This might be due to masking effects exerted by NGF-mediated stimulation of ChAT activity in the remaining functional neurons, as discussed elsewhere (Hellweg et al., 1990, 1992a,b).

ICV STZ treatment as a possible model for age-related cognitive dysfunction: pathophysiological and therapeutic implications

In contrast to the aged rat brain, where the pathophysiology of aging has probably started long before the investigation is started, our current approach provides the possibility of studying the early events leading to the kind of brain function that is operational in the presence of severe learning impairment (Hellweg et al., 1992c). The nature of the relationship between our observations and the ICV STZ-induced impairment of cerebral glucose metabolism causing pathobiochemical alterations comparable to those known to occur in incipient early-onset and incipient late-onset AD, remains to be investigated (for further discussion see Hoyer et al., 1991). It may be suggested that the critical event for memory impairment may be a central disturbance of glucose utilization, which ultimately affects NGF-responsive central cholinergic neurons (Hellweg et al., 1992b), although other transmitter systems (see Ding et al., 1992) and growth factors may also be affected. For the peripheral nervous system, we have recently shown that successful pancreatic islet transplantation reverses decreased NGF levels that are related to experimental diabetes mellitus (DM) and may ultimately account for some neuronal dysfunctions associated with DM (Hellweg et al., 1991). It remains to be investigated whether there is a similar situation in the central nervous system.

In conclusion, our results obtained so far suggest that the animal model used might permit studies of the early metabolic events in brain aging and neurodegenerative disorders, which may well lead to a better understanding of the underlying pathophysiological mechanisms. Alterations in central NGF and ChAT activity levels observed after ICV STZ treatment resemble those in behaviorally impaired aged rats in many respects (Hellweg et al., 1992b,c). Our results indicate that the impairment of passive avoidance learning observed in adult rats after short-term ICV treatment with STZ (Mayer et al., 1990) may be related to a perturbation of the forebrain cholinergic system. Such perturbation may be due to a diminished capacity of central NGF-responsive neurons to bind and/or transport NGF both after experimental impairment of central glucose metabolism and during aging (Hellweg et al., 1992c). Our results are consistent with the concept that — at least early — deprivation of target-derived retrograde messenger(s) such as NGF (and possibly related neurotrophins such as BDNF) accounts for some of the functional deficits known to occur in age-related dementia disorders. Thus (early) prevention of neurotrophin deprivation in neurotrophin-dependent neurons might play an important role in future therapy of disorders associated with degeneration of cholinergic neurons in the basal forebrain. The availability of sufficient quantities of recombinant human neurotrophins should allow comprehensive research programs in future (Vantini, 1992).

Acknowledgments

We like to thank Dr. Jockers-Scherübl for linguistic revision of the manuscript.

References

Alderson RF, Alterman AL, Barde YA, Lindsay RM (1990) Brain-derived neurotrophic factor increases survival and differentiated functions of rat septal cholinergic neurons in culture. Neuron 5: 297–306

Allen SJ, MacGowan SH, Treanor JJS, Feeney R, Wilcock GK, Dawbarn D (1991) Normal β-NGF content in Alzheimer's disease cerebral cortex and hippocampus. Neurosci Lett 131: 135–139

Angelucci L, Ramacci MT, Taglialatela G, Hulsebosch C, Morgan B, Werrbach-Perez K, Perez-Polo R (1988) Nerve growth factor binding in aged rat central nervous system: effect of acetyl-L-carnitine. J Neurosci Res 20: 491–496

Arendt T, Brückner M (1992) Perisomatic sprouts immunoreactive for nerve growth factor receptor and neurofibrillary degeneration affect different neuronal populations in the basal nucleus in patients with Alzheimer's disease. Neurosci Lett 148: 63–66

Auburger G, Heumann R, Hellweg R, Korsching S, Thoenen H (1987) Developmental changes of nerve growth factor and its mRNA in the rat hippocampus: comparison with choline acetyltransferase. Dev Biol 120: 322–328

Crutcher KA, Scott SA, Liang S, Everson WV, Weingartner J (1993) Detection of NGF-like activity in human brain tissue: increased levels in Alzheimer's disease. J Neurosci 13: 2540–2550

Ding A, Nitsch R, Hoyer S (1992) Changes in brain monoaminergic neurotransmitter concentrations in rat after intracerebroventricular injection of streptozotocin. J Cereb Blood Flow Metab 12: 103–109

Ebendal T (1989) NGF in CNS: experimental data and clinical implications. Prog Growth Factor Res 1: 143–159

Fischer W, Gage FH, Björklund A (1989) Degenerative changes in forebrain cholinergic nuclei correlate with cognitive impairments in aged rats. Eur J Neurosci 1: 34–45

Gage FH, Kelly PAT, Björklund A (1984) Regional changes in brain glucose metabolism reflect cognitive impairments in aged rats. J Neurosci 4: 2856–2865

Goedert M, Fine A, Hunt SP, Ullrich A (1986) Nerve growth factor mRNA in peripheral and central rat tissue and in the human central nervous system: lesion effects in the rat brain and levels in Alzheimer's disease. Mol Brain Res 1: 85–92

Hellweg R, Hock C, Hartung HD (1989) An improved rapid and highly sensitive enzyme immunoassay for nerve growth factor. Technique 1: 43–48

Hellweg R, Fischer W, Hock C, Gage FH, Björklund A, Thoenen H (1990) Nerve growth factor levels and choline acetyltransferase activity in the brain of aged rats with spatial memory impairments. Brain Res 537: 123–130

Hellweg R, Wöhrle M, Hartung HD, Stracke H, Hock C, Federlin K (1991) Diabetes mellitus-associated decrease in nerve growth factor levels is reversed by allogeneic pancreatic islet transplantation. Neurosci Lett 125: 1–4

Hellweg R, Hock C, Hartung HD (1992a) The physiological role of β-nerve growth factor and its possible pathophysiological implication in the central nervous system. In: Emrich HM, Wiegand M (eds) Integrative biological psychiatry. Springer, Berlin Heidelberg New York Tokyo, pp 105–122

Hellweg R, Nitsch R, Hock C, Jaksch M, Hoyer S (1992b) Nerve growth factor and choline acetyltransferase activity levels in the rat brain following experimental impairment of cerebral glucose and energy metabolism. J Neurosci Res 31: 479–486

Hellweg R, Hock C, Nitsch R, Hartung HD, Mayer G, Hoyer S (1992c) Impaired cerebral glucose and energy metabolism alters nerve growth factor levels and choline acetyltransferase activity in the rat forebrain cholinergic system: a model for age-related cognitive impairment? In: Ruiz-Torres A, Hofecker G (eds) Modification of the rate of aging. Facultas, Wien, pp 163–169

Hoyer S, Nitsch R, Oesterreich K (1991) Predominant abnormality in cerebral glucose utilization in late-onset dementia of the Alzheimer type: a cross-sectional comparison against advanced late-onset and incipient early-onset cases. J Neural Transm [PD Sect] 3: 1–14

Kesner RP (1988) Reevaluation of the contribution of the basal forebrain cholinergic system to memory. Neurobiol Aging 9: 609–616

Knusel B, Winslow JW, Rosenthal A, Burton LE, Seid DP, Nikolics K, Hefti F (1991) Promotion of central cholinergic and dopaminergic neuron differentiation by brain-derived neurotrophic factor but not neurotrophic-3. Proc Natl Acad Sci USA 88: 961–965

Koh S, Chang P, Collier TJ, Loy R (1989) Loss of NGF receptor immunoreactivity in basal forebrain neurons of aged rats: correlation with spatial memory impairment. Brain Res 498: 397–404

Lorigados L, Söderström S, Ebendal T (1992) Two-site enzyme immunoassay for βNGF applied to human patient sera. J Neurosci Res 32: 329–339

Mayer G, Nitsch R, Hoyer S (1990) Effects of changes in peripheral and cerebral glucose metabolism on locomotor activity, learning and memory in adult male rats. Brain Res 532: 95–100

Murase K, Nabeshima T, Robitaille Y, Quirion R, Ogawa M, Hayashi K (1993) NGF level is not decreased in the serum, brain-spinal fluid, hippocampus, or parietal cortex of individuals with Alzheimer's disease. Biochem Biophys Res Comm 193: 198–203

Nitsch R, Mayer G, Hoyer S (1989) The intracerebroventricularly streptozotocin-treated rat: impairment of cerebral glucose metabolism resembles the alterations of carbohydrate metabolism of the brain in Alzheimer's disease. J Neural Transm [PD Sect] 1: 109–110

Nitsch R, Hoyer S (1991) Local action of the diabetogenic drug, streptozotocin, on glucose and energy metabolism in rat brain cortex. Neurosci Lett 128: 199–202

Otten U, Lorez HP, Kettler R, Keller HH, Gallati H (1989) Age-related changes in nerve growth factor, neurotransmitters, and monoamine oxidase-B in rat and mouse brains. In: Kewitz H, Thomsen T, Bickel U (eds) Pharmacological interventions on central cholinergic mechanisms in senile dementia (Alzheimer's disease). Zuckschwerdt, München, pp 176–177 (Klinische Pharmakologie, vol 2)

Philips HS, Hains JM, Armanini M, Laramee GR, Johnson SA, Winslow JW (1991) BDNF mRNA is decreased in the hippocampus of individuals with Alzheimer's disease. Neuron 7: 695–702

Raivich G, Hellweg R, Graeber MB, Kreutzberg GW (1990) The expression of growth factor receptors during nerve regeneration. Restor Neurol Neurosci 1: 217–223

Smith GS, de Leon MJ, George AE, Kluger A, Volkow ND, McRae T, Golomb J, Ferris SH, Reisberg B, Ciaravino J, La Regina ME (1992) Topography of cross-sectional and longitudinal glucose metabolic deficits in Alzheimer's disease — pathophysiologic implications. Arch Neurol 49: 1142–1150

Strada O, Hirsch EC, Javoy-Agid F, Lehéricy S, Ruberg M, Hauw JJ, Agid Y (1992) Does loss of nerve growth factor receptors precede loss of cholinergic neurons in Alzheimer's disease? An autoradiographic study in the human striatum and basal forebrain. J Neurosci 12: 4766–4774

Thoenen H, Bandtlow C, Heumann R (1987) The physiological function of nerve growth factor in the central nervous system: comparison with the periphery. Rev Physiol Biochem Pharmacol 109: 145–178

Thoenen H (1991) The changing scene of neurotrophic factors. Trends Neurosci 14: 165–170

Vantini G (1992) The pharmacological potential of neurotrophins: a perspective. Psychoneuroendocrinology 17: 401–410

Wenk GL (1989) An hypothesis on the role of glucose in the mechanism of action of cognitive enhancers. Psychopharmacology 99: 431–438

Author's address: Dr. R. Hellweg, Psychiatrische Klinik und Poliklinik, Freie Universität Berlin, Eschenallee 3, D-14050 Berlin, Federal Republic of Germany.

J Neural Transm (1994) [Suppl] 44: 219–230

Alzheimer's disease and transgenic mice

Ch. Czech[1], **C. Masters**[2], and **K. Beyreuther**[1]

[1] Center for Molecular Biology Heidelberg (ZMBH), University of Heidelberg, Heidelberg, Federal Republic of Germany
[2] Department of Pathologie, University of Melbourne, Parkville, Victoria, Australia

Summary. Transgenic mice overexpressing the three major neuronal isoforms of the human amyloid precursor protein (APP), APP695, APP751, APP770 may provide an animal model for the analysis of the mechanisms and risk factors leading to amyloid deposition in Alzheimer's disease (AD) and Downs syndrome (DS). We have therefore generated transgenic mice expressing these isoforms under the control of the strong metallothionin promoter. Although we can demonstrate expression of transgenic APP in several tissues including brain, expression levels never exceeded those of the endogenous mouse APP. So far we have not been able to detect pathological changes resembling those of AD and DS. However we could demonstrate significant changes in spatial navigation tasks and motor behavior in the transgenic mice. The question remains open whether overexpression of APP is sufficient to induce Alzheimer pathology.

Introduction

Alzheimer's disease (AD) is an irreversible neurodegenerative disorder of the human central nervous system. The neuronal dysfunction is correlated with massive deposition of fibrillary aggregates at distinct locations: intracellularly as neurofibrillary tangles, extracellularly as amyloid plaques and perivaskular amyloid (Glenner and Wong, 1984; Masters et al., 1985; Müller-Hill and Beyreuther, 1989). Brains from patients with Down's syndrome (DS) show the same pathology, but with a fifty-year earlier onset than the normal karyotype, that is otherwise characteristic for AD (Rumble et al., 1989). The major proteinaceous component of the corresponding fibrils isolated from brains of patients with AD and DS is a peptide of maximally 39–43 amino acid residues called the βA4 peptide. The βA4 peptide is derived from a larger precursor called amyloid precursor protein (APP) (Kang et al., 1987). The corresponding APP gene maps to the middle of the long arm of Chromosome 21 in band 21q21, close to the obligate Down's syndrome trisomic region (Patterson et al., 1988). The APP gene consists of 19 exons that are alternatively spliced into a variety of different products named to their length in amino acids (Kang et al., 1987; König et al., 1992; Ponte et al., 1988; Sandbrink et al., 1994; Tanzi et al., 1988). APP695, the isoform that lacks exon 7 and 8, is the most abundant isoform in the brain (Kang et al., 1990; König et al., 1991; Neve et al., 1988). In peripheral organs, but also in microcoglia and astrocytes the major primary translation product is APP751 and APP770 that include exon 7 and 8 respectively (Mönning et al., 1992; Sandbrink et al., 1994; Tanzi et al., 1988; Weidemann et al., 1989). Exon 7 encodes a 57-amino acid domain with strong homology to a serine protease inhibitor of the Kunitz type (KPI) (Kitaguchi et al., 1988). Exon 8 encodes a 19-amino acid domain with homology to the MRC Ox-2 antigen found on rat thymocytes (Beyreuther and Müller-Hill, 1989; Clark et al., 1985).

Recently, a third alternative used splice site in the APP gene was identified involving exon 15 of the APP gene (König et al., 1992; Mönning et al., 1992; Sandbrink et al., 1994). This exon code for 18 amino acids that precede the βA4 region of APP by 16 amino acids. It was therefore suggested that alternative splicing of exon 15 may influence the pathway leading to the generation of the βA4 peptide. The corresponding splice forms were first discovered in peripheral leukocytes, therefore denoted leukocyte-derived APP (L-APP). In the meantime splice forms lacking exon 15 have been detected in a variety of tissues and cells. In the brain substantial amounts of L-APP were detected in astrocytes, microgia cells and in blood vessels (Banati et al., 1993; Sandbrink et al., 1994). Neurons express mainly APP isoforms and, if at all, very low amounts of L-APP (Sandbrink et al., 1994).

The generation of the βA4 subunit with 39 to 40 residues from APP has been shown to be a normal processing event in different cell types (Busciglio et al., 1993; Haass et al., 1992; Shoji et al., 1992) and this βA4 subunit can be detected in cerebrospinal fluid and plasma of both normal and AD patients (Seubert et al., 1992; Shoji et al., 1992). However the process which leads to accumulation and aggregation of βA4 in the brains of AD patients is still unclear, although soluble βA4 could be transformed into aggregating molecules in vitro by metal-catalyzed oxidation (Dyrks et al., 1993). It also remains obscure weather neuronal cells are able to release the βA4 subunit of the parenchymal deposits which are composed of βA4 molecules of up to 42/43 residues (Kang et al., 1987; Masters et al., 1985). So far the majority of the βA4 proteins that are generated normally by cells, express the 39/40-residue subunit of vascular amyloid deposits (Haass et al., 1992; Shoji et al., 1992).

Several mutations in the APP gene are linked to familial forms of AD (for Review see Mullan et al., 1993). Recently it has been shown that in comparison to wild-type APP, one of these mutations increases the amount of βA4 released by transfected cells, by a factor of six (Citron et al., 1992). However the mechanisms underlying the processing of APP into βA4 are not understood.

Expression of APP in cell culture systems did not lead to characteristic aggregates of βA4 observed in Alzheimer's disease (Weidemann et al., 1989). The absence of cell-cell interactions in cell culture and use of cells maintained in an altered state of differentiation may not be sufficient to simulate the processing of APP in an Alzheimer's brain.

Similar deposits as observed in AD and DS have been detected in aged monkeys, dogs and polar bears (von Braunmühl, 1956; Price, 1986; Selkoe et al., 1987; Stuble et al., 1985; Wisniewski et al., 1970). The APP gene in mouse and rat is highly conserved (Shivers et al., 1988; Yamada et al., 1987), but the pathological changes similar to AD have so far not been found to occur spontaneously in rodents. This may be due to the short life span of these animals and due to the three amino acid exchanges found in the βA4 sequence of murine and rat APP. Although in vitro experiments with synthetic βA4 peptide of mouse and man revealed no difference in the tendency to form fibrils (Hilbich et al., 1991), there is the growing body of evidence that rodent βA4 sequences exhibit reduced amyloidogenicity in vitro and in vivo (Dyrks et al., 1993).

An animal model of AD in laboratory rodents would greatly facilitate analysis of the mechanisms and the identification of risks factors leading to βA4 aggregation and amyloid formation. It would also facilitate to study the effect of amyloid deposits in brain function with diagnostic and therapeutic implications. We choose to develop transgenic mice for human APP as an in vivo approach to analyze whether overexpression of APP would result in βA4 pathology. Here we report the generation of transgenic mice expressing the APP isoforms 695, 751 and 770 under the control of the strong human metallothyonin promoter (Karin and Richard, 1982). This approach was chosen since there is direct evidence that overexpression of APP is sufficient to generate Alzheimer pathology:

- Overexpression of APP at mRNA and protein level in DS is correlated with earlier onset of βA4 protein deposition. In these patients this onset precedes that of normal individuals by about 50 years (Davies et al., 1988; Rumble et al., 1989).
- Transplants of hippocampal tissue from fetal mice with trisomy 16 into forebrain locations of 6-week old normal mice showed βA4 immunoreactivity. In situ hybridization of mouse embryos with trisomy of chromosome 16, the equivalent of the human chromosome 21, showed increased amount of murine APP mRNA. This increase and the factors induced by the transplantation are presumably responsible for the βA4 immunoreactivity in the transplants since control transplants with tissues from normal mice did not develop this pathology (Richards et al., 1991).

Material and methods

Construction of the transgenic expression vectors

For the construction of APP695 vector, a 0.8 kb BamHI/HindIII fragment of the hMTII promoter was fused with a 3 kb SmaI fragment of the human APP695 cDNA which comprises the full length coding region and the 3′ untranslated part up to the first polyadenylation site. We used pUC 19 as a vector. The DNA used for microinjection was purified over a CsCl gradient and linearized with the SspI site in pUC 19. For the construction of the APP751/770 vectors, the β-globin intron was inserted into the BamHI site of the 5′ untranslated part of the APP cDNA. The SV40 polyA and large T-antigen Intron were inserted into the PvuI site of the 3′ untranslated part of the APP cDNA. The DNA used for microinjection was purified over a CsCl gradient and linearized with a unique NotI site 3′ of the SV40 polyadenylation sequence.

Generation of transgenic mice

The transgenic MTA4 line was prepared at the Jackson Laboratory, USA by microinjecting the APP695 construct into (C57BL6/J × LT/Sv) F1 mice. Breeding pairs were transported to ZMBH laboratory in Heidelberg. The transgenic lines 106/2 for APP770 and 133/1 for APP751 were prepared in collaboration with D. Ganten at the University of Heidelberg, Institute for Pharmacology, by microinjection of the APP751/770 constructs into NMRI mice. Offspring of the transgenic mice were tested by standard southern-blot analysis, using a 1.6 kb BamHI fragment of the APP cDNA, for the presence of the transgene.

Expression analysis

S1 protection analysis of total RNA was done as described previously by König (1991). For immunohistochemical staining the monoclonal antibody Mab 22C11 and the polyclonal antibody fd42 were raised against a fusion protein, which was expressed in E. coli. The construct contained the sequence of APP695 and the VH domain of a murine IgM heavy chain. The epitope of the monoclonal Antibody is near the N-terminus of the cystein-rich domain of APP. Another polyclonal Antibody was generated in rabbits immunized with synthetic peptide which represents the 2–43 amino acid residues of βA4. The immunohistochemical studies were performed on frozen brain sections from transgenic and non transgenic animals that were processed in parallel. Analysis of brain protein extracts were performed by homogenizing transgenic and non transgenic brains in a buffer containing 0.32 M Saccharose and several proteinase inhibitors. Protein concentration of these extract were determined and same amounts fractionated by SDS-PAGE (Laemmli, 1970) and transferred to nitrocellulose (Tobwin et al., 1979). The nitrocellulose sheet was soaked in PBS containing 1% bovine serum albumin and then incubated at room temperature for 2 h with monoclonal antibody 22C11 or polyclonal antibody fd42. The nitrocellulose sheet was washed three times with Tris buffered saline (10 mM Tris, pH 8.0, 150 mM NaCl, 0.05% Tween 20). Specific binding was visualized with the protoblot alkaline phosphatase system (Promega, Heidelberg).

Behavioral testing

Behavioral testing on transgenic animals was done in Cambridge, UK as described previously by Yamaguchi (1991).

Results

Generation of transgenic mice

For the first generation of transgenic mice we chose the APP695 isoform that is predominantly expressed in brain and spinal cord of humans and rodents, to construct an transgenic expression vector. We selected the promoter hMTII of the human metallothionin gene. This promoter is inducible by heavy metal ions and glucocorticoids (Karin and Richard, 1982). Although the constitutive expression driven by this promoter is considerably high, it has been shown that application of zinc(II) or cadmium(II) to the mice can induce transcription several fold (Sandgren et al., 1990). The transgenic mouse line MTA4 positive for this construct contained 2–3 copies of the APP695 transgene and was used for the further experiments described (Richards et al., 1991).

For developing the second generation of transgenic mice we used the APP751 and the APP770 cDNA. In order to increase efficiency of expression in these constructs, we introduced an additional beta-globin intron 5′ to the APP cDNA and the intron and polyadenylation site of the SV40 large T-antigen 3′ to the APP cDNA, which has been shown to increase the efficiency of expression in transgenic mice considerably (Brinster et al., 1988; Palmiter et al., 1991). Two transgenic lines named 106/2 for APP770 and 133/1 for APP751 respectively, were established. Approximately 25–30 copies of the transgene were integrated into the genome of both lines.

Analysis for expression of the APP transgene

The expression of the transgene on the mRNA level was determined by S1-nuclease protection. This method allows not only to distinguish between the transcripts of the transgene and the endogenous APP transcripts but also to quantify the relative amount of mRNA.

The human APP695 transcript was found to be constitutively expressed in a number of tissues. Short term induction of the hMTII promoter was achieved by treating the animals with 10 mM Cd(II) for 8 hours. We did not find any enhancement of transgenic activity in the brain. This is probably due to the induction of the hMTII promoter by endogenous brain glucocorticoid levels through hormone responsive elements. In kidney we were able to detect a 5-fold increase in transgene activity. In heart and lung the activity was slightly increased whereas spleen, muscle, thymus and testis did not response to that treatment with stronger transcriptional activity. Compared to the endogenous expression of APP, the amount of transgenic APP in the brains of the MTA4 mice was at or below 10%. Long term induction experiments were performed by giving the animals drinking water supplemented with $ZnCl_2$, which is far less toxic then the $CdCl_2$ that was used in the short term experiments. Analysis of the mice after a six month exposure to zinc(II) revealed a 2–3 fold increase of transgenic mRNA in the brain.

Zinc-ions are believed to cross the blood brain barrier by using the transferrin receptor and would thus be available to activate the hMTII promoter in brain cells.

Analysis of brain protein extracts from the transgenic MTA4 mice as well as from control mice showed only a slight increase of APP protein. Unfortunately after some generation of breeding the MTA4 mice, the amount of expression of the transgene decreased considerably. Although at the DNA level they still contained the functional transcription unit.

Expression analysis of the lines 106/2 transgenic for APP700 and 133/1 transgenic for APP751 was also performed by S1-nuclease protection. Despite the fact that the expression vectors for these constructs were optimized for high levels of expression; we were unable to detect high transgene expression. Even after long term induction of the hMTII promoter with 25 mM zinc(II), which was supplied in the drinking water for over 12 months, the expression level in the brain of the mice never exceeded one percent of the endogenous APP expression.

Immunohistochemical analysis of the transgenic mice

In order to examine the cells that express the transgene in the mouse brain, we used the immunoperoxydase method for the localization of the transgenic product. In these immunohistochemical studies we used monoclonal and polyclonal antibodies raised against different epitopes of APP. These antibodies are able to detect all membrane bound isoforms of APP since the epitopes are present in these proteins.

In the brain of a normal mouse the APP staining is strongest for pyramidal cells of the hippocampus as well as the purkinje cells in the cerebellum. We estimated that these cells contain more than hundred thousand but less than a million APP molecules per cell. After inducing the transgene with zinc(II), we could show a clear enhancement of APP expression in pyramidal and purkinje cells in the MTA4 mice. In the brains of the APP751 and APP770 transgenic mice, we did not see any enhancement of APP expression even after zinc(II) induction of 12 months, probably due to the low levels of transgenic mRNA expressed in these mice.

We used a polyclonal antibody raised against a synthetic peptide representing amino acid residues 2–43 of the human βA4 sequence to analyze sections of brain tissue of APP695, APP751 and APP770 heterozygous and homozygous transgenic mice with and without zinc(II) treatment. So far we have been unable to demonstrate any βA4 immunoreactivity in the brains of these animals, some of which have been exposed to zinc(II) for 24 months.

Behavioral testing of transgenic animals

Alzheimer's disease is not only manifested by pathological changes like βA4 immunoreactivity in the brains. It is also the most common cause of late-life intellectual impairment in men. In order to assay more subtile changes in the brains of the transgenic mice, we investigated in collaboration with S.-J. Richard and S. B. Dunnet, Cambridge, whether these mice develop functional deficits in spatial learning tasks. This method has been widely used to evaluate several different animal models for aging and dementia (Gage et al., 1984; Morris, 1981; Whishaw et al., 1985).

The mice transgenic for APP695 showed moderate impairment in a spatial navigation task, used to assess learning and memory deficits in rodents (Yamaguchi et al., 1991). The mice transgenic for APP751 although they did not express amounts of human APP transcripts, exceeding 1% of the endogenous APP, showed significant deficits in motor behavior compared with littermate controls (Richards et al., manuscript in preparation).

Discussion

The aim of this study was to analyze whether transgenic expression of APP can lead to deposition of βA4 in the brains of transgenic mice expressing the three major neuronal isoforms of APP. Since the generation of transgenic mice overexpressing a protein that is already present in high levels of 10^5-10^6 copies per cell, is not without conceptual and technical problems, we have chosen a strong promoter that has been shown to achieve high levels of transgene expression in mice. The hMTII promoter we used, can induce the level of expression several fold when applying zinc(II) to the drinking water of the animals (Sandgren et al., 1990). Toxic effects of overexpression of APP during embryogenesis, resulting in prenatal lethality of the embryo, would thus be avoided.

Several APP mRNA isoforms are generated by alternative splicing of the APP gene. In mouse brain APP695 is predominantly expressed in neurons whereas APP751/770 and the corresponding L-APP733/752 isoforms are the major splice products in glia cells. In contrast to glia, neurons do mainly express APP isoforms which include the 18 amino acid residues on exon 15. But it is still unclear which APP or L-APP isoform is preferably processed into βA4. The ratio of the isoforms APP695 to APP751/770 are changing during the development and aging of the human but not of the rodent brain. There is no clear answer whether there is a different distribution of APP splice forms in AD patients, compared to age-matched controls. This issue is still under controversial discussion (König et al., 1991).

For the generation of the transgenic mice we used the three major neuronal isoforms APP695, APP751 and APP770. The ectopic expression of these isoforms under the control of the strong hMTII promoter should provide insight into the physiological function of APP and its processing into βA4 in different cell types and tissues.

With the mouse line MTA4 expressing the human APP695 we could show expression of human APP in several tissues, including brain. We have also been able to show induction of the hMTII promoter in the brain of the MTA4 mice, after long term induction with zinc(II). Despite the fact of using a strong promoter driving the expression of the transgene we have not been able to detect clear overexpression of APP protein and mRNA in any of our transgenic mouse lines. It might well be that the constitutive expression levels of the hMTII promoter induced by endogenous glucocorticoids during embryogenesis resulted in prenatal lethality of the overexpressing mice. This in turn would suggest to use a different promoter that is silent during embryogenesis of the mouse.

The expression levels of human APP in our transgenic mice were not sufficient to develop any βA4 immunoreactivity. On the other hand it is also possible that human APP in mouse is not metabolized in the same way as it is in human brain, and does not lead to the same amounts of amyloidogenic APP fragments.

The mouse lines transgenic for APP751 and APP770 showed only low levels of transgene mRNA expression, despite the fact that the number of copies of the transgene integrated into the genome of these mice was very high (25–30 copies per mouse genome). It has been shown that methylation of transgenes increases with copy number (Methali et al., 1990). This could be one reason why the expression levels were thus low. Another possibility, that could also apply for the MTA4 mice, is negative selection.

The human APP695 transgenic mice, expressing substantial amounts of transgenic mRNA and protein, lost expression almost completely after several generations of breeding. Although this phenomenon remains obscure, high expression of APP isoforms, during embryonic development of the mice could lead to prenatal lethality or result in a reduced fertility. The latter was indeed observed.

Behavioral testing of our transgenic mice showed a significant difference in spatial navigation tasks and motor behavior. (Richard's et al., in preparation; Yamaguchi et al., 1991). However, it is not clear whether this is due to the expression of the transgene or to position effect of transgene integration.

Similar transgenic approaches by other investigators have also been unsuccessful in reproducing the βA4 deposition's characteristic for AD and DS. For mice expressing the human βA4 peptide under the control of the human APP promoter transgene expression at the protein level was not shown (Wirak et al., 1991). Transgenic mice expressing the βA4 region within the C-terminal hundred residues of APP have also been generated by several groups (Kammesheit et al., 1992; Kawabata et al., 1992; Sandhu et al., 1991). Expression of this construct which consists of the βA4 sequence at the N-terminus and which is believed to represent an intermediate in the pathway of amyloid release (Dyrks et al., 1993) did not produce the expected βA4 deposits, and if at all, did not lead to marginal histopathological changes. Expression of the entire coding region of APP in transgenic mice, by us and several other groups, has so far not resulted in neuronal histopathological changes resembling these observed in AD and DS (Quon et al., 1991; Yamaguchi et al., 1991). A common feature of all these studies is the low level of transgene expression at the protein level compared with that of endogenous mouse APP protein, suggesting that the direct approach of generating transgenic mice, by introducing amyloidogenic constructs may not be sufficient to overexpress βA4 containing molecules.

Using a different approach, Lamb et al. (1993) introduced a yeast artifical chromosome harboring the entire human APP gene into embryonic stem cells. The generation of transgenic mice resulted in an expression of human APP mRNA and proteins at levels comparable to that of the endogenous murine APP. Up to now it has not been demonstrated that this approach leads to the desired pathological changes in the brains of such "transgenomic" mice. The question remains open whether overexpression of APP alone is sufficient to induce Alzheimer pathology.

Acknowledgements

We acknowledge essential contributions and collaborations by J. M. Saalbaum, J. Beer, G. M. Carlson, T. Dyrks, S. B. Dunnet, S-J. Richards, E. Schlichtmann, Y. H. Suh and F. Zimmermann. We thank P. J. Tienari for critically reading the manuscript. This work was supported by grant from the Deutsche Forschungsgemeinschaft through SFB 317 and 258 and the Bundesminister für Forschung und Technologie of Germany.

References

Banati RB, Gehrmann J, Czech C, Mönning U, Jones LL, König G, Masters CL, Beyreuther K, Kreutzberg GW (1993) Early and rapid denovo synthesis of Alzheimer βA4 amyloid precursor protein (APP) in activated mircoglia. Glia 9:199–210

Brinster RL, Allen JM, Behringer RE, Gelinas RE, Palmiter RD (1988) Introns increase transcriptional efficency in transgenic mice. Proc Natl Acad Sci USA 85: 836–840

Busciglio J, Gabuzda DH, Matsudaira P, Yankner BA (1993) Generation of beta-amyloid in the secretory pathway in neuronal and nonneuronal cells. Proc Natl Acad Sci USA 90: 2092–2096

Citron M, Oltersdorf T, Haass C, McConlogue L, Hung AY, Seubert P, Vigo PC, Lieberburg I, Selkoe DJ (1992) Mutation of the beta-amyloid precursor protein in familial Alzheimer's disease increases beta-protein production. Nature 360: 672–674

Clark M, Gagnon J, William AF, Barclay AN (1985) MRC Ox-2 antigen: a lymphoid/neuronal membrane glycoprotein with a structure like a single immunoglobuline light chain. EMBO J 4: 113–118

Davies L, Wolska B, Hilbich C, Multhaup G, Martins R, Simms G, Beyreuther K, Masters CL (1988) A4 amyloid protein deposition and the diagnosis of Alzheimer's disease: prevalence in aged brains determined by immunocytochemistry compared with conventional neuropathologic techniques. Neurology 38: 1688–1693

Dyrks T, Dyrks E, Masters CL, Beyreuther K (1993) Amyloidogenicity of rodent and human βA4 sequences. FEBS 324: 231–235

Dyrks T, Dyrks E, Hartmann T, Masters C, Beyreuther K (1992) Amyloidogenicity of beta A4 and beta A4-bearing amyloid protein precursor fragments by metal-catalyzed oxidation. J Biol Chem 267: 18210–18217

Gage FH, Björklund A, Stenevi U, Dunnett SB, Kelly PAT (1984) Intrahippocampal septal grafts ameliorate learning impairments in aged rats. Science 225: 533–536

Glenner GG, Wong (1984) Alzheimer's disease: inital report of the purification and characterisation of a novel cerebrovascular amyloid protein. C.M. Biochem Biophys Res Commun 120: 885

Haass C, Koo EH, Mellon A, Hung AY, Selkoe DJ (1992) Targeting of cell-surface beta-amyloid precursor protein to lysosomes: alternative processing into amyloid-bearing fragments. Nature 357: 500–503

Hilbich C, Kisters WB, Reed J, Masters CL, Beyreuther K (1991) Human and rodent sequence analogs of Alzheimer's amyloid beta A4 share similar properties and can be solubilized in buffers of pH 7.4. Eur J Biochem 201: 61–69

Kammesheidt A, Boyce FM, Spanoyannis AF, Cummings BJ, Ortegon M, Cotman C, Vaught JL, Neve RL (1992) Deposition of beta/A4 immunoreactivity and neuronal pathology in transgenic mice expressing the carboxyl-terminal fragment of the Alzheimer amyloid precursor in the brain. Proc Natl Acad Sci USA 89: 10857–10861

Kang J, Muller HB (1990) Differential splicing of Alzheimer's disease amyloid A4 precursor RNA in rat tissues: PreA4(695) mRNA is predominantly produced in rat and human brain. Biochem Biophys Res Commun 166: 1192–1200

Kang J, Lemaire HG, Unterbeck A, Salbaum JM, Masters CL, Grzeschik KH, Multhaup G, Beyreuther K, Muller HB (1987) The precursor of Alzheimer's disease amyloid A4 protein resembles a cell-surface receptor. Nature 325: 733–736

Karin M, Richards RI (1982) Human metallothionine genes-primary structure of metallothionine-II gene and a related processed gene. Nature 299: 797–802

Kawabata S, Higgins GA, Gordon JW (1991) Amyloid plaques, neurofibrillary tangles and neuronal loss in brains of transgenic mice overexpressing a C-terminal fragment of human amyloid precursor protein [retracted by Kawabata S, Higgins GA, Gordon JW. In: Nature 356: 23 (1992) and Nature 356: 265 (1992) [see comments]. Nature 354: 476–478

Kitaguchi N, Takahashi Y, Tokushima Y, Shiojiri S, Ito H (1988) Novel precursor of Alzheimer's disease amyloid protein shows protease inhibitory activity. Nature 331: 530–532

König G, Mönning U, Czech C, Prior R, Banati R, Schreiter GU, Bauer J, Masters CL, Beyreuther K (1992) Identification and differential expression of a novel alternative splice isoform of the beta A4 amyloid precursor protein (APP) mRNA in leukocytes and brain microglial cells. J Biol Chem 267: 10804–10809

König G, Salbaum JM, Wiestler O, Lang W, Schmitt HP, Masters CL, Beyreuther K (1991) Alternative splicing of the beta A4 amyloid gene of Alzheimer's disease in cortex of control and Alzheimer's disease patients. Brain Res Mol Brain Res 9: 259–262

Laemmli UK (1970) Cleavage of structural proteins during the assembly of the head of bacteriophage T4. Nature 227: 680

Lamp BT, Sisodia SS, Lawler AM, Slunt HH, Kitt CA, Kearns WG, Pearson PL, Price DL, Gearhart JD (1993) Introduction and expression of the 400 kilobase precursor amyloid protein gene in transgenic mice. Nature Gene 5: 22–29

Masters CL, Multhaup G, Simms G, Pottgiesser J, Martins RN, Beyreuther K (1985) Neuronal origin of a cerebral amyloid: neurofibrillary tangles of Alzheimer's disease contain the same protein as the amyloid of plaque cores and blood vessels. EMBO J 4: 2757–2763

Masters CL, Simms G, Weinman NA, Multhaup G, McDonald BL, Beyreuther K (1985) Amyloid plaque core protein in Alzheimer's disease and Down syndrome. Proc Natl Acad Sci USA 82: 4245–4249

Methali M, LeMeur M, Lathe R (1990) The methylation-free status of a housekeeping transgene is lost at high copy number. Gene 91: 179–184

Monning U, Konig G, Banati RB, Mechler H, Czech C, Gehrmann J, Schreiter GU, Masters CL, Beyreuther K (1992) Alzheimer beta A4-amyloid protein precursor in immunocompetent cells. J Biol Chem 267: 23950–23956

Morris RGM (1981) Learn Motiv 12: 239–249

Mullan M, Crawford F (1993) Genetic and molecular advances in Alzheimer's disease. TINS 16: 398–403

Müller HB, Beyreuther K (1989) Molecular biology of Alzheimer's disease. Annu Rev Biochem 58: 287–307

Neve RL, Finch EA, Bird ED, Benowitz LI (1988) Growth-associated protein GAP-43 is expressed selectively in associative regions of the adult human brain. Proc Natl Acad Sci USA 85: 638–642

Palmiter RD, Sandgren EP, Avarbock MR, Allen DD, Brinster RL (1991) Heterologous introns can enhance expression of transgenes in mice. Proc Natl Acad Sci USA 88: 478–482

Patterson D, Gardiner K, Kao FT, Tanzi R, Watkins P, Gusella JF (1988) Mapping of the gene encoding the beta-amyloid precursor protein and its relationship to the Down syndrome region of chromosome 21. Proc Natl Acad Sci USA 85: 8266–8270

Ponte P, Gonzalez DP, Schilling J, Miller J, Hsu D, Greenberg B, Davis K, Wallace W, Lieberburg I, Fuller F (1988) A new A4 amyloid mRNA contains a domain homologous to serine proteinase inhibitors. Nature 331: 525–527

Price DL (1986) New perspectives on Alzheimer's disease. Annu Rev Neurosci 9: 489–512

Quon D, Wang Y, Catalano R, Scardina JM, Murakami K, Cordell B (1991) Formation of beta-amyloid protein deposits in brains of transgenic mice. Nature 352: 239–241

Richards SJ, Waters JJ, Beyreuther K, Masters CL, Wischik CM, Sparkman DR, White CI, Abraham CR, Dunnett SB (1991) Transplants of mouse trisomy 16 hippocampus provide a model of Alzheimer's disease neuropathology. Embo J 10: 297–303

Rumble B, Retallack R, Hilbich C, Simms G, Multhaup G, Martins R, Hockey A, Montgomery P, Beyreuther K, Masters CL (1989) Amyloid A4 protein and its precursor in Down's syndrome and Alzheimer's disease. N Engl J Med 320: 1446–1452

Sandbrink R, Master CL, Beyreuther K (1994) βA4-Amyloid protein precursor mRNA isoforms without exon 15 (L-APP mRNAs) are ubiquitously expressed in rat tissues including brain, but not in neurons. J Cell Biol 261: 1510–1517

Sandgren EP, Luetteke NC, Palmiter RD, Brinster RL, Lee DC (1990) Overexpression of TGFa in transgenic mice: induction of epithelial hyperplasia, pancreatic metaplasia, and carcinoma of the breast. Cell 61: 1121–1135

Sandhu FA, Salim M, Zain SB (1991) Expression of the human beta-amyloid protein of Alzheimer's disease specifically in the brains of transgenic mice. J Biol Chem 266: 21331–21334

Selkoe DJ, Bell DS, Podlisny MB, Price DL, Cork LC (1987) Conservation of brain amyloid proteins in aged mammals and humans with Alzheimer's disease. Science 235: 873–877

Seubert P, Vigo PC, Esch F, Lee M, Dovey H, Davis D, Sinha S, Schlossmacher M, Whaley J, Swindlehurst C, et al (1992) Isolation and quantification of soluble Alzheimer's beta-peptide from biological fluids [see comments]. Nature 359: 325–327

Shivers BD, Hilbich C, Multhaup G, Salbaum M, Beyreuther K, Seeburg PH (1988) Alzheimer's disease amyloidogenic glycoprotein: expression pattern in rat brain suggests a role in cell contact. Embo J 7: 1365–1370

Shoji M, Golde TE, Ghiso J, Cheung TT, Estus S, Shaffer LM, Cai XD, McKay DM, Tintner R, Frangione B, et al (1992) Production of the Alzheimer amyloid beta protein by normal proteolytic processing. Science 258: 126–129

Struble RG, Price DL Jr, Cork LC, Price DL (1985) Senile plaques in cortex of aged normal monkeys. Brain Res 361: 267–275

Tanzi RE, McClatchey AI, Lamperti ED, Villa KL, Gusella JF, Neve RL (1988) Protease inhibitor domain encoded by an amyloid protein precursor mRNA associated with Alzheimer's disease. Nature 331: 528–530

von Braunmühl A (1956) Arch Psychiat Z Neurol 194: 396–414

Weidemann A, Konig G, Bunke D, Fischer P, Salbaum JM, Masters CL, Beyreuther K (1989) Identification, biogenesis, and localization of precursors of Alzheimer's disease A4 amyloid protein. Cell 57: 115–126

Whishaw IQ, O'Conner WT, Dunnett SB (1985) Behav Brain Res 17: 103–115

Wirak DO, Bayney R, Ramabhadran TV, Fracasso RP, Hart JT, Hauer PE, Hsiau P, Pekar SK, Scangos GA, Trapp BD, et al (1991) Deposits of amyloid beta protein in the central nervous system of transgenic mice [see comments]. Science 253: 323–325

Wisniewski HM, Johnson AB, Raine CS, Kay WJ, Terry RD (1970) Lab Invest 23: 287–296

Yamada T, Sasaki H, Furuya H, Miyata T, Goto I, Sakaki Y (1987) Complementary DNA for the mouse homolog of the human amyloid beta protein precursor. Biochem Biophys Res Commun 149: 665–671

Yamaguchi F, Richards SJ, Beyreuther K, Salbaum M, Carlson GA, Dunnett SB (1991) Transgenic mice for the amyloid precursor protein 695 isoform have impaired spatial memory. Neuroreport 2: 781–784

Authors' address: Dr. Ch. Czech, Center for Molecular Biology Heidelberg (ZMBH), University of Heidelberg, Im Neuenheimer Feld 282, D-69120 Heidelberg, Federal Republic of Germany.

J Neural Transm (1994) [Suppl] 44: 231–243

An aspect of Alzheimer neuropathology after suicide transport damage

I. P. Chessell[1], P. T. Francis[1], M-T. Webster[1], A. W. Procter[1], P. R. Heath[2], R. C. A. Pearson[2], and D. M. Bowen[1]

[1] Miriam Marks Department of Neurochemistry, Institute of Neurology, London, and [2] Department of Biomedical Sciences, University of Sheffield, Sheffield, United Kingdom

Summary. Concentrations of APP-like immunoreactivity have been determined by western blotting in a soluble fraction and two membrane fractions of brain cortex from demented patients (14 with Alzheimer's disease and 8 with other diagnoses). The concentration of APP in the soluble fraction correlated with the number of pyramidal neurones but not astrocytes or indices of interneurones. Experimental lesions in rats and quantitative autoradiography were used to investigate the cellular localisation of receptors. Lesions were produced by intrastriatal or intracortical injections of volkensin to destroy corticofugal and corticortical pyramidal neurones respectively. Volkensin treatment caused significant loss of pyramidal neurones which was accompanied by reduced binding to muscarinic cholinergic m_1 receptors. [^{3}H] 8-OH-DPAT (serotonin 1A receptors) binding was reduced only following intrastriatal volkensin. Results from the human and rat investigations are discussed in terms of the biology of cortical pyramidal neurones and drugs for the treatment of Alzheimer's disease.

Introduction

The regional and laminar distribution of receptors (including many subtypes) in the rat cerebral cortex are well known, yet the cellular localization of many is not clearly understood. The excitability of pyramidal neurones of the neocortex is affected by many neurotransmitters, based on in vitro electrophysiological studies (as reviewed by McCormick and Williamson, 1989). However, such studies have consistently shown that only a proportion of cells respond to a particular drug, which suggested that there was a substantial degree of heterogeneity with respect to the neurotransmitter receptors on pyramidal neurones.

It is well established that volkensin, a retrogradely transported toxic lectin from the kilyambiti plant, can be used to selectively destroy subpopulations of cortical and hippocampal pyramidal neurones distant from the site of injection (for a review see Wiley, 1992). By this mechanism, unilateral intrastriatal volkensin injections lead to loss of large, infragranular, pyramidal neurones of the rat frontal cortex with preservation of glutamic acid decarboxylase mRNA-positive interneurones (Pangalos et al., 1991, 1992). Francis and colleagues (Francis et al., 1992; Pangalos et al., 1991) have used receptor autoradiography with sections of rat brain from volkensin-injected animals to establish whether a subpopulation of cortical pyramidal cells (layer V corticostriatal projection neurones) are enriched with a particular receptor type. Corticocortical and corticostriatal pyramidal neurones are probably subject to qualitatively similar regulation, irrespective of species (McCormick and Williamson, 1989), so it is hoped that the approach will provide better understanding of the neurotransmitter receptor regulation of cortical pyramidal neurones in general. To date, serotonin $(5\text{-HT})_{1A}$ receptors, appear to be present in greatest numbers on corticostriatal neurones, by comparison with 5-HT_2 receptors and the N-methyl-D-aspartate (NMDA) receptor complex.

These studies may have particular relevance for treatment and improved diagnosis of neurodegenerative disorders such as Alzheimer's disease (AD), where loss of cortical pyramidal neurones underlies the pathology of the condition. This is because some receptors may provide the basis for development of novel ligands for in vivo brain imaging (as reviewed by Francis et al., 1993b, 1994) as well as targets for drug therapy. A substantial presynaptic cholinergic abnormality, accompanied by adequate coupling (Pearce and Potter, 1991) of the appropriate (Francis et al., 1993b) receptor (muscarinic m_1) in the brains of Alzheimer patients has suggested the basis of a rational palliative treatment (e.g. acetylcholinesterase inhibitors).

Evidence from cultured neurone cell lines and human ventricular CSF of an additional effect of such neurotransmitter replacement is now emerging. It is well known that β-amyloid precursor protein (APP) is subject to several rare dementia-producing genetic mutations and that this has been interpreted as the strongest evidence that the common non-familial form of AD results from APP mismetabolism. APP metabolism may be regulated by neuronal activity, in particular transmitter transduction mechanisms. These include m_1 and other receptors which use products of phosphatidylinositol bisphosphate (PIP_2) hydrolysis as second messengers and, following activation, increase APP secretion by the α-secretase pathway (Buxbaum et al., 1992; Clarke et al., 1993; Nitsch et al., 1992, 1993; Sinha and Lieberburg, 1992).

Neurobiological research has emphasized the role of amyloid deposition, exemplified by senile plaque formation, and loss of, or tangle formation within, subpopulations of cortical pyramidal neurones though it has not yet been possible to establish any direct link in AD between these cardinal features. However, pyramidal neurones of both rat and human brain appear to contain most of the APP and this is localised in granular structures. At the electron microscope level these appear to be the Golgi apparatus and lysosomes or other secretory granules (as reviewed by Shigematsu et al., 1992). These data are in keeping with the predominant localisation of APP mRNAs to pyramidal neurons (Bahmanyar et al., 1987). Glial cells in culture have also been shown to express APP but do not appear to secrete large amounts of the protein (as reviewed Haass et al., 1991).

This study has attempted to link data concerning receptor enrichment on pyramidal neurones using an animal paradigm with other results from humans to advance a hypothesis regarding APP, pyramidal neurones and the treatment of AD. Postmortem brain tissue from demented patients has been collected using techniques designed to minimize autolysis (Procter et al., 1990). Concentrations of the major species of APP, in a soluble fraction and two membrane fractions, have been determined. We have also investigated any intercorrelations of these measures with numbers of pyramidal neurones and astrocytes as well as with putative biochemical indices of interneurones. These were γ-aminobutyric acid (GABA) and the serotonin (5-HT) 2 receptor (Procter et al., 1988).

In volkensin treated animals we have previously reported on only one receptor that uses PIP_2 hydrolysis, the $5HT_2$ receptor, the present study extends the receptors investigated to other phospholipase C-linked receptors, m_1 and adrenoceptors (α_{1total} and the subtype α_{1b}). We have used [^{3}H] pirenzepine to label muscarinic receptors (predominantly the m_1 subtype; Buckley and Burnstock, 1986; Hulme et al., 1990) and [^{3}H] prazosin to label α_1 receptors (α_{1total}), and in the presence of 5-methyl urapadil to label the α_{1b} subtype (Zilles et al., 1991). We also report on the localisation of m_1 to corticocortical neurones.

Materials and methods

Human samples

Postmortem examinations were made, within 3.5 h of death, on fifteen patients who had a clinical diagnosis of dementia. The brains were removed and bisected sagitally. One hemisphere was preserved in 10% buffered formalin for histological examination, the other sliced coronally into 1-cm thick slices; these were placed into an ice-cold Krebs-Ringer phosphate buffer and transferred to the neurochemistry laboratory in an insulated box as previously described (Procter et al., 1988, 1990). Subjects with AD had clinical features, together with a degree of senile plaque formation and neurofibrillary degeneration, consistent with a diagnosis of AD. The other subjects did not have neuropathological features of AD (Harrison et al., 1994; Webster et al., 1994). Neocortex samples were processed as described (Arai et al., 1991) with minor modification for the measurement of APP-like immunoreactivity (Clarke et al., 1993; Webster et al., 1994). For determination of $5HT_2$ receptors [^{3}H] ketanserin was used (Procter et al., 1990) and total tissue GABA content was measured by HPLC (Procter et al., 1988).

From the tissue fixed in 10% neutral formalin, paraffin blocks were prepared by routine methods. For counts of pyramidal neurones, five sections of 6 µm thickness from BA 9 were cut and stained with cresyl violet and the number of nucleolated pyramidal neurones counted in cortical layers III and V in 10 consecutive non-overlapping microscope fields using a Weibel graticule at a magnification of ×250, from which the mean number of such cells per mm^2 was calculated as previously described (Neary et al., 1986). The sum of these two mean values is given as the "total number of pyramidal neurones". Sections of 8 μm thickness from BA 20 were stained with a monoclonal antibody against glial fibrillary acidic protein (Dako, High Wycombe, U.K.) and an avidin biotin peroxidase technique with a diaminobenzidine chromogen, previously described (Pangalos et al., 1991). Astrocytes were counted in alternate contiguous visual fields across all laminae of the cerebral cortex of one section and repeated four times at 0.12 mm intervals along the length the cortex at a magnification of ×1,000. From these data the mean number of astrocytes per mm^3 was determined using the optical dissector method (Coggeshall, 1992). These values were expressed per mm^2 of a hypothetical 6 μm thick section, equivalent to that used for pyramidal neurone counts.

Lesion and histological procedures in rats

There were only a limited number of sections remaining from volkensin-injected rats prepared previosly by injection of 2 ng of the toxin (Pangalos et al., 1991); hereafter referred to as the first group, n = 6). Therefore, second (n = 6) and third (n = 5) groups of volkensin-injected rats were prepared. Destruction of the cells of origin of the corticostriatal pathway, by low-dose intrastriatal volkensin injection, was as described for the first group (Pangalos et al., 1991). Cell numbers and size were determined as described (Pangalos et al., 1991), in medial areas, cortical regions Fr1/Fr2 and part of HL as described (Pangalos et al., 1991), and in parietal areas (Par1/Par2). In order to count pyramidal cells and exclude most smaller neurones and glia, a size constraint was set at 80 μm^2 for infragranular neurones and at 70 μm^2 for supragranular. Sections from ricin (10 ng) injected animals (n = 4), and high dose (6 ng) volkensin-injected rats (n = 4) were as previously described (Pangalos et al., 1991). An additional group of ricin (10 ng) injected, animals (n = 5) was studied, prepared as described (Pangalos et al., 1991).

For intracortical volkensin injection, male Wistar rats (200–250 g) were anaesthetised and placed in a stereotaxic frame. Following the removal of a bone flap and dural exposure, five 1 μl injections, each containing 0.5 ng volkensin or 2.5 ng ricin in phosphate buffered saline (PBS), were made into the left dorso-rostral cortex. Following a 28 day postoperative survival period, animals were terminally anaesthetised and perfused transcardially with PBS followed by PBS containing 30% sucrose. The brains were then processed as previously described (Pangalos et al., 1991). All procedures were performed under a Home Office licence, Animals (Scientific Procedures) Act 1986.

Receptor autoradiography

Receptor autoradiography and quantitation were performed using published methods as summarised in Table 1 (see also Chessell et al., 1993). [^{3}H] 8-OH-DPAT, [^{3}H] L-689,560 and [^{3}H] DPCPX were used to label $5HT_{1A}$, glycine (NMDA complex) and adenosine A_1 receptors.

Results

Immunoblots with antibody 22C11 revealed the same 4 bands as previously described in human cerebral cortex (Webster et al., 1994). The band in the soluble fraction with the characteristics of APP containing the KPI domain

Table 1. Incubation conditions for the labelling of 5-HT_{1A}, 5-HT_2, NMDA, m_1 $GABA_A$, A_1, nicotine and kainate receptors and adrenoceptors

Ligand	Displacer	Incubation buffer*	Pre-incubation	Incubation	Wash
^{3}[H] 8-OH-DPAT (1 nM)	1 μM 5-HT	0.17 M Tris-HCl 4 mM $CaCl_2$ 0.01% ascorbate, pH 7.6	30 min, 25°C	60 min, 25°C	2 × 5 min, 4°C
^{3}[H] Ketanserin (2 nM)	1 μM methysergide	0.17 M Tris-HCl, pH 7.7	15 min, 25°C	120 min, 25°C	2 × 10 min, 4°C
^{3}[H] L-689,560 (5 nM)	1 mM glycine	50 mM Tris-Acetate, pH 7.0	60 min, 4°C	120 min, 4°C	4 × 1 min, 4°C
^{3}[H] Pirenzepine (2.5 nM)	10 μM atropine sulphate	Phosphate buffered Saline	20 min, 25°C	90 min, 25°C	1 × 5 min, 4°C
^{3}[H] GABA (20 nM)	100 μM isoquvacine	50 mM Tris-HCl, 100 μM baclofen, pH 7.7	30 min, 25°C	45 min, 25°C	2 × 25 sec, 4°C
^{3}[H] DPCPX (2 nM)	100 μM R-PIA	170 μM Tris-HCl 10 mM Gpp (NH) P, pH 7.4	30 min, 37°C	80 min, 37°C	2 × 5 min, 4°C
^{3}[H] Nicotine (6 nM)	100 μM carbachol	50 mM Tris-HCl, 8 mM $CaCl_2$, pH 7.4	30 min, 25°C	90 min, 25°C	2 × 10 sec, 4°C
^{3}[H] Kainate (30 nM)	10 μM kainate	50 mM Tris-citrate, pH 7.0	45 min, 2°C	45 min, 2°C	2 × 20 sec, 4°C dip in fixative
^{3}[H] Prazosin (0.4 nM)	10 μM phentolamine	50 mM Tris-HCl, pH 7.4	30 min, 30°C	45 min, 30°C	2 × 5 min, 2°C

* Pre-incubation buffers are identical except for ^{3}[H] DPCPX (no $GPPNH_P$ but contains 2 U/ml of adenosine deaminase and 1 mM EDTA) and ^{3}[H] prazosine (no 5-methylurapidil); R-PIA, (−)-N^6-(2-phenylisopropyl) adenosine; ^{3}[H] prazosin conditions are for $\alpha_{1\ total}$, buffer contained 30 nM 5-methyl urapidil for α_{1b}

(APP_{KPI}) correlated with only the number of pyramidal neurones in cerebral cortex (Fig. 1). A similar correlation was observed with antibody 10D5 (epitope mapped to β-AP 1–16).

Intrastriatal injection of volkensin produced a significant loss of infragranular pyramidal neurones with preservation of (GAD mRNA positive) interneurones (Tables 2–4). This was also a feature of intracortical injections where a subpopulation of contralateral transcallosally-projecting pyramidal neurones were selectively destroyed (R. C. A. Pearson and P. R. Heath, unpublished observations). Tables 2–5 show the results for receptor binding changes in volkensin injected rats. Animals injected with a non-retrogradely transported toxin (ricin) did not show significant alterations in receptor binding (data not shown).

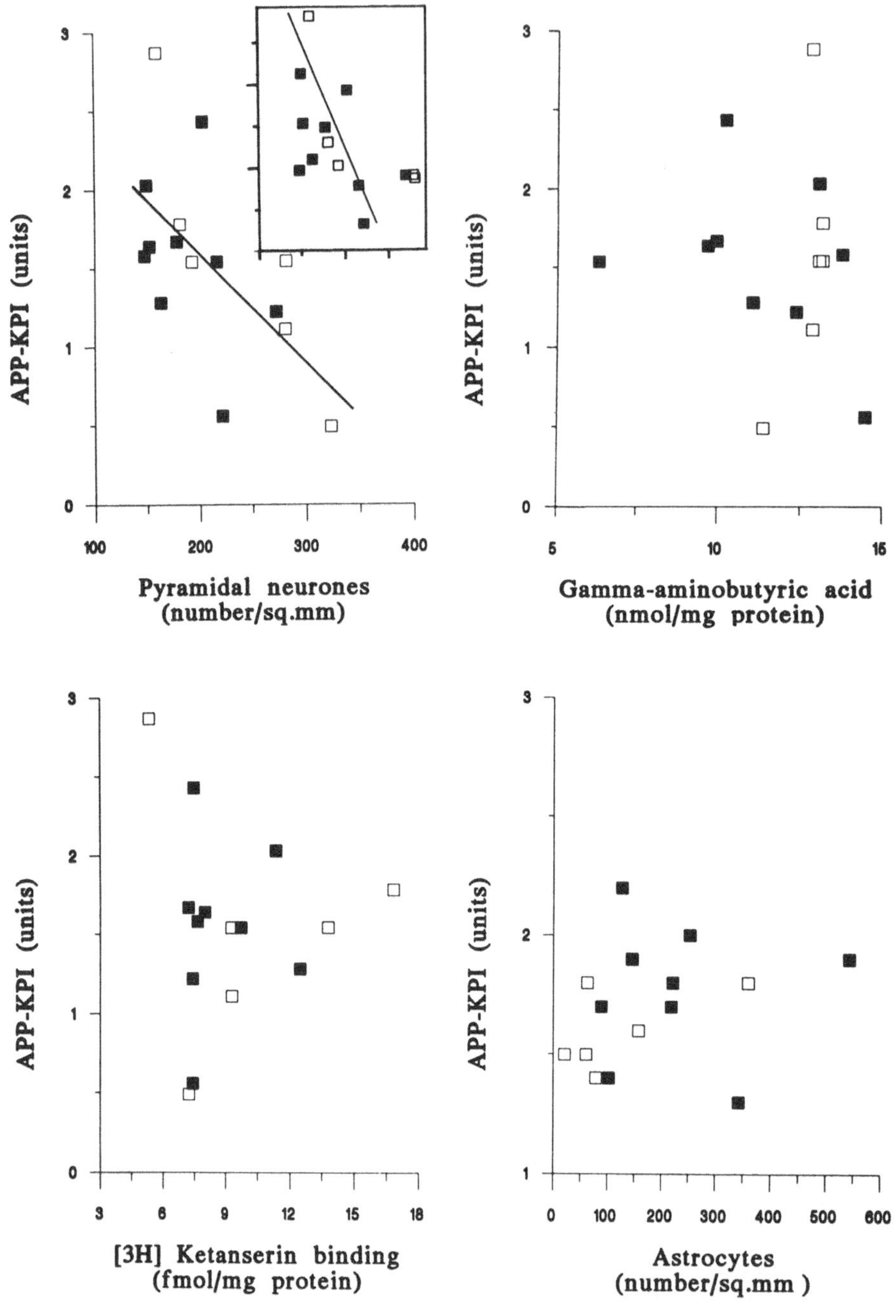

Fig. 1. Selective associations of APP_{KPI} values. For soluble fraction and number of pyramidal neurones in brain cortex (*top left* $r_s = -0.66$, $p = 0.008$, using antibody 22C11; box, $r_s = -0.68$, $p = 0.008$, using antibody 10D5). None of the values for other cellular markers were significantly correlated with APP_{KPI} values: GABA (*top right* $r_s = -0.16$, $p = 0.577$, 22C11; $r_s = -0.26$, $p = 0.366$, 10D5, data points not shown); ^{3}H ketanserin binding, 0.5 nM ligand (*bottom left* $r_s = 0.10$, $p = 0.718$, 22C11; $r_s = -0.27$, $p = 0.476$, 10D5, data points not shown), ^{3}H-ketanserin binding, 2 nM ligand (similar to previous values, data not shown); astrocytes (*bottom right* $r_s = 0.355$, $p = 0.19$, 22C11). Data is for superior frontal cortex, except for bottom right (inferior temporal cortex). Open and filled in data points identify non-Alzheimer and Alzheimer dementia cases respectively, correlation coefficients for these latter cases alone were $r_s = -0.51$, $p = 0.166$ (22C11) and $r_s = -0.95$ (10D5)

Table 2. Corticofugal neurones in medial area: overview of the effect on cortical neurone counts and receptor binding of unilateral striatal volkensin (2 ng) injections

Parameter measured	Cortical layer	Experiment (each n = 5 −6) 1	2	3
Pyramidal	Upper	81*, +	98	94
Neurones	Lower	71*	78*	60*
GAD mRNA	Upper	93	–	–
Neurones	Lower	95	–	–
5-HT_{1A}	Upper	85	98	86
Receptor	Lower	77*	79*	81*
m_1	Upper	98	–	98
Receptor	Lower	89*	–	84*
Kainate	Upper	88	98	–
Receptor	Lower	89*	82*	–
NMDA	Upper	92	–	–
Receptor	Lower	89*	–	–
GABA	Upper	–	96	–
Receptor	Lower	–	113	–
Nicotinic	Superficial	–	–	61*
Receptor	Middle	–	–	56*
	Deep	–	–	65*
A_1	Superficial	–	–	89
Receptor	Middle	–	–	90
	Deep	–	–	93
5-HT_2	Superficial	100	–	–
Receptor	Middle	94	–	–
	Deep	90	–	–
$\alpha_{1\ total}$	Superficial	–	107*	–
Receptor	Middle	–	96	–
	Deep	–	115*	–
α_{1b}	Superficial	–	115*	–
Receptor	Middle	–	100	–
	Deep	–	103	–

Values are ipsilateral side as percentage of the contralateral side. * Significant effect of volkensin; + indicative of cell shrinkage rather than neuronal loss; – not determined

Discussion

The study of volkensin treated animals has layed emphasis on the importance of m_1 receptors on regulating the activity of both corticofugal and cortico-cortical pyramidal neurones (Table 6). This is based on the small but *consistent* reductions in binding of [^{3}H] pirenzipine to m_1 receptors in areas of neocortex that showed pyramidal neurone loss. Reductions in binding were observed in the medial area following intrastriatal injections of 2 ng (Table 2) and 6 ng (Table 3) of volkensin and following cortical injection (Table 5). Changes in the lateral area (Table 4) were less convincing.

Table 3. Corticofugal neurones in medial area: overview of the effect on cortical neurone counts and receptor binding of unilateral striatal volkensin (6 ng) injections

Parameter measured	Cortical layers	Receptor binding
Pyramidal	Upper	93
Neurones	Lower	57*
GAD mRNA	Upper	102
Neurones	Lower	106
$5\text{-}HT_{1A}$	Upper	78
Receptors	Lower	68*
m_1	Upper	90
Receptors	Lower	76*
NMDA	Upper	91
Receptor	Lower	85*
$5\text{-}HT_2$	Superficial	86
Receptor	Middle	82
	Deep	85

Values are ipsilateral side as percentage of contralateral side.
* Significant effect of volkensin

Table 4. Corticofugal neurones in lateral area: overview of the effect on cortical neurones counts and receptor binding of unilateral striatal volkensin injections

Parameter measured	Cortical layers	Receptor binding	
		2 ng volkensin	6 ng volkensin
Pyramidal	Upper	96	—
Neurones	Lower	72*	—
$5\text{-}HT_{1A}$	Upper	78, 69*	69
Receptor	Lower	67, 66*	60*
m_1	Upper	98, 100	95
Receptor	Lower	83*, 89	80
NMDA	Upper	72*	69*
Receptor	Lower	80	71*
Nicotine	Superficial	80	—
Receptor	Middle	64*	—
	Deep	86	—

Values are ipsilateral side as percentage of the contralateral side. * Significant effect of volkensin; — not determined. Kainate, $GABA_A$, $\alpha_{1\ total}$, α_{1b}, A_1 and $5\text{-}HT_2$ receptors were not significantly affected except for two small (>90%) changes in $\alpha_{1\ total}$ and $5\text{-}HT_2$ binding

Table 5. Corticocortical neurones: receptor binding in neocortex ipsilateral to volkensin and ricin injections

Receptor layers	Volkensin (n = 7)	Ricin (n = 5)	No toxin (n = 8–13)
m_1			
Upper	43 ± 3**	55 ± 3	57 ± 2
Lower	30 ± 2*	38 ± 2	44 ± 2
$5\text{-}HT_{1A}$			
Upper	17 ± 1	16 ± 1	19 ± 3
Lower	26 ± 2	24 ± 1	30 ± 3
$5\text{-}HT_2$			
Superficial	26 ± 3	28 ± 1	32 ± 3
Middle	51 ± 7	55 ± 2	55 ± 6
Deep	30 ± 4	31 ± 2	31 ± 2
A_1			
Superficial	245 ± 11	243 ± 10	247 ± 7
Middle	278 ± 6	272 ± 8	277 ± 8
Deep	263 ± 6	259 ± 6	259 ± 5

Values are fmol/mg. Mean ± SEM. * Significantly different from no toxin; ** significantly different from ricin and no toxin. There are no other significant differences between treatment groups

Table 6. Summary of receptor localization

	Pyramidal neurones		
	Corticofugal Area		
Receptor	medial	lateral	Corticocortical
$5\text{-}HT_{1A}$	++	++	–
m_1	+	±	+
Nicotinic	++	±	nd
Kainate	+	–	nd
NMDA	+	–	nd
$GABA_A$	–	–	nd
$5\text{-}HT_2$	–	–	–
A_1	–	–	–
$\alpha_{1\ total}$	–	–	nd
α_{1b}	–	–	nd

++ obvious enrichment; + enriched; ± some evidence of enrichment; – no enrichment; *nd* not determined

From this work the most selective marker of the subpopulation of corticofugal neurones is [^{3}H] 8-OH-DPAT binding, the index of the $5HT_{1A}$ receptor (Table 5). Although this does not appear to be true for corticortical neurones in the rat, an important species difference should be emphasised. In rats neocortical $5HT_{1A}$ receptors are found in the lower layers (Table 5) whilst in primates (including man) there is an additional high density in upper layers. This observation suggests that the receptor is enriched on ipsilateral projecting corticocortical pyramidal neurones (Francis et al., 1993b). Receptors that are present in greatest numbers are obvious as targets for drugs to normalise activity of these neurones. The most suitable receptors by other criteria are those that are phospholipase C-linked and may avoid the possibility of excitotoxicity. Hence, positive modulation of impulse flow by the m_1 receptor, perhaps in combination with an antagonist of $5\text{-}HT_{1A}$ receptor induced hyperpolarisation, is considered to represent a most promising approach. A further approach suggested by others (Court and Perry, 1991) is the use of nicotinic agonists. This is supported by the present study as these receptors appear to be enriched at least on corticofugal pyramidal neurones. Work is in progress to establish if these receptors are also enriched on corticocortical neurones.

Positron and single photon emission tomography at present visualize only loss of overall brain substance, with a few exceptions (see Chessell et al., 1993). This has improved diagnostic accuracy in the clinic but further improvements could be made. By using toxins, such as volkensin, brain tissues can be produced that are deficient in various subpopulations of cortical pyramidal neurone, which may be used to help select receptors for visualizing the type and extent of neuronal loss in patients presenting in the clinic.

This study also provides further evidence that secretion of APP from pyramidal neurones may be regulated by the state of receptor activation, since the amount of an $APP_{751/770}$ species in frontal cortex correlated inversely with the number of pyramidal neurones (Francis et al., 1993c); confirmed with antibody 10D5 but not with biochemical markers of either interneurones (GABA and serotonin 2 receptor content) or astrocyte cell counts. Similarly, an APP_{695} species was increased in temporal cortex of patients with AD compared with those with other dementias (Holmes et al., 1993; Webster et al., 1994). Hence, a putative accumulation of APP within remaining pyramidal neurones has been linked to reduced receptor activation consequent upon degeneration of other (pyramidal) neurones (Francis et al., 1993c). Almost certainly this would be exacerbated in AD by corticopetal pathology (e.g. loss of excitatory cholinergic modulation; Chessell et al., 1993; Francis et al., 1993b).

It is well known that assessment of treatment to slow progression of pathology in AD will require study of many subjects over a long time period. Likewise, the prominent strategy relies on developing novel inhibitors for correcting suspected abnormal proteolysis of APP which is at present very poorly understood. Therefore it is important to discover any role that currently understood receptors and available drugs may have in regulating metabolism of APP. The present data suggests that traditional pharmacological approaches being developed for the symptomatic treatment of AD (e.g. enhancing cholinergic neurotransmission with agonists or cholinesterase inhibitors (Davis et al., 1992) are likely to affect selectively and positively the α-secretase processing pathway for APP which cannot lead to senile plaque formation. This treatment is likely to restore secretion (Francis et al., 1993c) and maintain the extracellular supply of a protein apparently necessary for normal brain function. However, whilst the present evidence suggests that secreted APP has a trophic effect (Roch et al., 1993) this does not rule out the discovery of a harmful action.

In conclusion, the use of volkensin as described here demonstrates that it is possible to selectively lesion subpopulations of cortical pyramidal neurones. This has allowed biochemical changes to be ascribed to the loss of a specific type of cortical neurone, which has previously been impossible. It is proposed that investigations of animals with such a loss provide novel information regarding the biology of cortical pyramidal cells, in particular addressing the issue as to whether a specific neurotransmitter receptor(s) is enriched upon these cells. They will also assist development of pharmaceutical agents both for treatment of memory loss and other disorders of higher mental function and improving scanning techniques for living human brain. Evidence presented here suggests that processing of APP may be regulated by receptor activity. Thus, knowledge of receptor localisation and enrichment may facilitate manipulation of APP secretion in human brain.

More studies are also needed on the neurobiology of APP, in particular of the extracellular function and role in transmitter transduction mechanisms (Nishimoto et al., 1993); see also (Blier et al., 1992; McLaughlin et al., 1991; O'Neill et al., 1991; Pearce and Potter, 1991). However, drugs to affect transmission (Bowen, 1990; Francis et al., 1993c) would now seem an important goal. These will be required for most patients to improve if not reverse functional cognitive disabilities already present as well as for sufferers during development of what are currently considered alternative non-transmitter related strategies. On the basis of the present study such drugs may also affect beneficially the metabolism of APP in the brain of patients with AD.

Acknowledgements

We are grateful for support from Astra Arcus, The Brain Research Trust ("Miriam Marks Department") and The Wellcome Trust.

References

Arai H, Lee VM-Y, Messinger ML, Greenberg BD, Lowery DE, Trojanowski JQ (1991) Expression patterns of β-amyloid precursor protein (β-APP) in neural and nonneural human tissues from Alzheimer's disease and control subjects. Ann Neurol 30: 686–693

Bahmanyar S, Higgins GA, Goldgaber D, Lewis DA, Morrison JH, Wilson MC, Shankar SK, Gajdusek DC (1987) Localization of amyloid β protein messenger RNA in brains from patients with Alzheimer's disease. Science 237: 77–80

Blier P, Lista A, Montigny CD (1992) Differential properties of pre-and postsynaptic 5-Hydroxytryptamine$_{1A}$ receptors in the dorsal raphe and hippocampus. II. Effect of pertussis and cholera toxins. J Pharmacol Exp Ther 265: 16–23

Bowen DM (1990) Treatment of Alzheimer's disease: molecular pathology versus neurotransmitter-based therapy. Br J Psychiatry 157: 327–330

Buckley NJ, Burnstock G (1986) Autoradiographic localisation of peripheral M1 muscarinic receptors using 3H pirenzepine. Brain Res 375: 83–91

Buxbaum JD, Oishi M, Chen HI, Pinkas-Kramarski P, Jaffe EA, Gandy SE, Greengard P (1992) Cholinergic agonists and interleukin 1 regulate processing and secretion of the Alzheimer β/A4 amyloid protein precursor. Proc Natl Acad Sci USA 89: 10075–10078

Chessell IP, Francis PT, Pangalos MN, Pearson RCA, Bowen DM (1993) Localisation of muscarinic (m_1) and other neurotransmitter receptors on corticofugal-projecting pyramidal neurones. Brain Res 632: 86–94

Clarke NA, Webster M-T, Francis PT, Procter AW, Hodgkiss AD, Bowen DM (1993) β-Amyloid precursor protein-like immunoreactivity can be altered in humans by drugs affecting neurotransmitter function. Neurodegeneration 2: 243–248

Coggeshall RE (1992) A consideration of neural counting methods. Trends Neurosci 15: 9–13

Court JA, Perry EK (1991) Dementia: the neurochemical basis of putative transmitter orientated therapy. Pharmacol Ther 52: 423–443

Davis KL, Thal LJ, Gamzu ER, Davis CS, Woolson RF, Gracon SI, Drachman DA, Schneider LS, Whitehouse PJ, Hoover TM, Morris JC, Kawas CH, Knopman DS, Earl NL, Kumar V, Doody RS (1992) A doule-blind, placebo-controlled multicenter study of tacrine for Alzheimer's disease. N Engl J Med 327: 1253–1259

Francis PT, Pangalos MN, Pearson RCA, Middlemiss DN, Stratmann GC, Bowen DM (1992) 5-HT_{1A} but not 5-HT_2 receptors are enriched on neocortical pyramidal neurones destroyed by intrastriatal volkensin. J Pharmacol Exp Ther 261: 1273–1281

Francis PT, Cross AJ, Bowen DM (1994) Neurotransmitters and neuropeptides. In: Terry RD, Katzman R, Bick K (eds) Alzheimer disease. Raven Press, New York, pp 247–262

Francis PT, Sims NR, Procter AW, Bowen DM (1993b) Cortical pyramidal neurone loss may cause glutamatergic hypoactivity and cognitive impairment in Alzheimer's disease: investigative and therapeutic perspectives. J Neurochem 60: 1589–1604

Francis PT, Webster M-T, Procter AW, Clarke NA, Bowen DM, Doshi R, Mann DMA, Neary D (1993c) Soluble β-amyloid precursor protein and pyramidal neurone loss. Lancet 341: 431

Haass C, Hung AY, Selkoe DJ (1991) Processing of β-amyloid precursor protein in microglia and astrocytes favors an internal localization over constitutive secretion. J Neurosci 11: 3783–3793

Harrison PJ, Barton AJL, Procter AW, Bowen DM, Pearson RCA (1994) The effects of Alzheimer's disease, other dementias and premortem course upon amyloid β precursor protein messenger RNA in frontal cortex. J Neurochem 62: 635–644

Holmes C, Webster M-T, Procter AW, Francis PT, Bowen DM (1993) Relationship between β-amyloid precursor protein, pyramidal neurones and astrocytes in human neocortex. Biochem Soc Trans 21: 238S

Hule EC, Birdsall NJM, Buckley NJ (1990) Muscarinic receptor subtypes. Ann Rev Pharmacol Toxicol 30: 633–673

McCormick DA, Williamson A (1989) Convergence and divergence of neurotransmitter action in human cerebral cortex. Proc Natl Acad Sci USA 86: 8098–8102

McLaughlin M, Ross BM, Milligan G, McCulloch J, Knowler JT (1991) Robustness of G proteins in Alzheimer's disease: an immunoblot study. J Neurochem 57: 9–14

Neary D, Snowden JS, Mann DM, Bowen DM, Sims NR, Northen B, Yates PO, Davison AN (1986) Alzheimer's disease: a correlative study. J Neurol Neurosurg Psychiatry 49: 229–237

Nishimoto I, Okamoto T, Matsuura Y, Takahashi S, Okamoto TO, Murayama Y, Ogata E (1993) Alzheimer amyloid protein precusor complexes with brain GTP-binding protein G_0. Nature 362: 75–79

Nitsch RM, Slack BE, Wurtman RJ, Growdon JH (1992) Release of Alzheimer amyloid precursor stimulated by activation of muscarinic acetylcholine receptors. Science 258: 304–307

Nitsch RM, Farber SA, Growdon JH, Wurtman RJ (1993) Release of amyloid β-protein precursor derivatives by electrical depolarization of rat hippocampal slices. Proc Natl Acad Sci USA 90: 5191–5193

O'Neill C, Cowburn RF, Wiehager B, Alafuzoff I, Winblad B, Fowler CJ (1991) Preservation of 5-hydroxytryptamine$_{1A}$ receptor-G protein interactions in the cerebral cortex of patients with Alzheimer's disease. Neurosci Lett 133: 15–19

Pangalos MN, Francis PT, Middlemiss DN, Pearson RCA, Bowen DM (1991) Selective destruction of a sub-population of cortical neurones by suicide transport of volkensin, a lectin from Adenia volkensii. J Neurosci Meth 40: 17–29

Pangalos MN, Francis PT, Foster AC, Pearson RCA, Middlemiss DN, Bowen DM (1992) NMDA receptors assessed by autoradiography with [^{3}H] L-689,560 are present but not enriched on corticofugal-projecting pyramidal neurones. Brain Res 596: 223–230

Pearce BD, Potter LT (1991) Coupling of m1 muscarinic receptors to G protein in Alzheimer disease. Alzheimer Dis Assoc Disord 5: 163–172

Procter AW, Lowe SL, Palmer AM, Francis PT, Esiri MM, Stratmann GC, Najlerahim A, Patel AJ, Hunt A, Bowen DM (1988) Topographical distribution of neurochemical changes in Alzheimer's disease. J Neurol Sci 84: 125–140

Procter AW, Doshi R, Bowen DM, Murphy E (1990) Rapid autopsy brains for biochemical research: experiences in establishing a programme. Int J Geriatr Psychiatry 5: 287–294

Roch J-M, Jin L-W, Ninomiya H, Schubert D, Saitoh T (1993) Biologically active domain of the secreted form of the amyloid β/A4 protein precursor. Ann N Y Acad Sci 695: 149–157

Shigematsu K, McGeer PL, McGeer EG (1992) Localization of amyloid precursor protein in selective postsynaptic densitics of rat cortical neurons. Brain Res 592: 353–357

Sinha S, Lieberburg I (1992) Normal metabolism of the amyloid precursor protein (APP). Neurodegeneration 1: 169–175

Webster M-T, Francis PT, Procter AW, Stratmann GC, Doshi R, Mann DMA, Bowen DM (1994) Postmortem brains reveal similar but not identical APP-like immunoreactivity in Alzheimer compared with other dementias: implication for the novel homologue APPH. Brain Res 644: 347–351

Wiley RG (1992) Neural lesioning with ribosome-inactivating proteins: suicide transport and immunolesioning. Trends Neurosci 15: 285–290

Zilles K, Gross G, Schleicher A, Schildgen S, Bauer A, Bahro M, Schwendemann G, Zeck K, Kolassa N (1991) Regional and laminar distributions of α_1-adrenoceptors and their subtypes in human and rat hippocampus. Neuroscience 40: 307–320

Authors' address: Prof. D. M. Bowen, Institute of Neurology, Department of Neurochemistry, 1, Wakefield Street, London WC1N 1PJ, United Kingdom

J Neural Transm (1994) [Suppl] 44: 245–257

AF64A-induced brain damage and its relation to dementia

H. Hörtnagl

Institute of Biochemical Pharmacology, University of Vienna, Austria

Summary. Several data obtained in the AF64A-model are of particular relevance for our understanding of the pathogenesis and progression of Alzheimer's disease. The AF64A-induced withdrawal of cholinergic function in the rat hippocampus was associated with reversible functional changes in other neurotransmitters, including noradrenaline, serotonin, somatostatin and glutamate, thereby mimicking changes in Alzheimer's disease. Identical changes in markers for synaptic vesicles were found in Alzheimer's disease and AF64A-model. A study on the role of gender revealed a higher susceptibility to the neurotoxic action of AF64A in female rats. The cholinergic deficit was also responsible for a disinhibition of the negative feedback regulation of glucocorticoids. Increased exposure to glucocorticoids, however, enhanced the vulnerability of hippocampal cholinergic neurons to AF64A. These data indicate that the AF64A-induced cholinergic deficit in the rat brain represents a reliable tool to study several mechanisms possibly involved in Alzheimer's disease.

Introduction

The development of a relevant experimental model for Alzheimer's disease is still of particular interest. Because of the complexity of the neurochemical and histological changes and the yet unresolved pathogenesis of this devastating neuronal disorder, uptodate it has not been achieved to find an experimental basis fulfilling all required criteria. Nevertheless, each of the available models may cover a certain aspect of the disease process and thereby finally contribute to our understanding of this disease. When in 1980 the cytotoxic choline analog ethylcholine aziridinium (AF64A) was designed as a possible tool for a relevant experimental model for Alzheimer's disease, the cholinergic hypothesis of dementia in Alzheimer's disease was a generally accepted highlight in the research of this disease. A hypofunction of the basal forebrain cholinergic systems has been consistently reported in brains of patients with Alzheimer's disease (Bowen et al., 1976; Coyle et al., 1983; Davies and Maloney, 1976; Perry et al., 1978). In the meantime, however, disturbances in several other neurotransmitter systems have been discovered, including the transmitters noradrenaline (NA), serotonin (5-

HT), glutamate, GABA, somatostatin, neuropeptide Y, corticotropin releasing factor, substance P (see for example Arai et al., 1984; Beal et al., 1986; Cross et al., 1983; Davies et al., 1980; Francis et al., 1985; Gottfries, 1985).

At present the main interest in Alzheimer's research focuses on the profound modifications of cytoskeletal proteins that occur in many Alzheimer neurons and on the possible involvement of the so-called amyloid cascade in the pathogenesis of this devastating disease (see, for example, Hardy and Allsop, 1991; Murphy, 1992; Selkoe, 1991). Therefore, the rationale for using an experimental model based mainly on a cholinergic hypofunction of the basal forebrain in the research of an otherwise complex disturbance in brain function, might be questioned at present. However, it has to be kept in mind that, although cerebral amyloid deposition may represent a critical pathogenetic event in Alzheimer's disease, the underlying deficits in the function of specific neurotransmitter systems remain the crucial step in the manifestation of clinical symptoms. Among the multiple neurotransmitter changes described in Alzheimer's disease, the cholinergic deficit still remains the most profound neurotransmitter abnormality, which is most closely related to the cognitive deficits (Katzman, 1986; Reinikainen et al., 1990). In this respect, the specific imitation of this representative neurotransmitter disturbance in a well defined animal model (for review, see Hanin et al., 1987) offers the possibility to study the contribution of the cholinergic deficit of the basal forebrain in the manifestation and progression of this disease. The present paper focuses on recent findings in the AF64A-induced cholinergic hypofunction model, which are of particular relevance for our understanding in the pathogenesis and progression of Alzheimer's disease. Furthermore, striking similarities observed in Alzheimer's disease and in the AF64A model are pointed out.

Methods

Animals

Male and female Sprague-Dawley rats (Charles River, Wilmington, MA, USA and Forschungsinstitut für Versuchstierzucht, Himberg, Austria) weighing 200–400 g were used. The animals were housed in groups of 3–4 per cage on a 12 h light/dark cycle and had free access to laboratory chow and tap water.

Experimental procedures

Rats anaesthetized with chloral hydrate (325 mg/kg i.p.) received stereotaxic infusions (0.5 µl/min) of 1 or 2 nmol of AF64A in 3 µl or of vehicle (3 µl) into each of both lateral ventricles. Stereotaxic coordinates were the following: from bregma, posterior 0.8 mm, lateral ±2.5 mm; ventral (from dura) 3.0 mm. A small stereotaxic frame (David Kopf Instruments) was used. AF64A was freshly prepared from acetylethylcholine mustard according to the procedure of Fisher et al. (1982). Treatment with glucocorticoids in sesame oil and the procedure for bilateral adrenalectomy or sham operation has been described previously (Hörtnagl et al., 1993a). In the female rats the estrous cycle was continuously monitored throughout the entire experiment by daily vaginal smears and microscopic analysis (Hörtnagl et al., 1993b). Rats were sacrificed at various time points after injection of AF64A by decapitation. The brains were rapidly removed and either chilled in ice water kept at −0.8°C with NaCl, or frozen at −80°C (for dissection, see Hörtnagl et al., 1991).

Biochemical analyses

Tissue samples were homogenized by ultrasonication in 20–40 volumes of N2-saturated deionized water. Immediately after sonication an aliquot of the homogenate (200–300 µl) was added to an equal volume of 0.2 M perchloric acid containing 0.8 mM $NaHSO_3$ and centrifuged at 25,000 × g at 4°C for 15 min. The methods used for the measurement of choline acetyltransferase (ChAT) activity and somatostatin for the HPLC analyses of acetylcholine (ACh) as well as monoamines and their metabolites and the putative neurotransmitter amino acids have been described in detail (Hörtnagl et al., 1987a, 1989, 1990, 1991).

In situ hybridization

The perfusion technique of the rats, the fixation of the brains and the in situ hybridization is described by Mahata et al. (1993). Synthetic complementary oligonucleotide probes to rat mRNA sequences were used (corticotropin releasing factor: corresponding to amino acids 23–37; secretogranin II: 48-mer corresponding to amino acids 25–40).

Results and discussion

Comparable changes of neurotransmitter functions in Alzheimer's disease and in the AF64A rat model

After intracerebroventricular injection of AF64A into the lateral ventricle of rats, the hippocampus is the brain area almost exclusively affected by the cholinotoxin. Within 4 days after treatment with AF64A various cholinergic parameters dose-dependently and irreversibly decrease in the hippocampus, including the levels of ACh, activities of ChAT and acetylcholinesterase (AChE), high affinity choline transport, acetylcholine release and hemicholinium-3 binding sites (for review, see Hörtnagl and Hanin, 1992). The levels of ACh and the activities of ChAT and AChE decrease to a comparable extent reaching a maximal reduction of up to about 70% in the hippocampus (Fisher et al., 1982; Hörtnagl et al., 1989; Leventer et al., 1987). A complete reduction in cholinergic function has not been achieved. This finding is in good agreement with the reported reduction in ChAT activity in brains of Alzheimer's patients. In Alzheimer hippocampi a 54 to maximally 87% decrease of ChAT activity has been described (see

Ransmayr et al., 1992). In the temporal cortex of severely demented patients ChAT activity decreases only by about 50% (Ruberg et al., 1990).

It has become evident that the withdrawal of cholinergic function induced by AF64A consequently leads to reversible functional changes in other neurotransmitter systems. Transient changes in the levels of various neurotransmitters and their metabolites have been described in detail (Eva et al., 1987; Hörtnagl et al., 1987a,b, 1989, 1990, 1991; Potter et al., 1986). The affected neurotransmitters include NA, 5-HT and their metabolites 3-methoxy-4-hydroxyphenylglycol (MHPG) and 5-hydroxyindoleacetic acid (5-HIAA), somatostatin, neuropeptide Y, glutamate (see Table 1). The transient nature of these changes and their reversibility by pharmacological manipulations (Hörtnagl et al., 1989) allow the conclusion that the effect of AF64A on non-cholinergic parameters is a secondary, functional sequel in response to the abrupt loss of the cholinergic input to the hippocampus. This phenomenon may be related to the close synaptic interactions of the affected neurotransmitter systems with the cholinergic nerve endings within the hippocampus. As indicated in Table 1 comparable neurotransmitter changes also develop in Alzheimer's disease. The similarity in the diseased human hippocampi and in the animal model suggests that in Alzheimer's disease the observed non-cholinergic neurotransmitter changes are, at least partly, the consequence of the largely reduced cholinergic function.

Table 1. Comparable changes in Alzheimer's disease and AF64A-treated rats

	Alzheimer's disease	References	AF64A treated rats	References
ChAT activity	↓	Bowen et al. (1976) Davies and Maloney (1976) Perry et al. (1978)	↓	Leventer et al. (1987) Hörtnagl et al. (1989, 1990)
ACh			↓	Walsh et al. (1984) Hörtnagl et al. (1987a,b)
5-HT	↓	Arai et al. (1984) Cross et al. (1983)	↓	Hörtnagl et al. (1987a) Potter et al. (1986)
5-HIAA	↓ =	Francis et al. (1985)	↑ ↓	Eva et al. (1987)
NA	↓	Arai et al. (1984) Francis et al. (1985)	↓	Eva et al. (1987) Hörtnagl et al. (1987b)
MHPG	↑ ↓	Gottfries, (1985)	↑	Hörtnagl et al. (1989)
glutamate	↓	Lowe et al. (1990) Maragos et al. (1987)	↓	Hörtnagl et al. (1991)
somatostatin	↓	Davies et al. (1980) Beal et al. (1986)	↓	Hörtnagl et al. (1990)
MAO-B activity	↑	Gottfries, (1985) Reinikainen et al. (1988)	↑	Jossan et al. (1989)
aluminum accumulation	+	Perl and Brody (1980) Candy et al. (1986)	+	Szerdahelyi and Kása (1988)
degeneration process	retrograde	Allen et al. (1988) Burke et al. (1988)	retrograde	Hörtnagl and Hanin (1992)

A further aspect concerning the toxicity of AF64A is a subregional variation in the susceptibility of the hippocampal cholinergic innervation, the CA3 subregion and the ventral portion of the hippocampus being most affected (Hörtnagl et al., 1991; Laganiere et al., 1990). Similar subregional differences in the affection of cholinergic nerve terminals may occur in Alzheimer's hippocampi, although immunohistochemical evidence indicates that at least in CA4, CA1 and the subiculum the densities of cholinergic terminals are homogenously diminished (32–33%; Ransmayr et al., 1992). A further common feature of experimental model and disease is the evidence for a retrograde degeneration process (Allen et al., 1988; Burke et al., 1988; Hörtnagl and Hanin, 1992).

Effect of AF64A treatment on several markers for synaptic and large dense core vesicles and its similarity to changes in Alzheimer's disease

Ongoing studies in Alzheimer's disease also focus on synaptic pathology. Markers for small synaptic vesicles as well as large dense core vesicles have been used to obtain more information on changes at the synaptic level. Recent findings refer to increased brain levels of chromogranin A, a secretory protein present in large dense core vesicles in the brain (Brion et al., 1991; Lassmann et al., 1992; Munoz, 1989; Weiler et al., 1990).

A characteristic feature of Alzheimer brains is a high ratio of chromogranin A to synaptophysin (Lassmann et al., 1992; Weiler et al., 1990). To find a better understanding for this synaptic pathology we attempted in a preliminary experiment (Weiler et al., unpublished results) to find out whether the AF64A-induced cholinergic lesion in the rat hippocampus is also associated with comparable changes in markers for synaptic and large dense core vesicles. The preliminary data indicate an increase in the level of chromogranin A (196 ± 44.9%; n = 5) and decrease in that of synaptophysin (80.6 ± 4.4%; n = 5). The changes in these synaptic markers as well as in their ratio were almost identical to those in the temporal cortex of Alzheimer brain (Lassmann et al., 1992). Therefore, the changes in these vesicle markers may be closely related to the cholinergic degeneration process or the reduction of cholinergic function. Further studies are essential to prove whether the increase in chromogranin A reflects increased synthesis.

Gender differences in the neurotoxicity of AF64A

Concerning the occurrence of Alzheimer's disease, epidemiological studies indicate a generally higher prevalence ratio among women (Rocca et al., 1986; Schoenberg, 1986; Zhang et al., 1990). The reason for this prevalence remains an unsolved question. Recent experience in the AF64A-model might provide an answer to this question. It has been demonstrated that adult female rats in general, and particularly females in proestrus are more susceptible to the neurotoxic action of AF64A than age-matched male rats

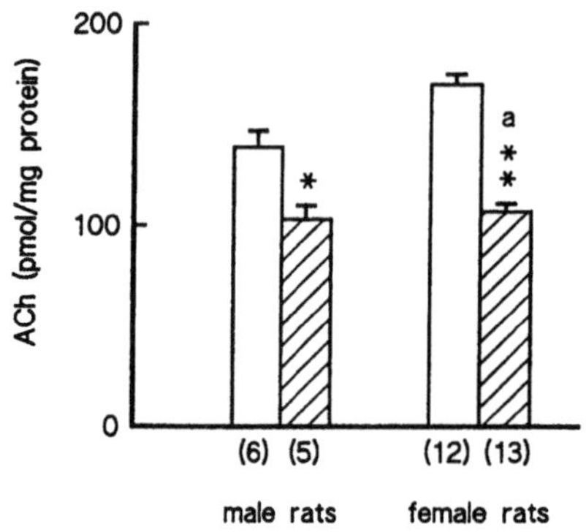

Fig. 1. Sex differences in the decrease of ACh in the rat hippocampus 8 to 10 days after bilateral i.c.v. injection of AF64A (1 nmol/ventricle). The female rats were pooled, irrespective of the stage of cycle at which AF64A or vehicle was injected. The open columns represent vehicle-injected rats, the shaded columns AF64A-injected rats. The number of rats in each group is given below the columns in parenthesis. The data are mean ± SEM (bar) values. The statistical significance of differences are indicated: $^{*}p < 0.05$; $^{**}p < 0.01$ versus the corresponding vehicle-injected rats (two-way ANOVA; Scheffe's test); $^{a}p < 0.03$ versus the effect of AF64A in male rats (two-way ANOVA)

(Hörtnagl et al., 1993b). Eight to 10 days after i.c.v. injection of AF64A (1 nmol/ventricle) the decrease of ACh in the hippocampus was larger in female rats than in male rats (Fig. 1), especially in female rats receiving AF64A at the stage of proestrus (42.7 ± 3.4% vs 25.9 ± 5.1% in male rats; $p < 0.05$). Across the estrous cycle the loss of ACh was most pronounced in female rats that had been injected at the stage of proestrus.

The sex difference in the vulnerability of the cholinergic septo-hippocampal pathway could conceivably be explained as follows: It is well established that the cholinergic systems in the brain are modulated by estrogens in various ways. Estradiol has e.g. the capacity to increase ChAT activity in specific cholinergic basal forebrain nuclei and projection areas (Kaufmann et al., 1983; Luine, 1984; Muth et al., 1980). The available data are suggestive for an enhanced cholinergic function in specific brain areas of female rats and its dependence on the stage of estrous cycle. AF64A, however, is known to avidly interact with the high affinity choline transport system, the velocity of which is dependent on the activity at the cholinergic synapse and the rate of ACh synthesis (for review, see Tuček, 1985). An increase in cholinergic activity, therefore, may be associated with an enhanced uptake of AF64A into the cholinergic nerve ending, thereby increasing its cytotoxicity. Independent of an increased accumulation of AF64A inside the cholinergic neuron, the higher vulnerability of the female hippocampus to the neurotoxin might additionally be related to the fact that a larger number of high affinity binding sites for glucocorticoids exists in the hippocampus of female than of male rats (Turner and Weaver, 1985). Glucocorticoids, however, appear to impair the ability of hippocampal neurons to survive insults (for review, see Sapolsky, 1987). Consequently the higher density of corticosteroid receptors in the female hippocampus might imply that in female rats the cholinergic innervation of the hippocampus has a reduced ability to resist noxious stimuli, such as the cytotoxic effect of AF64A, thereby resulting in a larger degree of cholinergic lesion. In fact, in male rats, glucocorticoids in supraphysiological concentrations aggrevate the cholinergic deficit in the rat hippocampus induced by AF64A, as will be discussed later. Similar mechanisms may also contribute to the higher incidence of Alzheimer's disease among women.

Influence of glucocorticoids on the AF64A induced cholinergic damage and its possible relation to Alzheimer's disease

The hippocampus appears to be an important inhibitory modulator of the negative feedback effects of glucocorticoids on the hypothalamo-pituitary-adrenal axis. Since 1960 evidence is available that damage to or isolation of the hippocampus as well as fimbria-fornix lesions are associated with hypersecretion of ACTH or glucocorticoids in various species (Feldman and Confronti, 1980; Fendler et al., 1961; Knigge, 1961; Magarinos et al., 1987; Sapolsky et al., 1989, 1991) or an increase in mRNA levels of corticotropin releasing factor in the paraventricular hypothalamic nucleus (Herman et al., 1989, 1992). In this respect it is of special interest that hypersecretion of ACTH or glucocorticoids as well as an abnormal glucocorticoid response occur in patients with Alzheimer's disease (DeLeon et al., 1988; Greenwald et al., 1986; Suemaru et al., 1991). The degree of hippocampal damage even parallels the magnitude of hypercortisolism (DeLeon et al., 1988). On the other hand, hypersecretion of glucocorticoids or prolonged exposure to glucocorticoids can damage hippocampal neurons. The hippocampus with the highest density of glucocorticoid receptors is the primary target tissue for glucocorticoids in the brain (for review, see Sapolsky, 1987).

Recent evidence from patients with Cushing's syndrome underlines the potential risk of hypercortisolemia. A clear association between reduced volume of hippocampal formation, memory deficits and elevated cortisol in patients with Cushing's syndrome has been found (Starkman et al., 1992). These data suggest a possible involvement of glucocorticoids in the pathogenesis of Alzheimer's disease. A vicious circle may develop, including the imbalance in the feedback regulation of the hypothalamo-pituitary-adrenal axis related to the hippocampal damage and the reinforcement of the hippocampal damage by exposure to glucocorticoids.

In the AF64A model we therefore tried to answer two important questions: (i) Is the cholinergic lesion in the hippocampus linked to a failure in the glucocorticoid feedback regulation? (ii) Do glucocorticoids increase the vulnerability of hippocampal cholinergic neurons to the neurotoxic effect of AF64A?

By use of in situ hybridization, we were able to demonstrate that 10 days after bilateral i.c.v. injection of 2 nmol AF64A the mRNA level for corticotropin releasing factor increased in the parvocellular neurons of the hypothalamic paraventricular nucleus by a factor of 2 as compared to control-injected rats (Mahata et al., 1993). This increase was associated with a comparable elevation in the message for secretogranin II, a secretory protein contained within large dense core vesicles, with a widespread distribution in the brain (Mahata et al., 1991). The increase in the mRNA expression of this peptide is a useful parameter for defining neurons under chronic stimulation (Mahata et al., 1992a,b). These data clearly indicate that the septo-hippocampal cholinergic system plays a vital role in glucocorticoid feedback regulation. A cholinergic deficit in the hippocampus, thus, could be associated with increased synthesis of corticotropin releasing factor and finally glucocorticoid hypersecretion. The consequences of increased exposure to glucocorticoids on the AF64A-induced lesion are summarized in Fig. 2. Long-term exposure (daily doses of 10 mg corticosterone in 1 ml sesame oil s.c. for 14 days) as well as short-term exposure (s.c. injections of 10 mg corticosterone in 1 ml sesame oil 24 h and 15 min before and 24 h after AF64A) resulted in a potentiation of the AF64A effect on the cholinergic system, as indicated in the significantly higher loss of ChAT activity in hippocampus. Similar results were obtained after long-term exposure to dexamethasone (Hörtnagl et al., 1993a). On the other hand, adrenalectomy 7 days previous to toxin injection significantly attenuated the cholinergic deficit. In the CA3-region of the hippocampus, the brain area most affected by AF64A, the ChAT activity declined to 42.9 ± 6.7% of controls ($n = 11$; $p < 0.01$) in sham-operated rats, whereas after adrenalectomy the AF64A

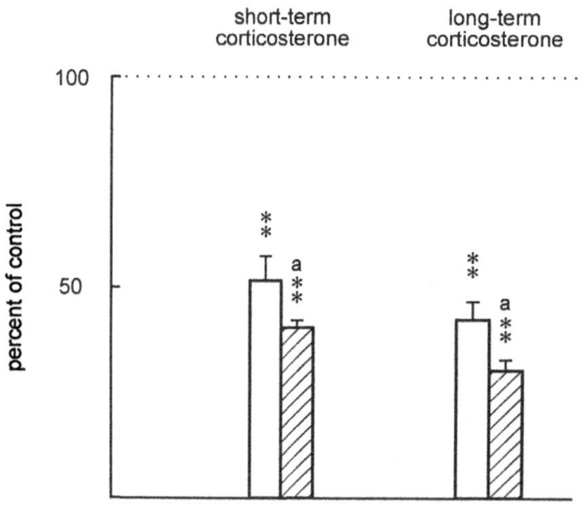

Fig. 2. Effect of short- or long-term exposure to corticosterone, expressed as percentage change, on the AF64A (1 nmol/ventricle)-induced reduction in ChAT activity in the rat hippocampus. The columns (open columns: AF64A/sesame oil, $n = 5$; shaded columns: AF64A/corticosterone, $n = 6$) represent mean percentage changes ± SEM (error bars) from the values of the respective vehicle-injected rats ($n = 6$). The significance of differences between the various groups as calculated from the actual values (one-way ANOVA followed by Newman-Keuls test) is indicated. $^{**}p < 0.01$ versus corresponding control group; $^{a}p < 0.05$ versus AF64A/sesame oil

(1 nmol/ventricle)-induced change in ChAT activity (78.7 ± 5.2% of controls; n = 7) did not reach the level of significance. Replacement of corticoids in adrenalectomized rats restored the toxicity of AF64A and ChAT activity again decreased to 49.5 ± 7.2% of the corresponding controls (n = 9). These data confirm the possibility that glucocorticoids play a significant role in the pathogenesis and progression of Alzheimer's disease.

Conclusions

Several functional and neurochemical correlates of Alzheimer's disease can be demonstrated in the AF64A hypocholinergic function-model. A striking similarity in the changes of certain neurochemical parameters can be achieved in the AF64A-model. This refers not only to the complexity of neurotransmitter changes but also to the specific changes of markers for synaptic vesicles. The observed sex-differences in the vulnerability of the septo-hippocampal cholinergic pathway to a neurotoxin might offer an explanation for the higher prevalence ratio of Alzheimer's disease among women. In addition, convincing evidence on a possible involvement of glucocorticoids in the degeneration process of Alzheimer's disease has been obtained in this model. These data underline the reliability of the AF64A model for future studies on further mechanisms possibly involved in the pathogenesis and progression of Alzheimer's disease.

References

Allen SJ, Dawbarn D, Wilcock GK (1988) Morphometric immunochemical analysis of neurons in the nucleus basalis of Meynert in Alzheimer's disease. Brain Res 454: 275–281

Arai H, Kosaka K, Jizuka R (1984) Changes of biogenic amines and their metabolites in post mortem brains from patients with Alzheimer's type dementia. J Neurochem 43: 388–393

Beal MF, Mazurek MF, Svendsen CN, Bird ED, Martin JB (1986) Widespread reduction of somatostatin-like immunoreactivity in the cerebral cortex in Alzheimer's disease. Ann Neurol 20: 489–495

Bowen DM, Smith CB, White P, Davison AN (1976) Neurotransmitter-related enzymes and indices of hypoxia in senile dementia and other abiotrophics. Brain 99: 459–496

Brion J-P, Couck A-M, Bruce M, Anderton B, Flament-Durand J (1991) Synaptophysin and chromogranin A immunoreactivities in senile plaques of Alzheimer's disease. Brain Res 539: 143–150

Burke WJ, Chung HD, Huang JS, Huang SS, Haring JH, Strong R, Marshall GL, Joh TH (1988) Evidence for retrograde degeneration of epinephrine neurons in Alzheimer's disease. Ann Neurol 24: 532–536

Candy JM, Klinowski J, Perry RH, Perry EK, Fairbairn A, Oakley AE, Carpenter TA, Atack JR, Blessed G, Edwardson JA (1986) Aluminosilicates and senile plaque formation in Alzheimer's disease. Lancet i: 354–357

Coyle JT, Price DL, DeLong MR (1983) Alzheimer's disease: a disorder of cortical cholinergic innervation. Science 219: 1184–1190

Cross AJ, Crow TJ, Johnson JA, Joseph MH, Perry EK, Perry RH, Blessed G, Tomlinson BE (1983) Monoamine metabolism in senile dementia of Alzheimer type. J Neurol Sci 60: 383–392

Davies P, Katzman R, Terry RD (1980) Reduced somatostatin-like immunoreactivity in cerebral cortex from cases of Alzheimer disease and Alzheimer senile dementia. Nature 288: 279–280

Davies P, Maloney AJF (1976) Selective loss of central cholinergic neurons in Alzheimer's disease. Lancet ii: 11–14

DeLeon MJ, McRae T, Tsai JR, George AE, Maraus DL, Freedman M, Wolf AP, McEwen B (1988) Abnormal cortisol response in Alzheimer's disease linked to hippocampal atrophy. Lancet ii: 391–392

Eva C, Fabrazzo M, Costa E (1987) Changes of cholinergic, noradrenergic and serotonergic synaptic transmission indices elicited by ethylcholine aziridinium ion (AF64A) infused intraventricularly. J Pharmacol Exp Ther 241: 181–186

Feldman S, Confronti N (1980) Participation of the dorsal hippocampus in the glucocorticoid feedback effect on adrenocortical activity. Neuroendocrinology 30: 52–61

Fendler K, Karmos G, Telegdy G (1961) The effect of hippocampal lesion on pituitary adrenal function. Acta Physiol (Budapest) 20: 283–297

Fisher A, Mantione CR, Abraham DJ, Hanin I (1982) Long-term central cholinergic hypofunction induced in mice by ethylcholine aziridinium ion (AF64A) in vivo. J Pharmacol Exp Ther 222: 140–145

Francis PT, Palmer AM, Sims NR, Bowen DM, Davison AN, Esiri MM, Neary D, Snowden JS, Wilcock GK (1985) Neurochemical studies of early-onset Alzheimer's disease. Possible influence on treatment. N Engl J Med 313: 7–11

Gottfries CG (1985) Alzheimer's disease and senile dementia: biochemical characteristics and aspects of treatment. Psychopharmacology 86: 245–252

Greenwald B, Mathe A, Mohs R, Levy M, Johns C, Davis K (1986) Cortisol in Alzheimer's disease. II. Dexamethasone suppression, dementia severity and affective symptoms. Am J Psychiatry 143: 442–449

Hanin I, Fisher A, Hörtnagl H, Leventer SM, Potter PE, Walsh TJ (1987) Ethylcholine aziridinium (AF64A; ECMA) and other potential cholinergic neuron-specific neurotoxins. In: Meltzer HY (ed) Psychopharmacology: the third generation of progress. Raven Press, New York, pp 341–349

Hardy J, Allsop D (1991) Amyloid deposition as the central event in the aetiology of Alzheimer's disease. Trends Pharmacol Sci 12: 383–388

Herman JP, Cullinan WE, Young EA, Akil H, Watson SJ (1992) Selective forebrain fiber tract lesions implicate ventral hippocampal structures in tonic regulation of paraventricular nucleus corticotropin-releasing hormone (CRH) and arginine vasopressin (AVP) mRNA expression. Brain Res 592: 228–238

Herman JP, Schäfer MK-H, Young EA, Thompson R, Douglass J, Akil H, Watson SJ (1989) Evidence for hippocampal regulation of neuroendocrine neurons of the hypothalamo-pituitary adrenocortical axis. J Neurosci 9: 3072–3082

Hörtnagl H, Berger ML, Hornykiewicz O (1991) Cholinergic deficit induced by ethylcholine aziridinium (AF64A) in rat hippocampus: effect on glutamatergic systems. Naunyn Schmiedebergs Arch Pharmacol 344: 213–219

Hörtnagl H, Berger ML, Hornykiewicz O (1993a) Role of glucocorticoids in the cholinergic degeneration in rat hippocampus induced by ethylcholine aziridinium (AF64A). J Neurosci 13: 2939–2945

Hörtnagl H, Hanin I (1992) Toxins affecting the cholinergic system. In: Herken H, Hucho F (eds) Selective neurotoxicity. Springer, Berlin Heidelberg New York Tokyo, pp 293–332 (Handbook of experimental pharmacology, vol 102)

Hörtnagl H, Hansen L, Kindel G, Schneider B, El Tamer A, Hanin I (1993b) Sex differences and estrous cycle-variations in the AF64A-induced cholinergic deficit in the rat hippocampus. Brain Res Bull 31: 129–134

Hörtnagl H, Potter PE, Hanin I (1987a) Effect of cholinergic deficit induced by ethylcholine aziridinium on serotonergic parameters in rat brain. Neuroscience 22: 203–213

Hörtnagl H, Potter PE, Hanin I (1987b) Effect of cholinergic deficit induced by ethylcholine aziridinium (AF64A) on noradrenergic and dopaminergic parameters in rat brain. Brain Res 421: 75–84

Hörtnagl H, Potter PE, Singer EA, Kindel G, Hanin I (1989) Clonidine prevents transient loss of noradrenaline in response to cholinergic hypofunction induced by ethylcholine aziridinium (AF64A). J Neurochem 52: 853–858

Hörtnagl H, Sperk G, Sobal G, Maas D (1990) Cholinergic deficit induced by ethylcholine aziridinium (AF64A) transiently affects somatostatin and neuropeptide Y levels in rat brain. J Neurochem 54: 1608–1613

Jossan SS, Hiraga Y, Oreland L (1989) The cholinergic neurotoxin ethylcholine mustard aziridinium (AF64A) induces an increase in MAO-B activity in the rat brain. Brain Res 476: 291–297

Katzman R (1986) Alzheimer's disease. N Engl J Med 314: 964–972

Kaufmann H, Vadasz C, Lajtha A (1983) Effect of estradiol on choline acetyltransferase activity in various brain regions. Brain Res 453: 389–392

Knigge KM (1961) Adrenocortical response to stress in rats with lesions of hippocampus and amygdala. Proc Soc Exp Biol 108: 67–69

Laganiere S, Marinko M, Corey J, Wülfert E, Hanin I (1990) Sector-dependent neurotoxicity of ethylcholine aziridinium (AF64A) in the rat hippocampus. Neuropharmacology 29: 961–964

Lassmann H, Weiler R, Fischer P, Bancher C, Jellinger K, Floor E, Danielczyk W, Seitelberger F, Winkler H (1992) Synaptic pathology in Alzheimer's disease: immunological large dense-core vesicles. Neuroscience 46: 1–8

Leventer SM, Wülfert E, Hanin I (1987) Time course of ethylcholine aziridinium ion (AF64A)-induced cholinotoxicity in vivo. Neuropharmacology 26: 361–365

Lowe SL, Bowen DM, Francis PT, Neary D (1990) Antemortem cerebral amino acid concentrations indicate selective degeneration of glutamate-enriched neurons in Alzheimer's disease. Neuroscience 38: 571–577

Luine VN (1984) Estradiol increases choline acetyltransferase activity in specific basal forebrain nuclei and projection areas of female rats. Exp Neurol 89: 484–490

Magarinos AM, Somoza G, DeNicola AF (1987) Glucocorticoid negative feedback and glucocorticoid receptors after hippocampectomy in rats. Horm Metabol Res 19: 105–109

Mahata M, Hörtnagl H, Mahata SK, Fischer-Colbrie R, Winkler H (1993) Messenger RNA levels of chromogranin B, secretogranin II and VGF in rat brain after AF64A-induced septo-hippocampal cholinergic lesions. J Neurochem 61: 1648–1656

Matha SK, Mahata M, Marksteiner J, Sperk G, Fischer-Colbrie R, Winkler H (1991) Distribution of mRNAs for chromogranin A and B and secretogranin II in rat brain. Eur J Neurosci 3: 895–904

Mahata SK, Mahata M, Steiner H-J, Fischer-Colbrie R, Winkler H (1992a) In situ hybridization: mRNA levels of secretogranin II, neuropeptides and carboxypeptidase H in brains of salt-loaded and Brattleboro rats. Neuroscience 48: 669–680

Mahata SK, Marksteiner J, Sperk G, Mahata M, Gruber B, Fischer-Colbrie R, Winkler H (1992b) Temporal lobe epilepsy of the rat: differential expression of mRNAs for chromogranin B, secretogranin II, synaptin/synaptophysin and p65 in subfields of the hippocampus. Mol Brain Res 16: 1–12

Maragos WF, Greenamyre JT, Penney JB, Young AB (1987) Glutamate dysfunction in Alzheimer's disease: a hypothesis. Trends Neurosci 10: 65–68

Munoz DG (1989) Chromogranin A immunoreactive peptides are major components of neocortical and limbic plaques in Alzheimer's disease. J Neuropathol Exp Neurol 48: 378 (abstract)

Murphy M (1992) The molecular pathogenesis of Alzheimer's disease: clinical prospects. Lancet 340: 1512–1515
Muth EA, Crowley WR, Jacobowitz DM (1980) Effect of gonadal hormones on luteinizing hormone in plasma and on choline acetyltransferase activity and acetylcholine levels in discrete nuclei of rat brain. Neuroendocrinology 30: 329–336
Perl DP, Brody AR (1980) Alzheimer's disease: X-ray spectrometric evidence of aluminium accumulation in neurofibrillary tangle-bearing neurons. Science 208: 297–298
Perry EK, Tomlinson BE, Blessed G, Bergmann K, Gibson PH, Perry RH (1978) Correlation of cholinergic abnormalities with senile plaques and mental test scores in senile dementia. Br Med J 2: 1457–1459
Potter PE, Hársing LG Jr, Kakucska I, Gaál G, Vizi ES (1986) Selective impairment of acetylcholine release and content in the central nervous system following intracerebroventricular administration of ethylcholine mustard aziridinium ion (AF64A) in the rat. Neurochem Int 8: 199–206
Ransmayr G, Cervera P, Hirsch EC, Berger W, Fischer W, Agid Y (1992) Alzheimer's disease: is the decrease of the cholinergic innervation of the hippocampus related to intrinsic hippocampal pathology? Neuroscience 47: 843–851
Reinikainen KJ, Paljärvil L, Halonen T, Malminen O, Kosma V-M, Laakso M, Riekkinen PJ (1988) Dopaminergic system and monoamine oxidase B activity in Alzheimer's disease. Neurobiol Aging 9: 245–252
Reinikainen KJ, Soininen H, Riekkinen PJ (1990) Neurotransmitter changes in Alzheimer's disease: implications to diagnostics and therapy. J Neurosci Res 27: 576–586
Rocca WA, Amaducci LA, Schoenberg BS (1986) Epidemiology of clinically diagnosed Alzheimer's disease. Ann Neurol 19: 415–424
Ruberg M, Mayo W, Brice A, Duyckaerts C, Hauw JJ, Simon H, LeMoal M, Agid Y (1990) Choline acetyltransferase activity and [^{3}H]vesamicol binding in the temporal cortex of patients with Alzheimer's disease, Parkinson's disease and rats with basal forebrain lesions. Neuroscience 35: 327–333
Sapolsky RM (1987) Glucocorticoids and hippocampal damage. Trends Neurosci 10: 346–349
Sapolsky RM, Armanini MP, Sutton SW, Plotsky PM (1989) Elevation of hypophysial portal concentrations of adrenocorticotropin secretagogues after fornix transection. Endocrinology 125: 2881–2887
Sapolsky RM, Zola-Morgan S, Squire LR (1991) Inhibition of glucocorticoid secretion by the hippocampal formation in the primate. J Neurosci 11: 3695–3704
Schoenberg BS (1986) Epidemiology of Alzheimer's disease and other dementing illnesses. J Chron Dis 39: 1095–1104
Selkoe DJ (1991) The molecular pathology of Alzheimer's disease. Neuron 6: 487–498
Starkman MN, Gebarski SS, Berent S, Schteingart DE (1992) Hippocampal formation volume, memory dysfunction, and cortisol levels in patients with Cushing's syndrome. Biol Psychiatry 32: 756–765
Suemaru S, Hashimoto K, Suemaru K, Maeba Y, Matsushita N, Zensuke O (1991) Hyperkinesia, plasma corticotropin releasing hormone and ACTH in senile dementia. Neuro-Report 2: 337–340
Szerdahelyi P, Kása P (1988) Intraventricular administration of the cholinotoxin AF64A increases the accumulation of aluminum in rat parietal cortex and hippocampus, but not in the frontal cortex. Brain Res 444: 356–360
Tuček S (1985) Regulation of acetylcholine synthesis in the brain. J Neurochem 44: 11–24
Turner BB, Weaver D (1985) Sexual dimorphism of glucocorticoid binding in rat brain. Brain Res 343: 16–23
Walsh TJ, Tilson HA, DeHaven DL, Mailman RB, Fisher A, Hanin I (1984) AF64A, a cholinergic neurotoxin, selectively depletes acetylcholine in hippocampus and

cortex and produces long-term passive avoidance and radial-arm maze deficits in the rat. Brain Res 321: 91–102

Weiler R, Lassmann H, Fischer P, Jellinger K, Winkler H (1990) A high ratio of chromogranin A to synaptin/synaptophysin is a common feature of brains in Alzheimer and Pick disease. Fed Eur Biochem Soc Lett 263: 337–339

Zhang M, Katzman R, Salmon D, Jin H, Cai G, Wang Z, Qu G, Grant I, Yu E, Levy P, Klauber MR, Liu WT (1990) The prevalence of dementia and Alzheimer's disease in Shanghai, China: impact of age, gender and education. Ann Neurol 27: 428–439

Author's address: Dr. H. Hörtnagl, Institute of Biochemical Pharmacology, University of Vienna, Borschkegasse 8a, A-1090 Wien, Austria.

J Neural Transm (1994) [Suppl] 44: 259–268

Desensitization of brain insulin receptor Effect on glucose/energy and related metabolism

S. Hoyer, D. Müller, and **K. Plaschke**

Department of Pathochemistry and General Neurochemistry, University of Heidelberg, Heidelberg, Federal Republic of Germany

Summary. The overall majority of cases of Alzheimer disease are not caused by genetic abnormalities. A pluricausal etiology is assumed, and the age factor may be of pivotal significance. Aging leads to inherent changes in basic metabolic principles, including the functionally most important cerebral glucose/energy metabolism. Experimentally induced perturbation of the neuronal control over the glucose metabolism by means of intracerebroventricular administration of streptozotocin leads to cascade-like abnormalities in glucose breakdown and energy formation and in membrane phospholipid and monoaminergic catecholamine metabolism, which closely resemble the disturbances found in sporadic Alzheimer disease. It is concluded that this model is a good tool for in vivo study of the cellular events characteristic for this human neurodegenerative disorder.

Introduction

It is a tried and tested principle in medicine that a differentiation be made between etiology and pathogenesis when the causation of any disease is sought. In the case of Alzheimer's disease (AD), this means that a distinction should necessarily be made between the genetically induced and the sporadically induced forms. Whereas for the former the single-cause etiology seems to be clear, the etiology of the latter is still unclear. However, some important pathogenetic factors have been defined, one of which is aging (Evans et al., 1989; Hoyer, 1992a; Storandt et al., 1988). Aging leads to inherent changes in fundamental metabolic principles, including the glucose/energy metabolism, which is central for brain function. In incipient sporadic AD, cerebral glucose utilization was found to be globally reduced (Mielke et al., 1992), the reduction ranging from 19% in mild cases to 40% and more in severely diseased patients (Hoyer et al., 1991; Kumar et al., 1991). This abnormality, along with the disproportionately slighter fall in cerebral oxygen utilization, led us to conclude that one major biochemical perturbation in incipient sporadic AD is the control over cerebral glucose metabolism. Clearly, this is in sharp contrast to the central dogma that amyloid deposition plays the major role in the etiology of AD (Beyreuther and Masters, 1991; Hardy and Allsop, 1991). With respect to the latter, much effort was formerly invested in attempts to mimic the abnormal formation of amyloid in transgenic animals. However, the results were rather disappointing (Czech et al., 1994; Quon et al., 1991).

Based on the findings in oxidative glucose metabolism, and related metabolism, in sporadic AD (Hoyer and Nitsch, 1989; Hoyer et al., 1988, 1990, 1991) it was proposed as a working hypothesis that these abnormalities come about subsequent to a signal transduction failure of the cerebral insulin receptor; in other words, that AD, and sporadic AD in particular, resembles non-insulin-dependent diabetes mellitus confined to the brain. There is evidence that insulin also controls the breakdown of glucose in central nervous tissue (Hoyer et al., 1993). Therefore, we damaged the control level of cerebral glucose/energy metabolism, and related metabolism, by intracerebroventricular administration of streptozotocin. The data from this animal model are compared with relevant findings in sporadic AD.

Material and methods

All studies were performed on 1-year-old male Wistar rats (Breeder: Zentralinstitut für Versuchstierzucht Hannover, FRG). The animals were randomly divided into control and experimentally damaged groups. Control animals received an intracerebroventricular (icv) injection of artificial cerebrospinal fluid with the following composition: 120 mM NaCl, 3 mM KCl, 1.15 mM $CaCl_2$, 0.8 mM $MgCl_2$, 0.33 mM NaH_2PO_4, and 27 mM $NaHCO_3$; pH 7.2, adjusted by CO_2 insufflation. Other animals were damaged by an icv injection of streptozotocin (STZ) in a subdiabetogenic dose (1.25 mg/kg body weight) dissolved in a final volume of 0.9 µl artificial cerebrospinal fluid. The icv injections were given into both lateral cerebral ventricles under stereotaxic guidance. Stereotaxical coordinates for the ventricular system were 2.00 mm lateral, 0.70 mm caudal and 4.00 mm ventral from the bregma at the level of the pia mater. In experiments lasting longer than 3 weeks, another injection was applied 3 weeks after the first one.

STZ is commonly used to induce experimental diabetes mellitus, which is achieved by means of a single intraperitoneal injection (Junod et al., 1968; Mordes and Rossini, 1981). The resulting diabetes mellitus is caused by way of both an inhibition of pancreatic insulin secretion (Bolaffi et al., 1987) and a decrease in insulin receptor signalling in target cells (Kadowaki et al., 1984). According to our working hypothesis, the function of the neuronal insulin receptor, which is densely distributed in various parts of the brain (Havrankova et al., 1978), was damaged by the icv administration of STZ.

Operation procedure

Bilateral icv injections were performed under anesthesia with chloral hydrate (240 mg/kg body weight i.p. in a 4% solution) with the head secured in a stereotaxic apparatus. The final experiment was started by induction of anesthesia with 3 vol% halothane in air in a glass box to allow performance of tracheotomy and intubation. Thereafter, anesthesia and respiration were continued with 1.5 vol% halothane and nitrous oxide/oxygen 70:30 by means of an animal respirator and immobilization with pancuronium bromide (2 mg/kg body weight). One femoral artery and one femoral vein were each exposed and cannulated to monitor mean arterial blood pressure, to sample arterial blood for analyses of blood gases, acid-base parameters, glucose, lactate, hemoglobin and hematocrit concentrations, and to administer bicarbonate if necessary (venous route). The galea was incised sagittally and removed from the skull. In one group of animals studied, the superior sagittal sinus was exposed and cannulated. A skin funnel was formed for later freezing of the brain. A warmed (37°C) wet sponge was placed in the funnel to avoid cooling of the brain. The animals were monitored over a 20-min steady state period of arterial normotension, arterial normocapnia, normoxemia and normoglycemia, while the body temperature was maintained close to 37°C. Thereafter, the brains were frozen in situ with liquid nitrogen poured into the skin funnel formed earlier. Each brain was chiselled away from the skull under liquid nitrogen. Different parts of the brain were dissected on a cold plate at −20°C and maintained at −80°C until biochemical analysis.

Biochemical analyses

The concentrations of glucose, glucose-6-phosphate (G6P), glucose-1-phosphate (G1P), fructose-6-phosphate (F6P), fructose-1,6-diphosphate ($F1,6P_2$), pyruvate, lactate, citrate, succinate, fumarate, adenosine triphosphate (ATP), adenosine diphosphate (ADP), adenosine monophosphate (AMP), creatine phosphate and glycogen were quantified photometrically. The activities of hexokinase (HK), phosphofructokinase (PFK), pyruvate kinase (PK), glyceraldehyde-3-phosphate dehydrogenase (GAPDH), phosphorylase a (PLa) and glucose-6-phosphatase (G6Pase) were determined by continuous optical tests at 340 nm and 30°C. The phospholipids phosphatidylcholine, phosphatidylethanolamine, phosphatidylserine, phosphatidylinositol and phosphatidic acid were measured according to Nitsch et al. (1992). Monoaminergic catecholamines were determined by means of HPLC.

Because of the diversity of the methods used, the reader is referred, for more details to Ding et al. (1992), Nitsch and Hoyer (1991), and Plaschke and Hoyer (1993).

Results

Cerebral utilization of substrates

As is demonstrated in Fig. 1, the cerebral arterio-venous glucose difference narrowed by nearly 40% 3 weeks after icv application of STZ, whereas the cerebral arterio-venous lactate difference doubled in the same period. There were no significant changes in the cerebral arterio-venous differences in oxygen and CO_2. Regional cerebral glucose utilization was diminished by 19% in parietal cortex, in frontal cortex by 14%, in the hippocampal CA1 subfield by 13% and in the entorhinal cortex by 10% (publication in preparation), but no changes were found in cerebral blood flow, either in these same areas or in the two hemispheres (unpublished data).

Substrate concentrations

In two different series of experiments (Nitsch and Hoyer, 1991; Plaschke and Hoyer, 1993), inconsistent results were found with respect to the

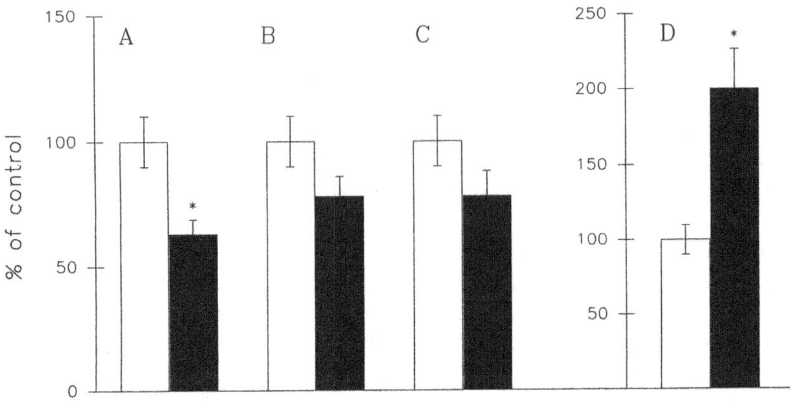

Fig. 1. Cerebral arterio-venous differences in glucose (A), oxygen (B), carbon dioxide (C) and lactate (D) three weeks after an intracerebroventricular injection of streptozotocin ■, as compared with controls □. Values are given as percentages of control values. $^*p \leq 0.05$

glucose concentration in cerebral parietotemporal cortex. First, we demonstrated a significant increase from 1.5 nmol/mg w.w. to 2.0 nmol/mg w.w. after STZ damage. However, this finding was not confirmed in the later study. While no changes were observed in the concentrations of glycolytic and TCA cycle substrates, glycogen concentration rose over time by up to 15% after STZ damage (Plaschke and Hoyer, 1993), and ATP and creatine phosphate fell by 12% and 17%, whereas ATP turnover increased by 47% (Nitsch and Hoyer, 1991).

Enzyme activities

The activities of enzymes that are active in glycolysis and glycogenolysis were studied in cerebral parietotemporal cortex and in hippocampus. As reported in detail elsewhere (Plaschke and Hoyer, 1993), the activities of hexokinase, phosphofructokinase and pyruvate kinase fell by 15% to 20% in cerebral parietotemporal cortex, and by 15% to 28% in hippocampus. There were reductions in GAPDH activity by between 10% and 30%, in the same areas. The activity of the gluconeogenetic enzyme G6Pase was also diminished by between 10% and 23% in the areas studied, as was the glycogenolytic enzyme PLa (11% to 30%).

Phospholipids

Preliminary data showed an STZ-induced reduction in phosphatidylethanolamine by 9% and in phosphatidylserine by 6% in the cerebral parietemporal cortex in 1-y-old animals. In 30-mth-old rats, the fall in phosphatidylserine was more pronounced with 12% (publication in preparation).

Monoaminergic catecholamines

As detailed elsewhere (Ding et al., 1992), noradrenaline levels were found to be decreased in the cerebral frontal cortex by 14%, in the entorhinal cortex by 18%, and in the striatum by 38%. 5-Hydroxytryptamine (5-HT) fell by 19% in the entorhinal cortex, but 5-HT turnover increased by 48%.

Discussion

In this section, prominent abnormalities related to glucose/energy metabolism in incipient AD are compared with correspondence changes after icv STZ administration. The data discussed are summarized in Table 1.

What became obvious from the studies in incipient-state AD patients was the disproportion between the cerebral utilization rates of oxygen and glucose and the release of lactate from the brain. This pattern was also found after icv STZ administration indicating the damaging and selective effect of STZ on cerebral glucose metabolism (Fig. 1). With respect to local cerebral glucose utilization, the most marked changes were found in frontal and temporo-parietal association areas, and these were more prominent in early-onset AD (−24% and −29%) than in late-onset AD (−9% and −18%) (Mielke et al., 1991; Small et al., 1989). After icv STZ administration, the changes in the local cerebral utilization rate were also most marked in frontal (−14%) and parietal cortex (−19%) (publication in preparation).

Ante- and post-mortem studies of brain tissue from patients suffering from sporadic AD clearly demonstrated reduced enzymatic activities in the glycolytic chain, whereas enzymes involved in the tricarboxylic acid cycle

Table 1. Metabolic changes in cerebral glucose/energy and related metabolism occurring in sporadic Alzheimer disease (AD) and after damage to the control of neuronal glucose metabolism by intracerebroventricular (icv) streptozotocin (STZ) application

Parameter	Sporadic AD	icv STZ
Glucose utilization	↓	↓
Lactate formation	↑	↑
Oxygen utilization	=	=
ATP concentration	↓	↓
Creatine phosphate concentration	↓	↓
Enzyme activity in glycolysis	↓	↓
Degradation of membrane phospholipids	↓	↓
Serotonin concentration	↓	↓
Noradrenaline concentration	↓	↓
Dopamine concentration	=	=

↓ reduction; ↑ enhancement; = no change

did not show corresponding changes in enzyme activities except for a drastic reduction in the activity of alpha-ketoglutarate dehydrogenase in obviously advanced AD (Bowen et al., 1979; Gibson et al., 1988; Liguri et al., 1990; Marcus et al., 1988, 1989; Perry et al., 1980; Sims et al., 1983; Sorbi et al., 1983; Yates et al., 1990). The reductions in enzyme activities of hexokinase and phosphofructokinase fluctuated between 20% and 40% in fresh autopsy material. Streptozotocin given icv diminished the activities of these two enzymes by 15% and 28%, respectively. The activity of another key enzyme in glycolysis, pyruvate kinase, was found to be decreased by the same amount. It may thus be concluded that icv STZ reduces the capacity of the glycolytic chain in parietotemporal cerebral cortex and in hippocampus (Plaschke and Hoyer, 1993).

In the mature, healthy, nonstarved mammalian brain, glucose is the only substrate for the formation of energy in the form of ATP (Erecinska and Silver, 1989). When the cerebral ATP formation rate was calculated from the rates of oxidized glucose and oxygen, a fall by at least 7% was observed in incipient early-onset AD, whereas the reduction was at least 20% in incipient late-onset sporadic DAT (Hoyer, 1992b). These data are congruent with the ATP formation rate in neocortical tissue samples removed at diagnostic craniotomy from AD brains (Sims et al., 1983) and from in vivo ^{31}P NMR profiles (Brown et al., 1989): reduction of available energy by 20% to 25% in AD brain. Otherwise, recent findings showed no changes in ATP content and an increase of creatine phosphate concentration by 28%, although the metabolic rate of glucose fell by around 15% (Murphey et al., 1993). The increase in creatine phosphate is hard to explain, because its formation is catalyzed by creatine kinase at elevated pH. In AD brain, however, no increase in pH, but rather a fall, was found (Yates et al., 1990). After STZ damage, the tissue levels of ATP were 12%, and those of creatine phosphate 17%, lower in parietal cerebral cortex than in controls (Nitsch and Hoyer, 1991). Thus, the reduction of cerebral availability of energy after icv STZ administration is equivalent to that found in incipient sporadic AD.

Glycogen accumulation was demonstrated in AD brain (Gertz et al., 1985; Mann et al., 1987), but was not quantified. After icv STZ damage, the glycogen content rose by 7% to 15% in parietotemporal cerebral cortex, indicating gluconeogenetic activity (Plaschke and Hoyer, 1993).

As yet unpublished data revealed a fall of phosphatidylethanolamine and of phosphatidylserine by 5% to 11% in parietotemporal cerebral cortex after STZ damage. In AD brain, there was a significant decrease in phosphatidylcholine and in phosphatidylethanolamine by 10% to 12% in parietal and frontal cerebral cortex (Nitsch et al., 1992). These changes provide evidence for a membrane defect occurring in both AD and after STZ damage. Furthermore, these data may indicate that the lacking glucose is substituted, at least in part, by fatty acids from degraded membrane phospholipids.

There are somewhat controversal data in the literature concerning in which cortical region the concentrations of monoaminergic catecholamines change in AD and to what extent. In this context, the reader is referred to Gottfries (1990), Palmer et al. (1988) and Reinikainen et al. (1988). When a rank order of abnormalities is formed in pathobiochemical terms, the most severe were those of 5-HT, followed by changes in NA, whereas DA was less involved or not at all in cerebral cortex of AD patients (Riederer et al., 1990). After icv STZ application, NA levels were found to be decreased in the frontal cortex by 14%, in the entorhinal cortex by 18%, and in the striatum by 38%. In the entorhinal cortex, 5-HT levels dropped by 19% and the 5-HT turnover rate increased by 48%. There were no detectable changes in DA metabolism (Ding et al., 1992).

Conclusion

The tentative conclusion that can be drawn from the present results obtained in the icv STZ model implies that the damage of the control of the neuronal glucose metabolism leads to metabolic abnormalities not only in cerebral glucose/energy metabolism itself but also in related metabolism, e.g. metabolism of membrane phospholipids and monoaminergic catecholamines. That is to say that a cascade-like perturbation is induced by the damage to the control of the neuronal metabolism. In this respect, the metabolic abnormalities found closely resemble, in both quality and quantity, those found in (incipient) sporadic AD. Therefore, the icv STZ model seems to be a good tool for further in vivo research into the cellular abnormalities of this age-related neurodegenerative disorder.

Acknowledgement

The work of D.M. and K.P. was supported by grants from the Hirnliga e.v. and from the Graduate Programme in Molecular and Cellular Neurobiology, University of Heidelberg.

References

Beyreuther K, Masters CL (1991) Amyloid precursor protein (APP) and A4 amyloid in the etiology of Alzheimer's disease: precursor-product relationships in the derangement of neuronal function. Brain Pathol 1: 241–251

Bolaffi JL, Nagamatsu S, Harris J, Grodsky GM (1987) Protection by thymidine, an inhibitor of polyadenosine diphosphate ribosylation, of streptozotocin inhibition of insulin secretion. Endocrinology 120: 2117–2122

Bowen DM, White P, Spillane JA, Goodhardt MJ, Curzon G, Iwangoff P, Meier-Ruge W, Davison AN (1979) Accelerated ageing or selective neuronal loss as an important cause of dementia? Lancet i: 11–14

Brown GG, Levine SR, Gorell JM, Pettegrew JW, Gdowski JW, Bueri JA, Helpern JA, Welch KMA (1989) In vivo ^{31}P NMR profiles of Alzheimer's disease and multiple subcortical infarct dementia. Neurology 39: 1423–1427

Czech C, Masters C, Beyreuther K (1994) Alzheimer's disease and transgenic mice. J Neural Transm [Suppl] 44 (this volume)

Ding A, Nitsch R, Hoyer S (1992) Changes in brain monoaminergic neurotransmitter concentrations in rat after intracerebroventricular injection of streptozotocin. J Cereb Blood Flow Metab 12: 103–109

Erecinska M, Silver IA (1989) ATP and brain function. J Cereb Blood Flow Metab 9: 2–19

Evans DA, Funkenstein H, Albert MS, Scherr PA, Cook NR, Chown MJ, Hebert LE, Hennekens CH, Taylor JO (1989) Prevalence of Alzheimer's disease in a community population of older persons. Higher than previously reported. J Am Med Ass 262: 2551–2556

Gertz HJ, Cervos-Navarro J, Frydl V, Schultz F (1985) Glycogen accumulation of the aging human brain. Mech Age Dev 31: 25–35

Gibson GE, Sheu KFR, Blass JP, Baker A, Carlson KC, Harding B, Perrino P (1988) Reduced activities of thiamine-dependent enzymes in the brains and peripheral tissues of patients with Alzheimer's disease. Arch Neurol 45: 836–840

Gottfries CG (1990) Disturbance of the 5-hydroxy-tryptamine metabolism in brains from patients with Alzheimer's dementia. J Neural Transm [Suppl] 30: 33–43

Hardy J, Allsop D (1991) Amyloid deposition as the central event in the aetiology of Alzheimer's disease. Trends Pharmacol Sci 12: 383–388

Havrankova J, Roth J, Brownstein MJ (1978) Insulin receptors are widely distributed in the central nervous system of rat. Nature 272: 827–829

Hoyer S (1988) Glucose and related brain metabolism in dementia of Alzheimer type and its morphological significance. Age 11: 158–166

Hoyer S (1992a) The biology of the ageing brain. Oxidative and related metabolism. Eur J Gerontol 1: 157–165

Hoyer S (1992b) Oxidative energy metabolism in Alzheimer brain. Studies in early-onset and late-onset cases. Mol Chem Neuropathol 16: 207–224

Hoyer S, Nitsch R (1989) Cerebral excess release of neurotransmitter amino acids subsequent to reduced cerebral glucose metabolism in early-onset dementia of Alzheimer type. J Neural Transm 75: 227–232

Hoyer S, Nitsch R, Oesterreich K (1990) Ammonia is endogenously generated in the brain in the presence of presumed and verified dementia of Alzheimer type. Neurosci Lett 117: 358–362

Hoyer S, Nitsch R, Oesterreich K (1991) Predominant abnormality in cerebral glucose utilization in late-onset dementia of the Alzheimer type: a cross-sectional comparison against advanced late-onset and incipient early-onset cases. J Neural Transm [P-D Sect] 3: 1–14

Hoyer S, Oesterreich K, Wagner O (1988) Glucose metabolism as the site of the primary abnormality in early-onset dementia of Alzheimer type? J Neurol 235: 143–148

Hoyer S, Prem L, Sorbi S, Amaducci L (1993) Stimulation of glycolytic key enzymes in cerebral cortex by insulin. NeuroReport 4: 991–993

Junod A, Lambert AE, Stauffacher W, Renold AE (1968) Diabetogenic action of streptozotocin: relationship of dose to metabolic response. J Clin Invest 48: 2129–2139

Kadowaki T, Kasuga M, Akanuma Y, Ezaki O, Takafu F (1984) Decreased autophosphorylation of the insulin receptor-kinase in streptozotocin diabetic rats. J Biol Chem 259: 14208–14216

Kumar A, Schapiro MB, Grady C, Haxby JV, Wagner E, Salerno JA, Friedland RP, Rapoport SI (1991) High-resolution PET studies in Alzheimer's disease. Neuropsychopharmacology 4: 35–46

Liguri G, Taddei N, Nassi P, Latorraca S, Nediani C, Sorbi S (1990) Changes in Na^+, K^+-ATPase, Ca^{2+}-ATPase and some soluble enzymes related to energy metabolism in brains of patients with Alzheimer's disease. Neurosci Lett 112: 338–342

Mann DMA, Sumpter PQ, Davies CA, Yates PO (1987) Glycogen accumulation in the cerebral cortex in Alzheimer's disease. Acta Neuropathol 73: 181–184

Marcus DL, deLeon MJ, Goldman J, Logan J, Christman DR, Wolf AP, Fowler JS, Hunter K, Tsai J, Pearson J, Freedman ML (1989) Altered glucose metabolism in microvessels from patients with Alzheimer's disease. Ann Neurol 26: 91–94

Marcus DL, deLeon M, Swerdlow R, Kaplan H, Tsai J, Freedman ML (1988) Decreased brain glucose metabolism in vivo and in microvessels in patients with Alzheimer's disease. Clin Res 36: 486A

Mielke R, Herholz K, Grond M, Kessler J, Heiss WD (1992) Differences of regional cerebral glucose metabolism between presenile and senile dementia of Alzheimer type. Neurobiol Aging 13: 93–98

Mordes JP, Rossini RA (1981) Animal model of diabetes. Am J Med 70: 353–360

Murphey DGM, Bottomley PA, Salerno JA, DeCarli C, Mentis MJ, Grady CL, Teichberg D, Giacometti KR, Rosenberg JM, Hardy CJ, Schapiro MB, Rapoport SI, Alger JR, Horwitz B (1993) An in vivo study of phophorus and glucose metabolism in Alzheimer's disease using magnetic resonance spectroscopy and PET. Arch Gen Psychiatry 50: 341–349

Nitsch RM, Blusztajn JK, Pittas AG, Slack BE, Growdon JH, Wurtman RJ (1992) Evidence for a membrane defect in Alzheimer disease brain. Proc Natl Acad Sci USA 89: 1671–1675

Nitsch R, Hoyer S (1991) Local action of the diabetogenic drug, streptozotocin, on glucose and energy metabolism in rat brain cortex. Neurosci Lett 128: 199–202

Palmer AM, Stratmann GC, Procter AW, Bowen DM (1988) Possible neurotransmitter basis of behavioral changes in Alzheimer's disease. Ann Neurol 23: 616–620

Perry EK, Perry RH, Tomlinson BE, Blessed G, Gibson PH (1980) Coenzyme A-acetylating enzymes in Alzheimer's disease: possible cholinergic "compartment" of pyruvate dehydrogenase. Neurosci Lett 18: 105–110

Plaschke K, Hoyer S (1993) Action of the diabetogenic drug streptozotocin on glycolytic and glycogenolytic metabolism in adult rat brain cortex and hippocampus. Int J Dev Neurosci 11: 477–483

Quon D, Wang Y, Catalano R, Scardina JM, Murakami K, Cordell B (1991) Formation of beta-amyloid protein deposits in brains of transgenic mice. Nature 352: 239–241

Reinikainen KJ, Paljärvi L, Halonen T, Malminen O, Kosma VM, Laakso M, Riekkinen PJ (1988) Dopaminergic system and monoamine oxidase-B activity in Alzheimer's disease. Neurobiol Aging 9: 245–252

Riederer P, Sofic E, Moll G, Freyberger A, Wichart I, Gsell W, Jellinger K, Hebenstreit G, Youdim MBH (1990) Senile dementia of Alzheimer's type and Parkinson's disease: neurochemical overlaps and specific differences. In: Dostert P, Riederer P (eds) Early markers in Parkinson's and Alzheimer diseases. Springer, Wien New York, pp 221–232 (New vistas in drug research, vol 1)

Sims NR, Blass JP, Murphy C, Bowen DM, Neary D (1987) Phosphofructokinase activity in the brain in Alzheimer's disease. Ann Neurol 21: 509–510

Sims NR, Bowen DM, Neary D, Davison AN (1983) Metabolic processes in Alzheimer's disease: adenine nucleotide content and production of $^{14}CO_2$ from (U-^{14}C) glucose in vitro in human neocortex. J Neurochem 41: 1329–1334

Small GW, Kuhl DE, Riege WH, Fujikawa DG, Ashford JW, Metter EJ, Mazziotta JC (1989) Cerebral glucose metabolic patterns in Alzheimer's disease. Effect of gender and age at dementia onset. Arch Gen Psychiatry 46: 527–532

Sorbi S, Bird ED, Blass JP (1983) Decreased pyruvate dehydrogenase complex activity in Huntington and Alzheimer brain. Ann Neurol 13: 72–78

Storandt M, Bäkman L, Baltes MM, Blass JP, Braak H, Gutzmann H, Hauw JJ, Hoyer S, Jorm AF, Kauss J, Kliegl R, Mountjoy CQ (1988) Relationships of normal aging and dementing diseases in later life. In: Henderson AS, Henderson JH (eds) Etiology of dementia of Alzheimer's type. Wiley, Chichester, pp 231–239

Yates CM, Butterworth J, Tennant MC, Gordon A (1990) Enzyme activities in relation to pH and lactate in postmortem brain in Alzheimer type and other dementias. J Neurochem 55: 1624–1630

Authors' address: Prof. Dr. S. Hoyer, Department of Pathochemistry and General Neurochemistry, University of Heidelberg, P.O.B. 104340, D-69033 Heidelberg, Federal Republic of Germany

Subject Index